Biology

Biology

A HISTORICAL PERSPECTIVE VOLUME I

JASON L. WORLEY
Patrick & Henry Community College

Bassim Hamadeh, CEO and Publisher
Anne Jones, Project Editor
Casey Hands, Production Editor
Jess Estrella, Senior Graphic Designer
Trey Soto, Licensing Specialist
Natalie Piccotti, Director of Marketing
Kassie Graves, Senior Vice President, Editorial
Jamie Giganti, Director of Academic Publishing

Printed in the United States of America.

BRIEF CONTENTS

DETAILED CONTENTS

INTRODUCTION

I do not believe studying history is a way of informing us of our past mistakes from which we are to learn to avoid repeating, but instead, as an opportunity to embrace the accomplishments of history to help us better prepare the trajectory of the future. There are times when figures instrumental in shaping the course of history are often forgotten or avoided to ensure that only those who were responsible for causing the mistakes receive recognition. For example, the amendments enacted during the time of Reconstruction that abolished slavery (13th), granted citizenship to all "born or naturalized in the United States," including former slaves (14th), and prohibited states from disenfranchising voters "on account of race, color, or previous condition of servitude," (15th) are often overshadowed by opposition from Southern states, vetoes implemented by President Andrew Johnson, and his impending impeachment. What is lost in all of this are those members of Congress who were instrumental in constructing these amendments and pushing them towards ratification despite the calls of opposition. These members include Lyman Trumbull of Illinois, Thaddeus Stevens of Pennsylvania, John Henderson of Missouri, Charles Sumner of Massachusetts, John A. Bingham of Ohio, Jacob Howard of Michigan, and William Stewart of Nevada. As someone once told me, these amendments, but more importantly, these Congressional members were responsible for legislation that helped shape the voice of advocates of the Civil Rights movement during the 1950s and 1960s.

Strangely enough, the same can be said about science, as the discipline is typically associated with the accomplishments of Newton, Leeuwenhoek, Darwin, or Watson and Crick. Please know that I am by no means diminishing the names or works of these great scientists, but oftentimes, textbooks and lectures resonate their accomplishments to the point of overshadowing and placing within the confines of historical obscurity the studies of their predecessors (note: sentence written as a means of self-accusation as well). Therefore, reflecting on the lack of recognition of certain scientists, such as Walther Flemming, Friedrich Miescher, Agnes Robertson Arber, or Rachel Carson, who were instrumental in helping shape the current knowledge of the concepts set forth within the scientific curriculum, I felt inspired to write a textbook that would present a *historical perspective* into the development of these concepts. Thus, each ... chapter begins with a brief biographical sketch of a scientist I believe deserves recognition for their accomplishments in helping shape the current understanding of a scientific concept, which is clearly defined thereafter. Although some of these biographical sketches were written on scientists highly touted in other textbooks or during a lecture, including Linnaeus, Mendel, and Woese, I felt it appropriate to include them for students to understand that even in science, "to embrace the accomplishments of history to help us better prepare the trajectory of the future."

CHAPTER

1

Science and Biology

PROFILES IN SCIENCE

FIGURE 1.1 Carolus Linnaeus.

Carolus Linnaeus is often called the father of modern taxonomy. Linnaeus was born in May 1707 in Småland, Sweden, and died at his Hammarby Estate (Sweden) in 1778. At an early age, Linnaeus learned Latin and geography, and he began receiving tutoring at age seven. He was enrolled at the Växjö Lower Grammar School (Sweden) at age nine; however, he rarely studied. Linnaeus spent a majority of his time collecting and organizing plants, and within several years, he became well-acquainted with the existing flora in the region. After spending only one year at Lund University (Sweden), he enrolled at Uppsala University (Sweden) in 1728, where he studied botany and began classifying the plants he collected by their number of stamens and pistils. It was in his 1732 publication *Flora Lapponica*, which detailed plants he encountered and discovered during his trip to Lapland (Sweden), that Linnaeus outlined for the first time his nomenclature and classification systems, along with a description of the species and their geographical distribution. His most famous work, *Systema Naturae*, was published in 1735 detailing a hierarchical classification system in which he placed minerals, plants, and animals in separate kingdoms, then subsequently divided each into classes, orders, genera, and species. Linnaeus published additional works in 1736 and 1737 summarizing both the principles and rules of the hierarchical classification system and binomial nomenclature, and how the principles were to be applied. Between the 1740s and 1760s, Linnaeus continued traveling across Sweden collecting, classifying, and naming minerals, animals, and plants using his newly created classification system. During those same years, he published *Philosophia Botanica* (1751), a compilation and expansion of his 1736 and 1737 publications, in which he gave his first full description of binomial nomenclature. Other notable publications during those same years included *Species Plantarum* (1753) and the tenth edition of *System Naturae* (1758), both of which are considered the beginning of botanical and zoological nomenclature. After becoming ennobled by Swedish King Adolf Frederick in 1761, ill health dominated Linnaeus until his death in 1778. Linnaeus has been memorialized through statues, awards, and street names, while his name has also been used to name moon craters and minerals.

Introduction to the Chapter

Science is the all-encompassing study of the natural world. The process of science begins with discovery and inquiry, which leads to a collection of facts, finally culminating in a better understanding of the general order of nature and the laws that govern it. In other words, without science, any gain in knowledge about how the natural world functions ceases to exist. Science simplifies even the most complex ideas and enables all who use it to find truth and objectivity. To use science to explain unique phenomena is a privilege and must be handled accordingly by means of an orderly system of methods. Any deviation from these methods can greatly disrupt the integrity and sanctity of the process. Science does not flaw the scientist, but rather, the scientist can flaw science. Biology is a subdiscipline of science and is the specified study of life and what it constitutes. Life is strictly defined and classified by biology. This chapter introduces science and its main components, including the scientific method, a detailed step-by-step process that helps answer complex questions, and outlines how science has recently come under scrutiny as some have taken advantage of the process. Finally, the chapter defines biology as the study of life, outlines what constitutes life, and introduces the scientists who were instrumental in classifying life and how the classification system has changed since it was first introduced.

Chapter Objectives

In this chapter, students will learn the following:

1-1. Science is a process of inquiry and discovery, involving a step-by-step scientific method, which leads to a better understanding of the natural world and how it works.

1-2. Biology is defined as the study of life and encompasses five main qualities: organization; energy and metabolism; homeostasis; reproduction, growth, and development; and environmental response.

1-3. Several scientists were responsible for contributing to the early history and development of the classification system of life.

1-4. The current classification system of life was developed using differences in ribosomal RNA (rRNA), leading to the modern-day three-domain, six-kingdom system.

1-1. What Is Science?

FIGURE 1.2 Earth, or the Blue Marble, provides scientists with the perfect laboratory to perform experiments to understand how the natural world works.

Are you a scientist? Well, of course you are! You may have never been specified by that title; however, if you have ever inquired about how something works and proceeded through a series of steps to find the answer, then you are a scientist! **Science** involves curiosity and the desire to learn more about the natural world and the phenomena within it. Science is about inquiry, discovery, and obtaining knowledge to have a better understanding of the natural world and how nature works, and Earth provides the laboratory for scientists to perform their experiments (Figure 1.2).

Performing the act of science can be a daunting task, because it is a considerably broad discipline, consisting of a multitude of different subdisciplines that require the attention of devoted and unique scientists. Many scientists prefer collaborating with a team of other scientists when performing scientific research because it incorporates different viewpoints, ideas, and contributions, leading to the overall success of the project. For example, the ... number of scientists preferring to collaborate with others increased slightly between 2007 and 2011 from 3.8 to 4.5 individuals. This incremental shift correlates to a recent universal trend toward larger team sizes occurring within the growing scientific community. The increase in team size has been attributed to a variety of factors, including more focus on specialized research and the need for multidisciplinary interactions to combat complex, modern issues. But how does team size influence the integrity of scientific research? Current research into this question has produced an interesting answer: small team sizes, consisting of five or fewer collaborating scientists, disrupts science through the introduction of novel ideas and concepts, while larger-sized teams tend to focus more resources on solving, developing, and elaborating on preexisting ideas and questions. Regardless of team size, however, contributions from both small and large teams are essential in maintaining the health and integrity of scientific research.

FIGURE 1.3 Galileo has been given credit for our current understanding of the scientific method.

Although the number of scientists collaborating as a team may greatly influence the course of scientific research, the common factor linking all team sizes is the scientific method. The scientific method is a step-by-step process all scientists follow in order to understand how the natural world works. Despite some historians giving credit to Sir Francis Bacon as the first to formally document the scientific method, other historians have credited Galileo's work as the foundation on which the scientific method is built. Galileo (Figure 1.3) was a renowned Italian astronomer and mathematician who famously challenged the idea that planetary objects, including

Earth, orbit the sun elliptically rather than circularly. This idea was met with great hostility and bitterness because it defiled the teaching of the church that the sun revolved around the earth, leading to continued conflict with the Inquisition. Galileo was condemned for heresy in 1633. In a recently discovered letter in August 2018, dated December 21, 1613, Galileo outlined his arguments against the church's doctrine and indicated that one should be free to practice science without theological influence. The letter, however, was discovered to contain multiple edits, suggesting Galileo was softening his claims. Nonetheless, Galileo's observations of facts leading to quantitative laws, resulting in the prediction of future facts, is a perfect example of the scientific method.

Figure 1.4 outlines the major main steps associated with the scientific method:

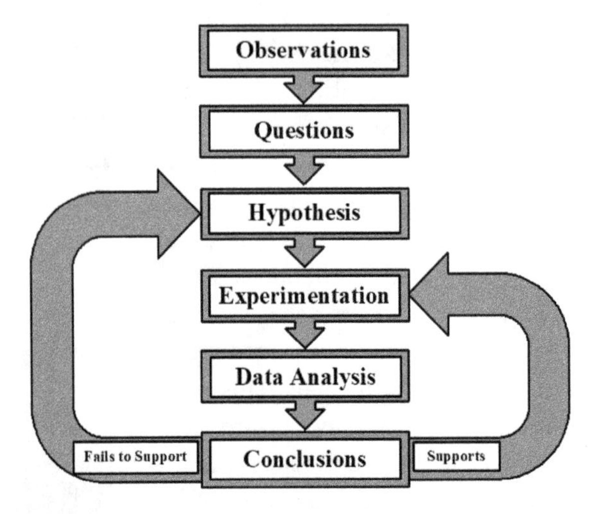

FIGURE 1.4 The scientific method is a step-by-step process that scientists use in order help explain unique phenomena and answer questions about the workings of the natural world.

FIGURE 1.5 Charles Darwin's voyage to the Galápagos Islands provided him with many different types of observations leading him to develop the theory of evolution.

Observations

Observations are made using a variety of different methods: Senses—hearing, sight, taste, smell, or touch—for example, Charles Darwin (Figure 1.5) made a startling discovery when observing the physical differences among the finch species he collected during his voyage to the Galápagos Island. Other methods include historical incidents or accidents—asbestos was a commonly used component of insulation materials for much of the 20th century in the construction of new homes and commercial buildings because of its extreme resistance to heat, electricity, and chemical corrosion. However, asbestos is also highly toxic and is now closely associated with an aggressive and malignant cancer called mesothelioma. Another method relates to prior scientific knowledge and experimentation—this may involve scientists exploring scientific journals and newspapers or having conversations with other scientists about similar experiments and collected data.

Hypothesis

Considered the heart of scientific inquiry, a **hypothesis** is a formulated explanation constructed based on information obtained from the observations and previous knowledge. A hypothesis involves the use of inductive reasoning, in which isolated facts are combined into a cohesive whole and is used to help explain a scientist's observations and questions. When formulating a hypothesis, the scientist must ensure that the hypothesis leads to testable predictions, while also ensuring that data collected during experimentation either supports or fails to support the hypothesis. A hypothesis does not prove anything but rather guides the scientist to an explanation for the observations made and questions asked.

Experimentation

This is the most important and critical step of the scientific method because it enables a scientist to test his or her hypothesis. Prior to conducting the experiment, scientists must carefully design the experiment to ensure the quality, quantity, and validity of their data. There are four common elements generally found in a good experimental design: **treatment**—an experimental condition applied to the research individuals; **experimental group**—research individuals receiving the treatment; **control group**—research individuals not exposed to the treatment, who oftentimes may receive a placebo; and **variables**—elements subject to change during the experiment.

There are two types of experimental designs a scientist may employ: a **blind experimental design**, in which the experimental group is uncertain as to which treatment they are receiving, if they receive the treatment at all; and a **double-blind experimental design**, a commonly used design in medical research, where neither the participants nor the scientists are aware of which treatment is being administered during the experiment.

Regardless of how a scientist designs his or her experiment, the design will generally include controls and variables. Controls are necessary in any experiment because they provide the basis with which to compare the experimental results. Depending on the experiment, a control may take one form or another. In some cases, a control may involve the use of a placebo.

Placebos are "fake pills," or substances given to individuals during experiments that lack an active ingredient. Placebos are generally sugar pills or some known treatment. It is relatively common for participants in the control group who receive a placebo to not show any effects from the inert substance. However, a United States physician in the mid-1950s discovered that in 15 studies he analyzed, roughly 33 percent of the patients showed a significant change in their pain levels when taking a placebo. This idea was supported with evidence from a 1978 study, where 33 percent of post-surgery patients reported significant pain reduction when administered saline instead of morphine. Another study in 2010, patients with irritable bowel syndrome (IBS) who knew they were administered a placebo reported a 59-percent reduction in symptoms, compared to 35-percent in a nontreatment group.

Controls are necessary in an experimental design because they remain constant and ensure the scientist can isolate a particular variable or factor resulting in an observed effect. Variables are an important component of an experiment design because they produce the information that the scientist is looking for to help verify a hypothesis. There are two main types of variables, including the **independent variable**, the variable manipulated by the scientist, which is evaluated based on its effect upon the phenomenon being studied, and the **dependent variable**, measured by the scientist to determine the subject's response to changes in the independent variable. It is important that the scientist carefully controls these variables to minimize disparities experienced between the control group and experimental group.

Finally, an important feature of an experimental design is to ensure its repeatability. Oftentimes, scientists will utilize the work of other scientists to validate the effectiveness of a treatment used during an experiment. In addition, a repeatable experiment helps prevent any unnecessary biases from influencing data collection and/or analysis, enabling the scientist to validate his or her conclusions.

Data Analysis and Conclusions

Once the experiment has concluded, the scientist compiles and analyzes the results of the experiment, known as **data**. The primary way in which a scientist can determine whether there was an effect on variables is through statistical analysis. Statistical analysis is an important mathematical tool for a scientist because it enables him or her to understand the data produced during the experiment, while also determining variations, relationships, or a lack thereof between the variables tested. Upon analysis, a **conclusion** will be reached by the scientist to determine whether the data supports or fails to support his or her hypothesis.

But what does a scientist do when the data supports his or her hypothesis? The scientist continues to build upon the idea through additional experimentation of the hypothesis. However, there are oftentimes when the conclusion of one experiment may lead to a hypothesis for another experiment that only verifies the findings of the original experiment. What if the data does not support a scientist's hypothesis? Does the scientist give up? Absolutely not! The scientist will revise his or her original hypothesis or formulate a new hypothesis and rerun the experiment.

There are several steps a scientist will take after a conclusion has been reached, including presenting his or her data in the form of graphs or tables, submitting the work for publication in journals, and/or having his or her work reviewed by peers within the scientific community. This last step is important, because it enables a scientist's work to be evaluated for validity, repeatability, and quality.

Recall that a hypothesis is a formulated explanation about an observed phenomenon; a **theory** is a hypothesis thoroughly supported by years of experimentation and data. The term *theory* is commonly misused during everyday conversation as a person states, "My theory is ..." when in fact the sentence should be rephrased as, "My hypothesis is ..." For example, in 2014, Malaysia Airlines Flight 370 left Kuala Lumpur International Airport on its way to Beijing Capital International Airport. Unfortunately, the flight mysteriously disappeared over the South China Sea, with 227 passengers and 12 crew members presumed dead. To this day, investigators have proposed different reasons as to why the flight disappeared, including a hypoxic event, hijacking, or communication failure on the part of the air traffic controllers to maintain contact after the disappearance. Regardless of the cause outlined by investigators or the media, each has been referenced as a "theory," a direct misuse of the term. Each reason would be considered a separate hypothesis rather than a theory.

There are several different theories in biology, including the cell, gene, and ecosystem theories. However, one of the more common theories is the theory of evolution, which states that all living organisms share a common ancestor, and each organism has adapted a specialized way of life. When does a theory become a *law*? A theory becomes a law when the theory is supported by years of experimental data and is accepted by most scientists. Examples of laws include the law of gravity and the law of thermodynamics.

Science and the methods employed by scientists to understand phenomena of the natural world have been under intense scrutiny in the past several years. Recent studies have revealed that ecologists and evolutionary biologists have been practicing unconventional data analysis and reporting, leading to irreproducible experiments, like those that have been reported in psychology. Of the 807 scientists surveyed in the study, 64 percent failed to report statistically insignificant data; 42 percent employed an act known as "p-hacking," which involves extending a study beyond its timeframe to collect additional data to reach statistical significance; and 51 percent reported unexpected findings but then claimed as if the findings had been hypothesized prior to the work being executed. These scientific malpractices have made science appear more flawed than in the past. It has been suggested that focusing on minute details during scientific studies has consequently made these issues more evident. However, better vigilance during scientific studies, along with education and preregistering of studies prior to their being conducted, may curb these issues, thereby improving the overall quality of science.

Statistics is an important tool used by all scientists to help analyze data generated during an experiment. What helps scientists conclude whether a hypothesis is supported is *statistical significance*. The most common statistical method scientists use to determine whether statistical significance exists within their data is to calculate what is known as a *P* value. The general rule is that upon statistical analysis of the data, if the *P* value is less than 0.05, scientists conclude that the data are statistically significant; therefore, an association exists between their data and their hypothesis. Conversely, a *P* value greater than 0.05 is considered statistically insignificant, suggesting that the hypothesis is not supported by the data. In recent years, several hundred biologists, statisticians, and medical researchers have called for better scientific practices and abandoning the use of *P* values as an arbitrary indicator of statistical significance. Advocates believe that the misuse of statistical significance is affecting the scientific community by promoting bias and encouraging researchers to manipulate their data and methodology to produce statistically significant and/or insignificant results. To rectify the issues, advocates believe that scientists should preregister their studies and publish all of their results. However, the idea of preregistering studies has been met with concern due to the potential for increased bias based on the methods employed to analyze the data. Although advocates believe implementing these scientific practices and ending the use of *P* values will eliminate apparent issues within the system, there are some opponents that believe otherwise.

Opponents agree that the use of statistical significance is a convenient and common-sense practice but can often be seen as an obstacle to scientific research as well. However, these opponents suggest that completely abandoning the practice altogether is impractical and could lead to more profound bias during research. Others opine that it is not the use of statistical significance that is the underlying issue but rather what constitutes sufficient statistical evidence to suggest that a correlation exists between the data and the hypothesis. It was proposed that a discussion among scientists is required to determine a better definition of "statistically significant" to alleviate any confusion. Additionally, some scientists believe there should be established rules in place that prevent data manipulation to reach statistical significance, proper education of those who misuse the term, and experiment designs in which multiple statistical tools are used for data analysis.

Another issue facing science today is the media's portrayal of "*scientific evidence*," the linking of two occurrences together when there is no scientific evidence supporting this link. One example of "scientific evidence" is **pseudoscience**, wherein something sounds scientific but lacks the appropriate evidence produced through scientific methods of correlating significance. A region of the world that has shown a significant increase in the use of pseudoscience is India. The rise of pseudoscience in India is rooted in ancient Hindu science and currently supported by the push towards Hindu nationalism by those in the highest levels of government. Scientists believe that pseudoscientific claims in India have threatened the quality of its science and have weakened its educational institutions, consequently affecting the country's development. Scientific funding has also been affected using pseudoscience in India. In 2017, the Indian government funded research validating the use of panchagavya, claiming that the mixture of cow urine and feces provided therapeutic relief for a variety of ailments. This claim, however, has been dismissed by several independent scientists.

Another example of "scientific evidence" is **anecdotal observations**, or observations that suggest there is or is not a link between two occurrences. This is especially evident in a study published in 1998, wherein researchers investigated a small sample size of carefully selected children with gastrointestinal disorders exhibiting autism. They reported that the link between these disorders could be attributed to MMR vaccinations received months prior to the autism diagnosis. Since this article was first published, multiple studies have been performed refuting the authors' claim suggesting that a link between MMR vaccines and autism exists. As a result, the original paper was retracted, and in 2011, the *British Medical Journal* declared the article "fraudulent."

1-2. Biology and "What Is Life?"

Biology is defined as the scientific study of life. Life on Earth is represented by several diverse forms (Figure 1.6) ranging from highly functional, single-celled microorganisms (e.g., bacteria and protists) to multicellular organisms exhibiting increased complexity (e.g., fungi, plants, and animals).

Science recognizes and defines life based on the combination of five qualities shared by all organisms. Although some of these qualities may be attributed to a nonliving object, such as a rock, an organism is considered living if it encompasses all five qualities.

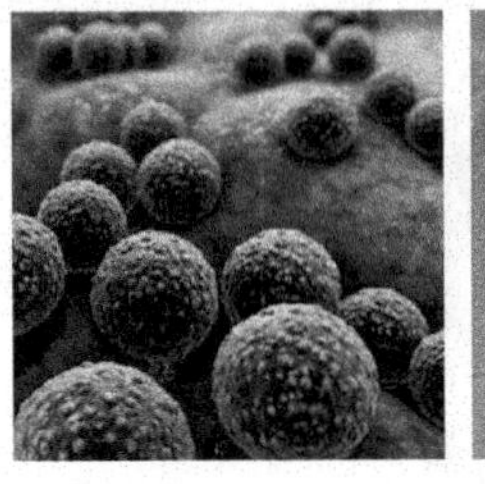

Bacteria

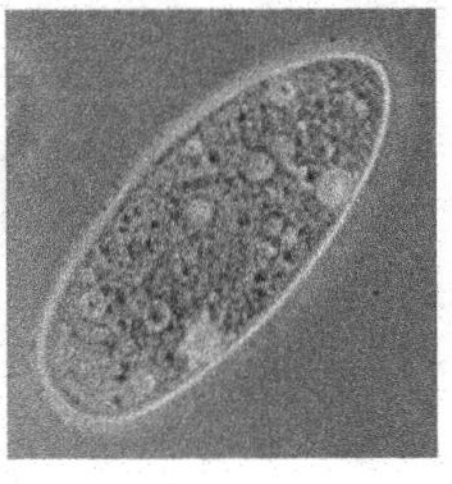

Paramecium

Fungi

Hyacinth

Tiger Swallowtail

FIGURE 1.6 Biology is defined as the scientific study of life, which comes in a variety of diverse forms.

Hierarchical Organization

All life exhibits organization, which occurs at different hierarchical levels (Figure 1.7). At each subsequent level within the hierarchy, the acquisition and aggregation of interactive properties, or **emergent properties**, exhibit distinct roles not evident in the previous level. For example, the heart is responsible for pumping blood through the body, but the individual cells comprising the organ do not possess this same property. Therefore, the pumping of blood only becomes an emergent property of the organ when the heart cells aggregate in a specific manner to facilitate the function.

The most basic level of organization begins with the **atom**. All pure substances, known as elements, are composed of atoms, which contain three main particles—protons, neutrons, and electrons. The essential atoms for life are carbon, hydrogen, oxygen, nitrogen, phosphorus, calcium, and sulfur. Atoms build **molecules**, which consist of two or more atoms bonded together into a larger structure. Common examples include water, glucose, and amino acids. **Macromolecules** are complex structures composed of two or more molecules joined together through chemical bonding. Proteins are an example of a complex macromolecule, composed of multiple amino acids joined by peptide bonds. DNA is another example of a macromolecule, consisting of several molecules, including sugar, phosphate groups, and nitrogenous bases. Groups of large macromolecules constitute **organelles**, structures that provide a functional purpose within a cell. Organelles include mitochondria, responsible for producing ATP within the cell and the nucleus, which consists of DNA and is responsible for all physiological processes within the cell.

A **cell** is considered the most basic unit of both structure and function of all organisms. Enclosed by a plasma membrane, the cell consists of different types of organelles, all surrounded by a jelly-like substance known as cytosol. Various types of cells exist, including bacteria, which are the simplest of all cells; plant cells; animal cells; and body cells, like muscle, nerve, and skin cells. Multiple cells combined together, all sharing a common function, constitute **tissues**. There are four basic types of tissues, including epithelial, connective, muscle, and nervous. Several tissues combined together that share a common, specific function are called **organs**. Organs have a recognizable shape and structure, and there are no two organs that look alike. The most common organs are the heart, lungs,

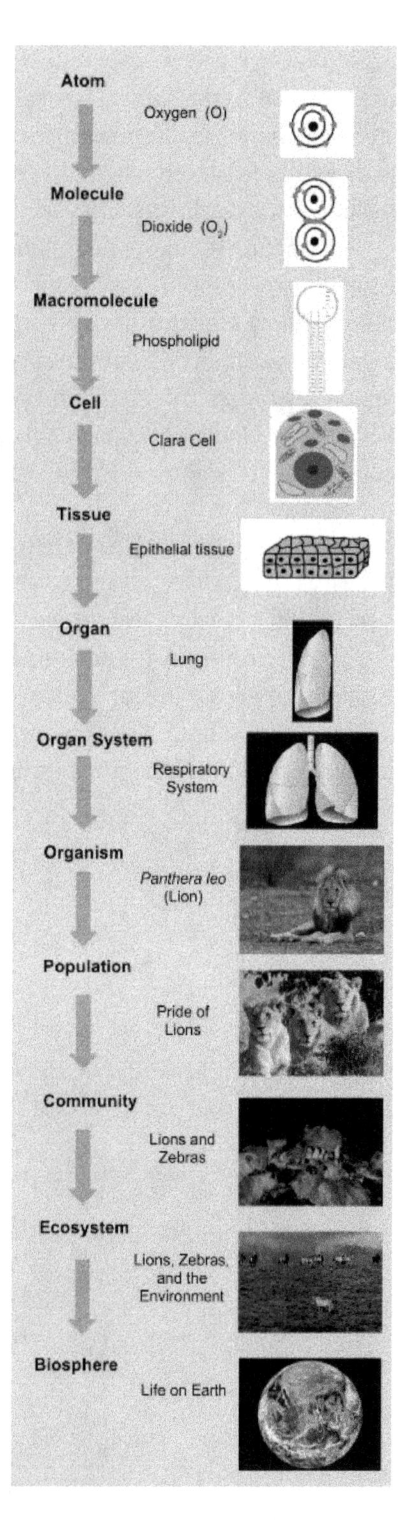

FIGURE 1.7 Hierarchical organization is an important feature of all living organisms.

and brain. **Organ systems** are a combination of related organs, all of which share a common function. Although each organ within the system may perform a specific function, the overall function of the system is the same. For example, each organ of the digestive system (e.g., tongue, esophagus, stomach, and large intestine) has a specialized function, but the ultimate function of the entire system is digesting food and extracting nutrients. The human body has eleven different organ systems, including the cardiovascular, skeletal, and nervous systems.

A complete set of organ systems that helps support life constitutes an **organism**. A **population** is when two or more organisms of the same species are interacting in the same place at the same time, while a **community** is the interaction among organisms of different species in a particular area. An example of a community would include not only the animals but also plant life as well. When organisms interact with each other, along with the nonliving environment, this is called an **ecosystem**. Nonliving components of an environment include soil, air, and water. Finally, a **biosphere** consists of all regions of Earth inhabited by organisms, which can sustain life.

Energy and Metabolism

Energy is defined as the ability to do work, and it is required in order to maintain organization and sustain physiological processes. The sun provides the ultimate source of energy on Earth. Solar energy is provided by the sun and is converted to chemical energy by plants (photosynthesis), which is then transferred through other various types of organisms, like animals and decomposers, upon consumption (Figure 1.8). This energy helps fuel chemical reactions that occur within the cell and is responsible for helping maintain life and is termed **metabolism**. Metabolism is responsible for

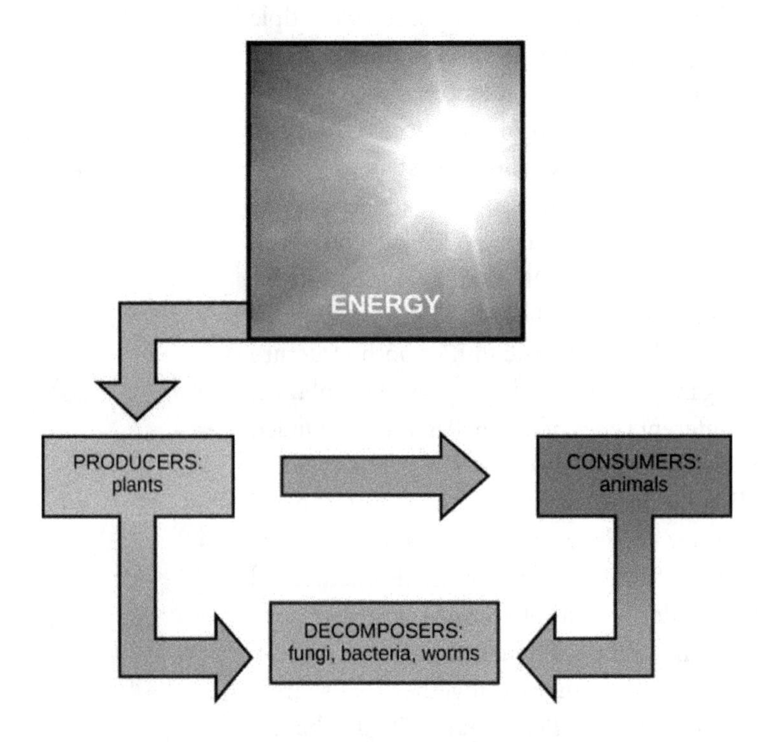

FIGURE 1.8 Energy flow in living organisms begins with the sun and is transferred through plants and animals to decomposers. Energy is important in fueling metabolic processes.

the breakdown of molecules (conversion of food into specific nutrients) and the synthesizing of molecules (building proteins from amino acids).

Homeostasis

Homeostasis is defined as the ability of an organism to maintain a constant, stable internal environment. Living organisms must maintain a consistent body temperature and water content in order to survive. Any fluctuations in an organism's internal environment can greatly jeopardize its survival, resulting in either illness or death. Homeostasis is highly dependent on feedback systems and requires a majority of the energy the organism acquires.

Reproduction, Growth, and Development

One of the main goals of every organism is **reproduction**. Through the process of reproduction, the organism ensures the survival of its population by being able to produce another organism similar to itself. Organisms accomplish reproduction in one of two ways. **Asexual reproduction** is defined as the process by which an organism produces offspring that is genetically identical to itself. This is a common reproductive method utilized by prokaryotes, as well as some plant and animal species. The other type of reproduction is **sexual reproduction,** and this type of reproduction involves the exchange of sex cells (gametes) from two distinct parents. This results in a mixing of genetic material, producing an offspring that is genetically different from the parents. After reproduction, the resulting offspring increases in body size and mass (**growth**) and begins the **development** of anatomical features and physiological functions, to the point of being able to reproduce (Figure 1.9).

FIGURE 1.9 Reproduction is the goal of every living organism because it helps sustain the population. Upon birth, offspring begin to grow and develop.

Environmental Response

The ability of an organism to respond to its environment and make changes to itself is key to its overall success. **Irritability** is the ability of an organism to respond to an environmental stimulus. This stimulus can involve an interaction with the nonliving environment, like a change in temperature, or an interaction with other living organisms. For example, heliotropism, in which a plant's growth is made in direct response to the position of the sun, is considered a type of irritability. Another example is the ability of a prey species to escape the advances of a predator (a mouse running away from a snake). Irritability ensures an organism survives and enables it to carry out necessary physiological functions. **Adaptation** is the modification of an organism to a particular environment, enhancing their survival. Adaptations are a by-product of the process of evolution, therefore occurring over long periods of time and within a population of organisms. One of the most common examples of adaptation is the modification of the beaks of the Galápagos finches (Figure 1.10). The change in beak size of the finches ensured their survival year after year, leading to reproductive success and growth of their populations.

FIGURE 1.10 A living organism must adapt to its environment to ensure survival as seen in different sizes of beaks of Galápagos Islands finches.

1-3. History of Classifying Life

The classification of life can be traced back to the Greek philosopher Aristotle. In 350 BCE, Aristotle introduced in his book, *History of Animals*, a hierarchical classification system wherein animals belonging to particular groups were arranged by increasing complexity in the form of a ladder or scale. This became known as *scala naturae* or "scale of nature." The *scala naturae* organized 14 groups of living things by increasing complexity, with each group being subdivided into further groups based on size (Figure 1.11). For example, all birds were organized into the same group based on distinguishing features, such as feathers, wings, beaks, and two bony legs.

The modern-day classification of organisms dates back to 1735 and was developed by a Swedish botanist Carolus Linnaeus. It was not until his book *Systema Naturae*, wherein Linnaeus outlined a comprehensive two-system classification system (Figure 1.12) of naming and grouping similar plant and animal **species** (currently defined as a group of organisms capable of producing viable offspring), that the modern foundation for organismal classification was established. Linnaeus grouped an organism according to six main categories: kingdom, phylum, class, order, genus, and species.

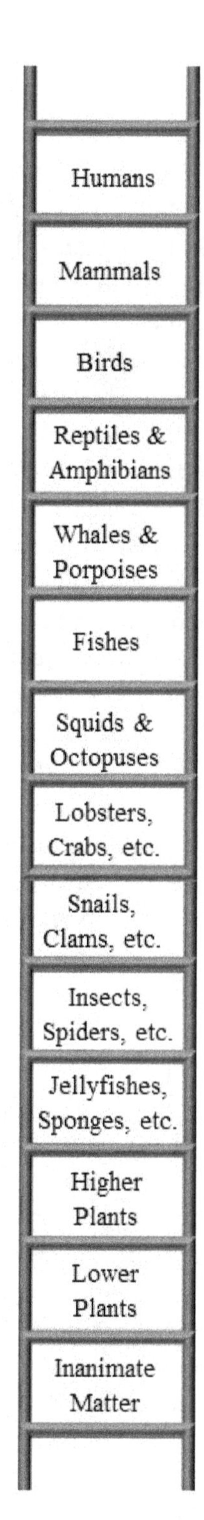

FIGURE 1.11 Aristotle's hierarchical classification system, known as *Scala Naturae*, organized all organisms based upon increased complexity.

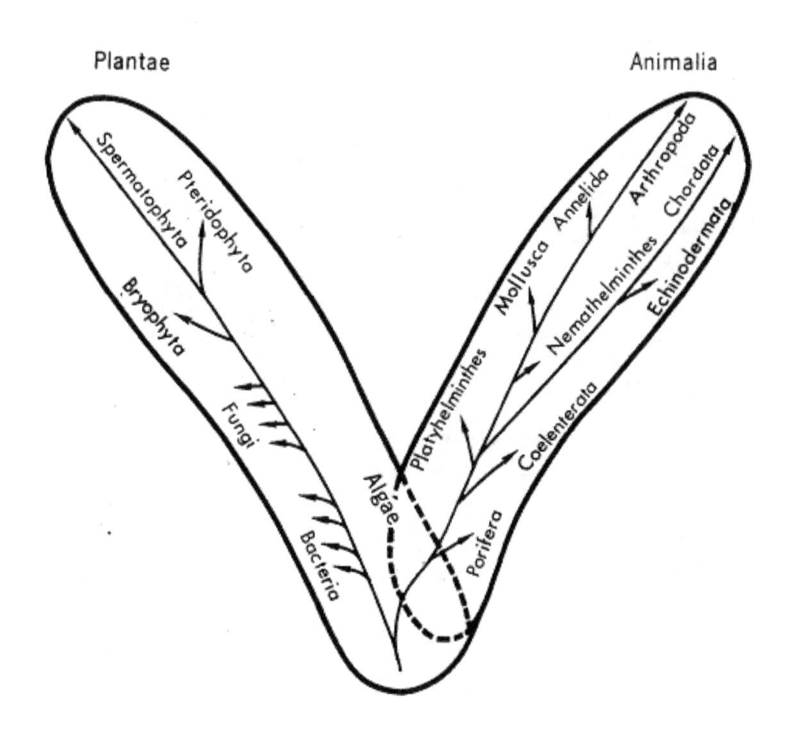

FIGURE 1.12 Carolus Linnaeus outlined a two-kingdom system in his 1735 publication, *Systema Naturae*, in which similar plant and animal species were organized.

In addition, Linnaeus also developed a system of naming organisms called **binomial nomenclature**, in which each organism is given a unique two-part name. The first part of the name indicates to which genus the organism belongs, while the second part of the name is called the **specific epithet** and provides a description of the organism and distinguishes the species within the genera. For example, *Sitta carolinensis* and *Poecile carolinensis* are two different species of birds commonly found in the eastern part of the United States. The first part of the binomial name outlines the genus into which the bird is categorized. The second part of the name is similar in both birds and provides a description of the geographic range in which the birds are commonly found. The first letter of the genus name is always capitalized, while the entire name is italicized. Oftentimes a scientist will shorten a binomial name by using the first letter of the genus name, for example *S. carolinensis* and *P. carolinensis*. The importance of establishing a binomial name for an organism is that it deters confusion and enables scientists to communicate information about a specific species in a language universally recognized and accepted.

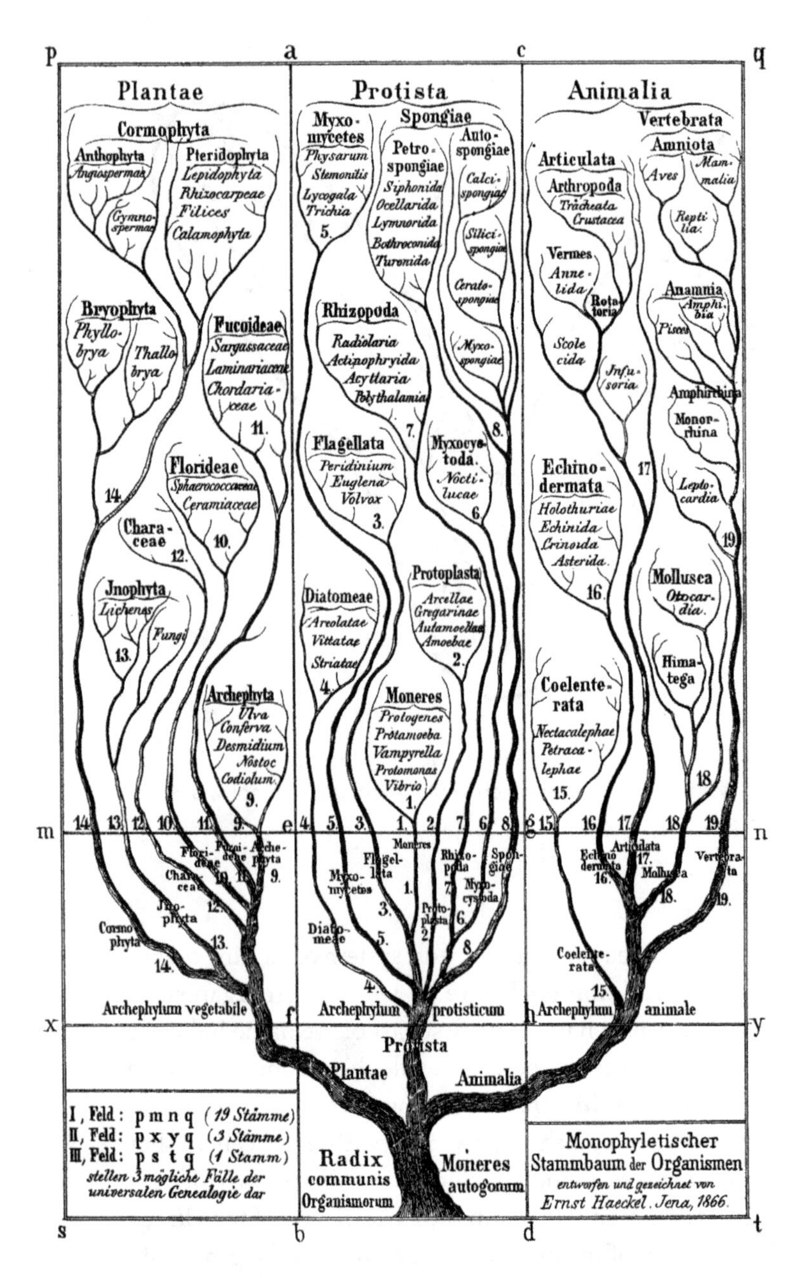

FIGURE 1.13 In 1866, Ernst Haeckel introduced a third kingdom to the classification system, known as Protista, to include organisms that were nonplant and nonanimal, including bacteria and protozoans.

The two-kingdom classification system came under scrutiny from German zoologist Ernst Haeckel in his book entitled *General Morphology of Organisms*. In 1866, Haeckel proposed the addition of a third kingdom, Protista, which included organisms considered to be nonplant and nonanimal (Figure 1.13). This new group included unicellular organisms, such as bacteria, protozoans, diatoms, and radiolarians. Unfortunately, Haeckel's system was met with controversy and not widely accepted. There were many opponents of Haeckel's three-kingdom system, including prominent figures in protozoological systematics, who argued that his new kingdom consisted of microorganisms whose phylogenetic interrelationships and taxonomic boundaries were unknown and vague. In addition, many of these opponents did not understand a necessity to place organisms into an additional kingdom based solely on their unicellularity or microscopic size when they could just as easily be placed in the existing plant and animal kingdoms.

In 1938, Herbert F. Copeland, a junior college biology professor, published a paper in which he proposed a four-kingdom system (Figure 1.14). Borrowing inspiration from Haeckel's ideas, better microscopy, and a more comprehensive understanding of microorganisms, Copeland established the kingdoms Monera, consisting of bacteria, and Protista, which included microorganisms containing a nucleus. At the time of his paper, Copeland did not delineate Fungi as separate but rather included this group in the kingdom Protista. In addition, other organisms such as green algae remained in the kingdom Plantae. Several years later, Copeland insisted on replacing the names of his newly created kingdoms with "Mychota" and "Protoctista." Copeland further solidified his four-kingdom system in his 1956 book, entitled *The Classification of Lower Organisms*.

In reaction to Herbert Copeland's book in 1956, another college professor, Robert Whittaker, decided on a different approach in categorizing organisms into kingdoms. Rather than using a morphological approach, Whittaker used ecology and the ways organisms obtain food in order to suggest a three-kingdom system. Each kingdom would consist of **producers**—organisms who make their own food (plants); **consumers**—organisms who consume other organisms (animals); and **decomposers**—organisms that consume dead or dying organisms (bacteria and fungi). However, after much thought and reflection, Whittaker proposed a four-kingdom system much different from that of Copeland in 1959. This new four-kingdom system consisted of Protista (including bacteria), Plantae, Animalia, and Fungi.

Whittaker revisited the four-kingdom system in 1969 and suggested instead a five-kingdom system (Figure 1.15), utilizing a dichotomous system, established by Roger Stanier and C. B. van Niel in 1962, of differentiating life as either prokaryotic (nonnucleated) or eukaryotic (nucleated). Whittaker removed bacteria from kingdom Protista because he realized that bacteria and single-celled protists were greatly different as protists had nuclei. Therefore, he established a new kingdom for bacteria he called Monera, borrowing the taxon name from work originally performed by Haeckel in the 1860s.

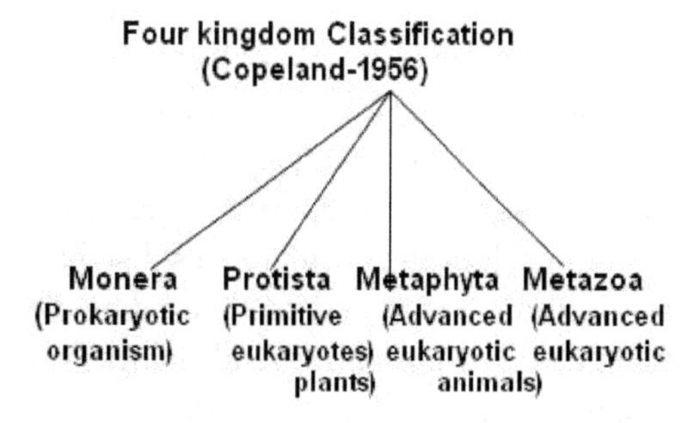

FIGURE 1.14 Herbert Copeland proposed a four-kingdom system in 1938, separating bacteria into their own kingdom, Monera, and placing nucleated microorganisms into the Kingdom Protista.

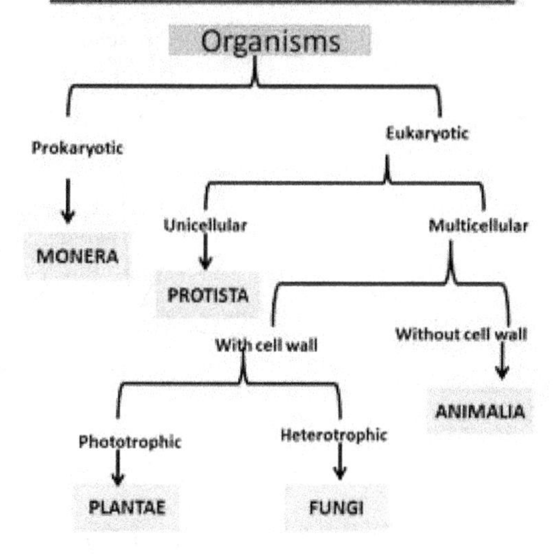

FIGURE 1.15 A five-kingdom system was proposed by Robert Whittaker in 1969, which included the kingdom Fungi and a reestablished Monera kingdom to include bacteria.

1-4. Current Classification System

Whittaker's five-kingdom system was about to drastically change, with work performed by two American scientists, Carl Woese and George E. Fox. In 1977, Woese and Fox proposed the existence of three taxonomic domains based on different cell types discovered through the sequencing of ribosomal RNA (rRNA). Their analysis of 16S rRNA, a type of ribosomal RNA found specifically in bacteria, revealed three subdivisions (blue-green bacteria, Gram-positive bacteria, and Gram-negative bacteria), which they labelled **Eubacteria**. The colleagues analyzed another type of 18S rRNA, a unique ribosomal RNA to eukaryotic cells. Woese and Fox labelled this group of organisms **Urkaryotes** (now referred to as eukaryotes), and this includes animals, plants, fungi, and protists. Finally, a third group of organisms, known as methanogenic bacteria, were identified using rRNA sequencing and represented their own separate group, **Archaebacteria**. This group of organisms is unique for several different reasons, including their ability to reduce carbon dioxide to methane, lack of peptidoglycan-based cell walls, and differences in both rRNA sequencing and base modification in comparison to rRNA found in the eubacteria group.

As a result of these differences discovered by Woese and Fox, a three-domain, six-kingdom classification system is currently used by biologists to organize life (Figure 1.16). The classification system has gone through several changes since being proposed by Woese and Fox, most notably in the names of each of the domains.

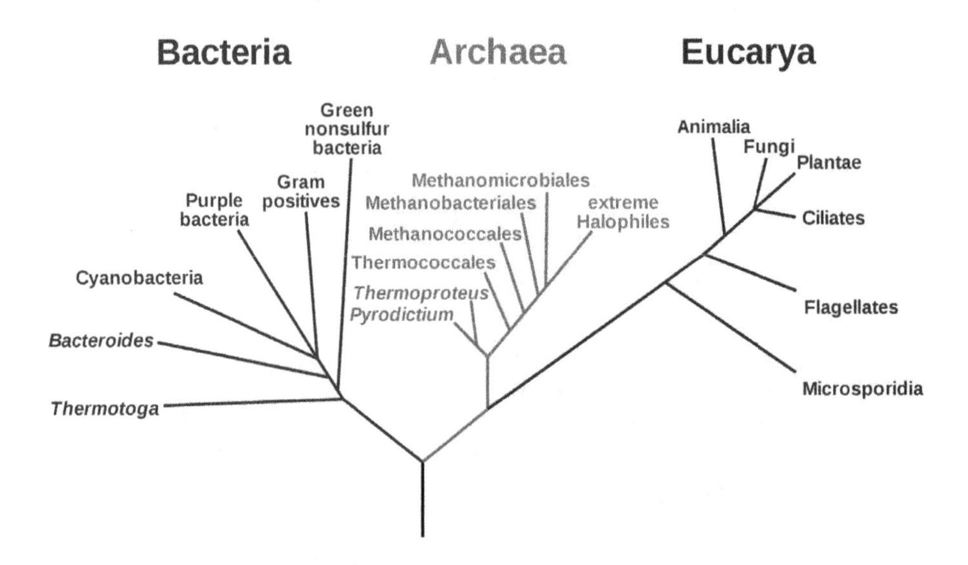

FIGURE 1.16 Upon analysis of ribosomal RNA in different cell types, Woese and Fox proposed a three-domain, six-kingdom system that is currently used by biologists to organize life.

1. Bacteria (domain and kingdom)
    - unicellular prokaryotes, lacking nuclei and membrane-bound organelles
    - consists of both autotrophic (producer) and heterotrophic (consumer) organisms
    - have distinct cell walls consisting of peptidoglycan
    - found almost anywhere (water, soil, atmosphere, in/on the human body, etc.)
2. Archaea (domain and kingdom)
    - unicellular prokaryotes, lacking nuclei and membrane-bound organelles
    - consists of both autotrophic (producer) and heterotrophic (consumer) organisms
    - have distinct cell walls lacking peptidoglycan
    - found in water devoid of oxygen and in harsh environmental conditions (temperatures, salinity, pH)

3. Eukarya (domain)
    - unicellular to multicellular eukaryotes, consisting of membrane-bound nucleus and membrane-bound organelles
    - consists of four kingdoms:
        1. Protists—mostly unicellular; some have cell walls; some are autotrophic, others are heterotrophic
        2. Fungi—mostly multicellular; chitin-based cell walls; heterotrophic
        3. Plants—multicellular; cellulose-based cell walls; autotrophic
        4. Animals—multicellular; no cell walls; heterotrophic

Chapter Summary

Below is a summary of the main ideas from each section of the chapter.

1-1. What Is Science?

- Science is a process of inquiry and discovery leading to a better understanding of the natural world and how it works.
- The scientific method is a multiple step process that enables scientists to understand and discover the natural workings of the world. Observations are made using the five senses, historical incidents or accidents, or prior scientific knowledge and experimentation. A hypothesis is a testable explanation that guides scientists to a better understanding of an observed phenomena. Experimentation is the most important and critical step, which enables the scientist to test his or her hypothesis. A good experiment must be carefully and intricately designed to include a treatment, experimental group, control group, and variables. Two different types of experimental designs are a blind experimental design (experimental group does not know which treatment is being applied) and double-blind experimental design (neither the experimental group nor the experimenter know which treatment is being applied). Placebos are inert substances given during an experiment that typically do not produce an effect, although some studies have proven otherwise. Controls allow scientists to isolate specific variables during the course of an experiment. Two different types of controls are the independent variable (variable available at the beginning of the experiment and controlled by the scientist) and the dependent variable (variable measured by the scientist in response to the independent variable). Data analysis is performed, and conclusions are reached once the experiment has concluded. Data are the results of the experiment, and upon their analysis by means of statistics, a conclusion is reached as to whether the data supports or does not support the proposed hypothesis. A scientist will continue to revise and build on an idea regardless of whether the data support or do not support his or her hypothesis. Scientists will typically display their data in graphs and tables and present this information in a peer-reviewed journal or publication. A theory is a hypothesis well supported by many years

of experimentation and data, while a law is a theory supported by years of experimental data that is accepted by most scientists.

- Science is currently facing many different types of issues, including unconventional data analysis and reporting (reporting statistically insignificant data, p-hacking, and reporting unexpected findings), pseudoscience, and anecdotal observations.

1-2. Biology and "What Is Life?"

- Biology is the scientific study of life.
- All living organisms share five common characteristics including organization, energy, homeostasis, reproduction and development, and environmental response. Organization is hierarchical consisting of thirteen different levels demonstrating distinct emergent properties. These hierarchical levels include an atom (defining structure of elements, composed of protons, neutrons, and electrons); molecule (two or more atoms bonded together); macromolecule (two or more molecules bonded together); organelle (large group of macromolecules); cell (the smallest unit of life, consisting of organelles surrounded by cytoplasm enclosed within a plasma membrane); tissue (two or more cells combined together sharing a common function); organ (two or more tissues combined together sharing a common, specific function); organ system (combination of related organs with a specific function); organism (individual entity composed of organ systems that constitute life); population (two or more organisms of the same species interacting with one another; community (two or more species interacting with one another); ecosystem (interaction of the living with the nonliving environment); and biosphere (all regions of Earth inhabited by organisms, which can sustain life). Energy is the ability to do work and fuels chemical reactions within a cell, while metabolism is the sum of all chemical reactions occurring in the cell. Homeostasis is the ability of an organism to maintain a constant, stable internal environment. Reproduction is the process by which an organism produces an organism similar to itself; growth is an increase in body size and mass; development is the formation of anatomical and physiological characteristics that enable the organism to reproduce. Environmental response is the ability of an organism to respond to a stimulus within its environment (irritability), leading to changes that improve its overall success and survival (adaptation).

1-3. History of Classifying Life

- A hierarchical classification system of life began with Aristotle, but the modern-day classification system dates back to Carolus Linnaeus. Plant and animal species (organisms who produce viable offspring) were classified into six categories by Linnaeus: kingdom, phylum, class, order, genus, and species. The binomial nomenclature is a unique two-part naming system developed by Linnaeus and enables scientists to communicate in a universally recognized and accepted language. A three-kingdom system was developed in 1866 by Ernst Haeckel, in which a new kingdom, Protista, was proposed to include unicellular organisms.

A four-kingdom system was proposed in 1938 by Herbert F. Copeland that split unicellular cells between kingdom Protista (cells consisting of a nucleus) and kingdom Monera (bacterial cells lacking a nucleus). In 1956, Robert Whittaker categorized organisms into three kingdoms based on their ability to obtain food: producers (organisms that make their own food), consumers (organisms that consume other organisms). and decomposers (organisms that consume dead or dying organisms). Whittaker changed his approach in 1959, proposing a four-kingdom system that included Protista (all unicellular organisms with or without a nucleus), Fungi, Plants, and Animals. In 1969, Whitaker once again modified the classification system by proposing a five-kingdom system, in which bacteria were placed in kingdom Monera. The other kingdoms were Protista, Fungi, Plants, and Animals.

1-4. Current Classification System

- In 1977, the current classification system was proposed based on work performed by Carl Woese and George Fox. The work performed by Woese and Fox demonstrated different cell types based on the analysis of ribosomal RNA (rRNA).
- The newly proposed classification system consisted of three domains and six kingdoms, based on the differences found in the ribosomal RNA (rRNA). Bacteria (domain and kingdom), consisting of unicellular prokaryotic cells, with distinct cell walls composed of peptidoglycan, found almost anywhere. Archaea (domain and kingdom), consisting of unicellular prokaryotic cells, with distinct cell walls lacking peptidoglycan, found in harsh environmental habitats. Eukarya (domain), consisting of unicellular and multicellular eukaryotic cells comprising four kingdoms, including Protists (unicellular organisms), Fungi, Plants, and Animals.

End-of-Chapter Activities and Questions

Directions: Please refer back to what you learned in this chapter to complete the following activities.

Define Each Term in Your Own Words

1. Science
2. Biology
3. Life
4. Binomial Nomenclature
5. Archaea

Chapter Review

1. What is the difference between science and biology?
2. Describe the four elements found in an experimental design.
3. What are the current issues facing science today?
4. Identify the scientists who were responsible for developing the classification system and describe the contributions each scientist made.
5. Describe the work performed by Carl Woese and George Fox using the steps of the scientific method.

Multiple Choice

1. Which of the following occurs in the correct order from simplest to most complex?
   - **a.** population, organism, community, biosphere
   - **b.** macromolecule, cell, organism, ecosystem
   - **c.** atom, macromolecule, molecule, population
   - **d.** molecule, organ system, tissue, organism

2. What is the best example of irritability?
   - **a.** rabbit escaping a pursuing fox
   - **b.** breakdown of starch into glucose
   - **c.** bacterial cells dividing to produce offspring
   - **d.** maple tree using sunlight to make food

3. All of the following are examples of common biological theories, except __________.
   - **a.** cell
   - **b.** thermodynamics
   - **c.** evolution
   - **d.** gene

4. Which of the following scientists is not correctly paired to the proper number of kingdoms in his proposed classification system?
   - **a.** Linnaeus—two kingdoms
   - **b.** Copeland—four kingdoms
   - **c.** Haeckel—five kingdoms
   - **d.** Woese—six kingdoms

5. What macromolecule is the basis for the currently used three-domain, six-kingdom classification system?
   - **a.** messenger RNA
   - **b.** mitochondrial RNA
   - **c.** transfer RNA
   - **d.** ribosomal RNA

Image Credits

Fig. 1.1: Copyright © by Alexander Roslin (CC BY-SA 3.0) at https://commons.wikimedia.org/wiki/File:Carolus_Linnaeus_(cleaned_up_version).jpg.

Fig. 1.2: Source: https://commons.wikimedia.org/wiki/File:The_Earth_seen_from_Apollo_17.jpg.

Fig. 1.3: Source: https://commons.wikimedia.org/wiki/File:-Justus_Sustermans_-_Portrait_of_Galileo_Galilei,_1636.jpg.

Fig. 1.5: Source: https://commons.wikimedia.org/wiki/File:Charles_Darwin_seated_crop.jpg.

Fig. 1.6a: Copyright © by www.scientificanimations.com (CC BY-SA 4.0) at https://commons.wikimedia.org/wiki/File:Staphylococcus_Bacteria.jpg.

Fig. 1.6b: Copyright © by Hämbörger (CC BY-SA 3.0) at https://commons.wikimedia.org/wiki/File:Paramecium_contractile_vacuoles.jpg.

Fig. 1.6c: Copyright © by Marymaid1 (CC BY-SA 3.0) at https://commons.wikimedia.org/wiki/File:Toadstool_Mushroom.JPG.

Fig. 1.6d: Source: https://commons.wikimedia.org/wiki/File:Hyacinth_Jacynthe1.JPG.

Fig 1.6e: Copyright © by BLM Nevada (CC BY 2.0) at https://commons.wikimedia.org/wiki/File:Tiger_Swallowtail_Butterfly_(19984085821).jpg.

Fig. 1.7: Copyright © by Mikala14 (CC BY-SA 3.0) at https://commons.wikimedia.org/wiki/File:Levels_of_Organization.svg.

Fig. 1.8: Copyright © by OpenStax (CC BY 4.0) at https://commons.wikimedia.org/wiki/File:Figure_06_01_01.jpg.

Fig. 1.9: Copyright © by Varadbansod (CC BY-SA 4.0) at https://commons.wikimedia.org/wiki/File:Feeding_Fawn.jpg.

Fig. 1.10: Source: https://commons.wikimedia.org/wiki/File:F-MIB_47321_Finches_from_Galapagos_Archipelago.jpeg.

Fig. 1.12: from "New Concepts of Kingdoms of Organisms," Science, vol. 163, no. 3863, p. 151. Copyright © 1969 by American Association for the Advancement of Science.

Fig. 1.13: Source: https://commons.wikimedia.org/wiki/File:Haeckel_arbol_bn.png.

Fig. 1.14: Source: https://www.toppr.com/ask/content/concept/basic-of-two-three-four-and-five-kingdom-classification-218669/.

Fig. 1.15: Source: https://bioisnotdifficult.blogspot.com/2016/07/kingdomclassification.html.

Fig. 1.16: Copyright © by Maulucioni; adapted by TilmannR (CC BY-SA 3.0) at https://commons.wikimedia.org/wiki/File:-PhylogeneticTree,_Woese_1990.svg.

CHAPTER

2

Chemistry

PROFILES IN SCIENCE

Dmitri Mendeleev, often referred to as the father of the periodic table, was born within the Russian province of Siberia on February 8, 1834, and died in Saint Petersburg, Russia, in 1907. His father was a teacher and graduate of Main Pedagogical Institute in Saint Petersburg. However, after Mendeleev's father's ill health and blindness forced his retirement in 1834, his mother took over her brother's former glassworks factory in order to support the family. Unfortunately, the factory burned down in 1848, forcing the Mendeleev family to travel and relocate to Saint Petersburg. In 1850, Mendeleev entered school at his father's alma mater, Main Pedagogical Institute, and graduated in 1855, having published his dissertation on isomorphism and other relationships between physical form and chemical composition, along with other postgraduate publications. After graduation, Mendeleev suffered from tuberculosis; however, after a short teaching tenure on the Crimean Peninsula and improvement in his health, he returned to Saint Petersburg. Upon his return, he received his master's degree in chemistry from the University of Saint Petersburg in 1856 for his dissertation on the relationships between the specific volumes of substances and their crystallographic and chemical properties. After studying abroad in Germany, Mendeleev resumed teaching at the University of Saint Petersburg in 1861, and by 1865, he had received his doctorate, and he became professor of general chemistry at the university in 1867. In 1869, while beginning

FIGURE 2.1 Dmitri Mendeleev.

work on the second volume of his textbook, he began examining the elements and, using an idea presented on atomic weights during an 1860 conference he attended, was convinced that an element's property was dependent upon this characteristic. He presented his first sketch of a periodic table to the Russian Chemistry Society (which he helped found) on March 6, 1869, and by 1871, a revised diagram was published that looks similar to the modern periodic table. Mendeleev was awarded the Davy Medal in 1882 and Copley Medal in 1905 by London's Royal Society and was nominated for the Nobel Prize in 1906. Although Mendeleev never won the Nobel Prize, he joined an exclusive club in 1955 when physicists at the University of California–Berkeley produced traces of element 101 from element 99 (einsteinium), which they officially named mendelevium.

Introduction to the Chapter

In 1977, Theodore L. Brown and H. Eugene LeMay stated in the first edition of their textbook that "chemistry ... is the *central science*" because it plays a central role in connecting a variety of diverse disciples. So what is the connection between chemistry and biology? Chemistry helps biologists better understand the basic chemical properties and materials that make the world work as it does. In addition, chemistry helps explain naturally occurring systems, such as a living organism. Recall that the organization of life begins on the simplest and most basic level—the atom. Atoms are the foundation of simple and complex molecules, both of which are important in participating in a variety of chemical reactions responsible for sustaining and supporting a living organism. Chemistry helps biologists answer important biological questions such as how a living organism is organized, how it interacts with its environment, and how it utilizes important biological molecules. In this chapter, we will outline the basic concepts of chemistry, including a description of matter and its composition, elemental organization, formation of molecules by means of chemical bonding, and the chemistry of water and its importance in supporting life.

Chapter Objectives

In this chapter, students will learn the following:

2-1. Matter is made of pure chemical substances known as elements that are composed of small particles called atoms.

2-2. Elements are arranged based on their atomic numbers and similar properties in a periodic table.

2-3. Chemical bonding between atoms forms molecules.

2-4. Water molecules are held together by hydrogen bonds, giving water unique properties important in sustaining life.

2-5. The properties of acids and bases are determined by how a substance dissolves in water.

2-1. Chemical Basis of the Living and Nonliving

If you look around, you will see chemistry everywhere—from the rocks on mountaintops to the organisms living in the ocean. Chemistry is an important component of both the living and the nonliving. These things are composed of **matter**, defined as anything that has mass and takes up space. Although matter can exist in many different forms, it exists in three defined states: solid, liquid, and gas.

Matter is composed of **elements**, pure chemical substances that cannot be chemically broken down into a simpler substance and are systematically organized into a periodic table. Elements are divided into two groups based on their requirement for life. Elements required in large amounts are called *macroelements* and make up about 20 to 25 percent of the elements essential for life. In the human body alone, there are 25 elements essential for life (Figure 2.2), with the most common of these making up over 96 percent, including oxygen (O), carbon (C), hydrogen (H), and nitrogen (N). Those elements required in smaller or trace amounts are called *microelements* and, in the human body, make up the remaining 4 percent. Examples of these elements include iron (Fe), potassium (K), sodium (Na), and calcium (Ca).

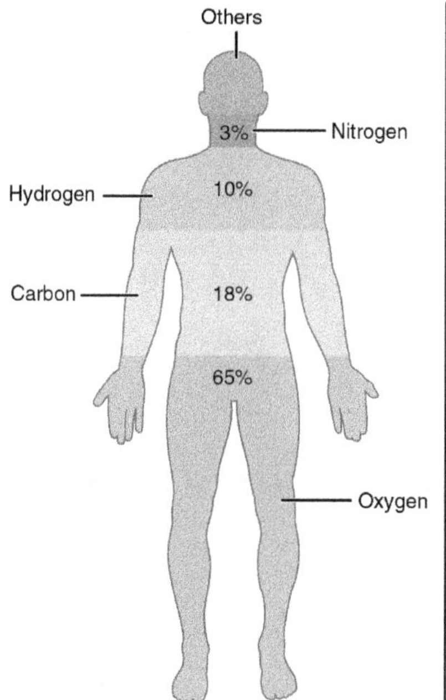

Element	Symbol	Percentage in Body
Oxygen	O	65.0
Carbon	C	18.5
Hydrogen	H	9.5
Nitrogen	N	3.2
Calcium	Ca	1.5
Phosphorus	P	1.0
Potassium	K	0.4
Sulfur	S	0.3
Sodium	Na	0.2
Chlorine	Cl	0.2
Magnesium	Mg	0.1
Trace elements include boron (B), chromium (Cr), cobalt (Co), copper (Cu), fluorine (F), iodine (I), iron (Fe), manganese (Mn), molybdenum (Mo), selenium (Se), silicon (Si), tin (Sn), vanadium (V), and zinc (Zn).		less than 1.0

FIGURE 2.2 About 25 elements are essential for life in the human body, including 11 different elements making up about 99%. Four elements of these 11 are most abundant—oxygen, carbon, hydrogen, and nitrogen—and make up about 96%.

Elements are composed of particles called **atoms**, tiny units that retain an element's properties. An atom exists in a uniform structure consisting of three types of subatomic particles (Figure 2.3). Within the center of an atom is an atomic nucleus composed of protons and neutrons. *Protons* are positively charged particles, and *neutrons* are neutrally charged. Surrounding this central core of subatomic particles is a diffuse region or "cloud" of negatively charged particles called *electrons,* which spin in orbitals around the atomic nucleus (Figure 2.3).

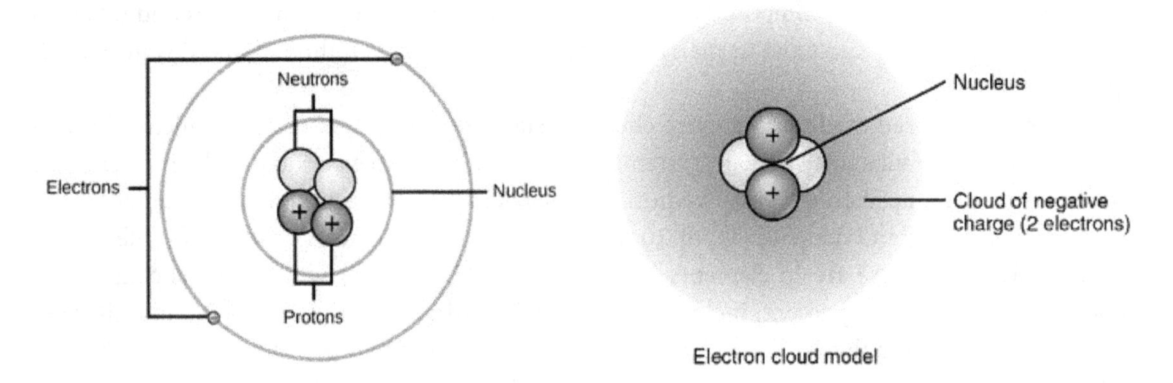

FIGURE 2.3 In the center of an atom is a nucleus, which consists of positively charged protons and neutrally charged neutrons, surrounded by negatively charged electrons orbiting around the nucleus.

Atoms of elements are arranged within the periodic table according to the number of protons that exist within the atomic nucleus. The total number of protons in an atom is called its **atomic number**. In addition, the **atomic mass** of an atom defines the overall mass of the atom, calculated as the sum of the number of protons and neutrons in the nucleus. The atomic number and atomic mass of an element are clearly defined using the element's chemical symbol and name. To the left of the symbol, the atomic number is typically written as a superscript and the atomic mass is indicated below the elemental symbol. (Figure 2.4).

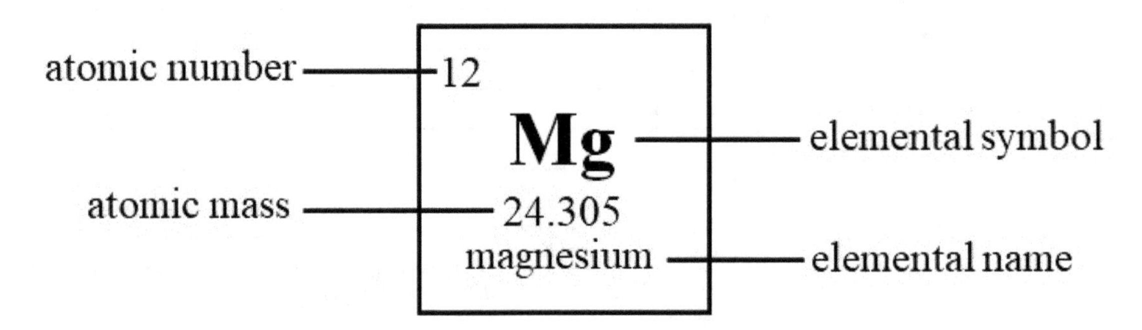

FIGURE 2.4 Magnesium is represented using the chemical symbol Mg. Magnesium has twelve protons, indicated by the atomic number and an atomic mass of 24.305, which represents the total of protons and neutrons.

In 1813, Jacob Berzelius, a Swedish chemist, developed the system of designating the names of the elements using symbols created by capitalizing the first letter of the element's name. For example, B for boron, O for oxygen, and C for carbon. However, if there was more than one element whose name started with the same letter, Berzelius assigned a second, lowercase letter to the symbol, generally the second letter in the element's name. Examples include Ba for barium, Os for osmium, and Ca for calcium.

Of the 26 letters in the alphabet, all are used in the periodic table except the letter J. The letter Q is not used in any official elemental names or elemental symbols; however, it is used within the temporary name and symbol of the element ununquadium (element 114). Some predict that the letter Q will be used in the temporary names and symbols of elements discovered past element 118 (ununoctium).

Isotopes are atoms of the same element consisting of the same number of protons but varying numbers of neutrons. Naturally occurring elements generally consist of a mixture of isotopes, as is evident in the element carbon (Figure 2.5). Carbon (C) consists of three different isotopes: Carbon-12 (^{12}C) = six protons and six neutrons (12 atomic mass), Carbon-13 (^{13}C) = six protons and seven neutrons (13 atomic mass), and Carbon-14 (^{14}C) = six protons and eight neutrons (14 atomic mass). Each of these isotopes has a different atomic mass; however, the number of protons remains the same. Carbon is primarily composed of Carbon12; however, about 1 percent of carbon consists of Carbon-13 and Carbon-14. It is interesting to note that some isotopes are radioactive, in that when they break apart to form different atoms, particles are emitted, and energy is released. Radioactive isotopes are an important biological tool used today for different reasons, including fossil dating and as a diagnostic tool in medicine.

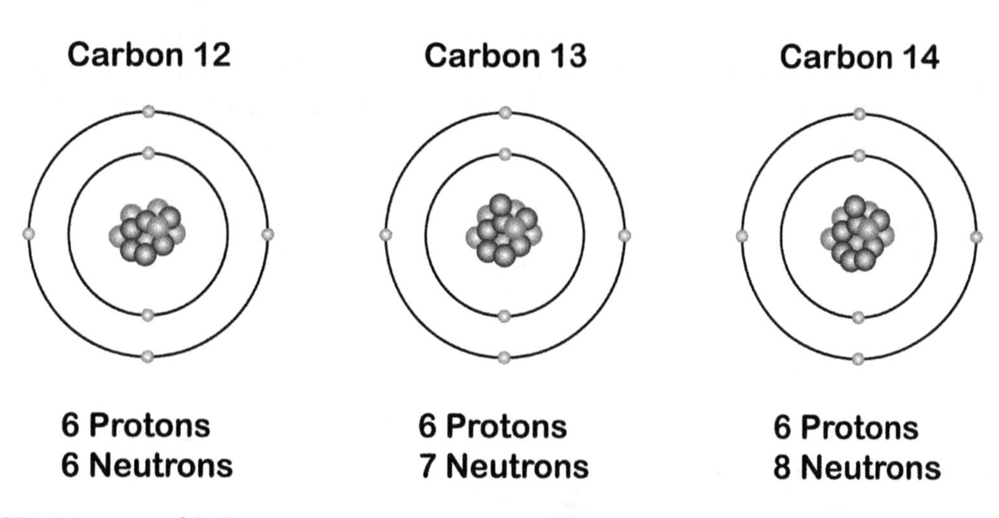

FIGURE 2.5 Isotopes of Carbon.

Oftentimes, an atom contains the same number of protons as electrons, resulting in a neutral atom. However, some atoms will either gain or lose electrons, upsetting the balance between these subatomic particles and making the atom charged. The resulting charged atom is called an **ion**. Any atom that gains one or more electrons is called an *anion* and has a negative charge, while an atom that loses one or more electrons is called a *cation* and has a positive charge.

2-2. The Periodic Table

The **periodic table** is an important tool for chemists, as it arranges the elements by their atomic number and similar chemical and physical properties. This particular order of columns and rows

ОПЫТЪ СИСТЕМЫ ЭЛЕМЕНТОВЪ.

ОСНОВАННОЙ НА ИХЪ АТОМНОМЪ ВѢСѢ И ХИМИЧЕСКОМЪ СХОДСТВѢ.

			Ti = 50	Zr = 90	? = 180.
			V = 51	Nb = 94	Ta = 182.
			Cr = 52	Mo = 96	W = 186.
			Mn = 55	Rh = 104,4	Pt = 197,1.
			Fe = 56	Rn = 104,4	Ir = 198.
			Ni = Co = 59	Pl = 106,6	O- = 199.
H = 1			Cu = 63,4	Ag = 108	Hg = 200.
	Be = 9,4	Mg = 24	Zn = 65,2	Cd = 112	
	B = 11	Al = 27,1	? = 68	Ur = 116	Au = 197?
	C = 12	Si = 28	? = 70	Sn = 118	
	N = 14	P = 31	As = 75	Sb = 122	Bi = 210?
	O = 16	S = 32	Se = 79,4	Te = 128?	
	F = 19	Cl = 35,5	Br = 80	I = 127	
Li = 7	Na = 23	K = 39	Rb = 85,4	Cs = 133	Tl = 204.
		Ca = 40	Sr = 87,6	Ba = 137	Pb = 207.
		? = 45	Ce = 92		
		?Er = 56	La = 94		
		?Yt = 60	Di = 95		
		?In = 75,6	Th = 118?		

Д. Менделѣевъ

FIGURE 2.6 Mendeleev's short-form periodic table published in his 1869 work, *An Attempt at a System of Elements, Based on Their Atomic Weight and Chemical Affinity.*

in the periodic table was developed and published by Russian chemist Dmitri Mendeleev in 1869 (Figure 2.6).

Although Mendeleev was not the first scientist to arrange the elements in a meaningful chart, his system gained widespread notoriety within the scientific community after the discovery of important elements such as gallium, scandium, and germanium, proving his tool was effective. Interestingly, Mendeleev's original chart from 1869 and his expanded 1904 periodic table (Figure 2.7) contained gaps, which he left for undiscovered elements.

By 2016, the periodic table was completely full (Figure 2.8), with 118 elements, the most recent four elements—nihonium (Nh), moscovium (Mc), tenness-ine (Ts), and oganesson (Og)—being added that year. However, some predict that elements 119 and 120 will be discovered within the next five years. It has been estimated that one element has been added to the periodic table every two to three years. Of the 118 elements in the periodic table, 92 of these are naturally occurring

Series	Zero Group	Group I	Group II	Group III	Group IV	Group V	Group VI	Group VII	Group VIII
0	x								
1	y	Hydrogen H=1·008							
2	Helium He=4·0	Lithium Li=7·03	Beryllium Be=9·1	Boron B=11·0	Carbon C=12·0	Nitrogen N=14·04	Oxygen O=16·00	Fluorine F=19·0	
3	Neon Ne=19·9	Sodium Na=23·05	Magnesium Mg=24·1	Aluminium Al=27·0	Silicon Si=28·4	Phosphorus P=31·0	Sulphur S=32·06	Chlorine Cl=35·45	
4	Argon Ar=38	Potassium K=39·1	Calcium Ca=40·1	Scandium Sc=44·1	Titanium Ti=48·1	Vanadium V=51·4	Chromium Cr=52·1	Manganese Mn=55·0	Iron Fe=55·9 Cobalt Co=59 Nickel Ni=59 (Cu)
5		Copper Cu=63·6	Zinc Zn=65·4	Gallium Ga=70·0	Germanium Ge=72·3	Arsenic As=75·0	Selenium Se=79	Bromine Br=79·95	
6	Krypton Kr=81·8	Rubidium Rb=85·4	Strontium Sr=87·6	Yttrium Y=89·0	Zirconium Zr=90·6	Niobium Nb=94·0	Molybdenum Mo=96·0	—	Ruthenium Ru=101·7 Rhodium Rh=103·0 Palladium Pd=106·5 (Ag)
7		Silver Ag=107·9	Cadmium Cd=112·4	Indium In=114·0	Tin Sn=119·0	Antimony Sb=120·0	Tellurium Te=127	Iodine I=127	
8	Xenon Xe=128	Cæsium Cs=132·9	Barium Ba=137·4	Lanthanum La=139	Cerium Ce=140	—	—		— — — (—)
9		—	—		—	—	—	—	
10	—	—	—	Ytterbium Yb=173	—	Tantalum Ta=183	Tungsten W=184	—	Osmium Os=191 Iridium Ir=193 Platinum Pt=194·9 (Au)
11		Gold Au=197·2	Mercury Hg=200·0	Thallium Tl=204·1	Lead Pb=206·9	Bismuth Bi=208	—	—	
12	—	—	Radium Rd=224	—	Thorium Th=232	—	Uranium U=239		

FIGURE 2.7 Mendeleev's periodic table published in 1904, which is like the periodic table used today. Although many differences exist between Mendeleev's original periodic table and the current periodic table, the main difference seen in this publication is the location of the noble gases, incorporated as a left-hand column. There are also empty boxes, purposely left blank for newly discovered elements.

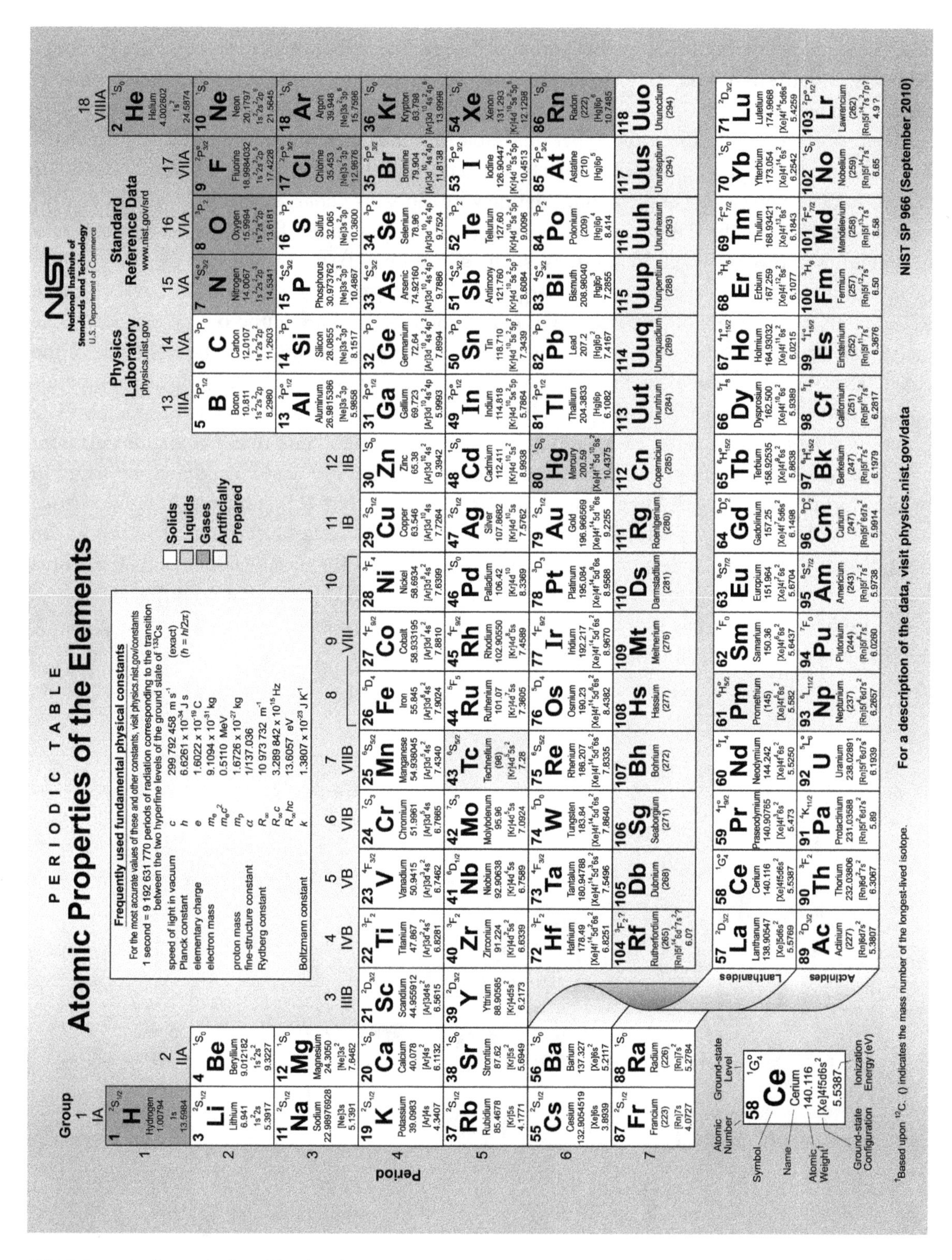

FIGURE 2.8 In the periodic table, the elements are arranged by their atomic number in groups and periods, with all elements within the same group having similar chemical properties.

and biologically important in the composition of matter. The remaining 26 elements are synthetic elements, with the first synthetic element—element 43, technetium (Tc)—being discovered via synthesis in a laboratory in 1937, while the first entirely synthetic element was curium, produced in 1944.

There are important aspects of the periodic table that should be noted when studying the chart. All of the elements are arranged in groups (vertical columns) and periods (horizontal rows). The groups indicate the charge of the atom, starting with a positive charge on the far left to negative charges on the far right. These charges are determined by whether the element is a metal or a nonmetal. Metals carry positive charges, such as lithium (Li), magnesium (Mg), and calcium (Ca), while nonmetals carry negative charges, such as oxygen (O), nitrogen (N), and chlorine (Cl). Metals make up about 75 percent of the periodic table and are found on the left side of the chart, and nonmetals make up the remaining 25 percent and are found on the right side of the chart. There is a group of elements called *metalloids* that divide the metals from the nonmetals and are called as such because they contain characteristics of both metals and nonmetals. The group numbers, 1–8, also indicate the number of **valence electrons** that a particular element contains. Valence electrons are found in the outermost orbital of the atom and are important in helping form specific bonds. For example, lithium (Li) is in group 1; therefore, it has one valence electron, while oxygen (O) is located in group 6 and has six valence electrons. There are seven periodic rows, and the corresponding row number indicates the number of orbitals the atoms have surrounding the atomic nucleus. For example, sodium (Na) is located in period 3, and therefore has three orbitals surrounding the nucleus, while francium (Fr) is found in period 7 and has seven orbitals around the nucleus.

FIGURE 2.9 Marie Curie in Her Lab.

Women's contributions to the periodic table—In 1898, Marie Curie, along with her husband, Pierre Curie, discovered two new elements, radium (element 88) and polonium (element 84). In 1911, Curie won her second Nobel Prize (the first Nobel Prize was a shared award with her husband, Pierre Curie, and Henri Becquerel in 1903 for discovering radioactivity) for her discoveries of the two elements and her continual isolation and study of radium. Curie was the first woman to win a Nobel Prize, the first woman to win the Nobel Prize twice, and the only person to win the Nobel Prize in two different scientific fields.

The number of orbitals a particular atom contains is defined by the period row the atom is found on within the periodic table. The orbitals are important because electrons are found within, spinning around the nucleus. For example, the sodium (Na) atom is located in period 3, which means that it has three orbitals surrounding the nucleus. Sodium (Na) has an atomic number of 11, and because this atom is neutral, the number 11 not only indicates the number of protons but the number of electrons as well. These 11 electrons are found within these three orbitals. The first orbital consists of a total of two electrons and can only hold this many. The second orbital contains eight electrons, and the remaining electron is found in the third orbital. When referring to the periodic table, sodium (Na) is located in group 1, which indicates the number of valence electrons this particular atom contains. The goal of every atom is to fill its outermost orbital. If an atom has only one orbital, such as hydrogen (H), the maximum number of electrons that will fill that orbital is two. However, for any atom with more than one orbital, like lithium (Li) or fluorine (F), the maximum number of electrons that will fill the outermost orbital is eight. When an atom reaches a maximum of eight electrons in its outermost energy level, the atom is said to have reached its **octet rule** (Figure 2.10). In other words, the maximum number of eight in the outermost orbital makes the atom stable. In the case of sodium (Na), when the atom loses that one additional electron in the outermost orbital, the sodium (Na) atom is said to be like neon (Ne), a noble gas.

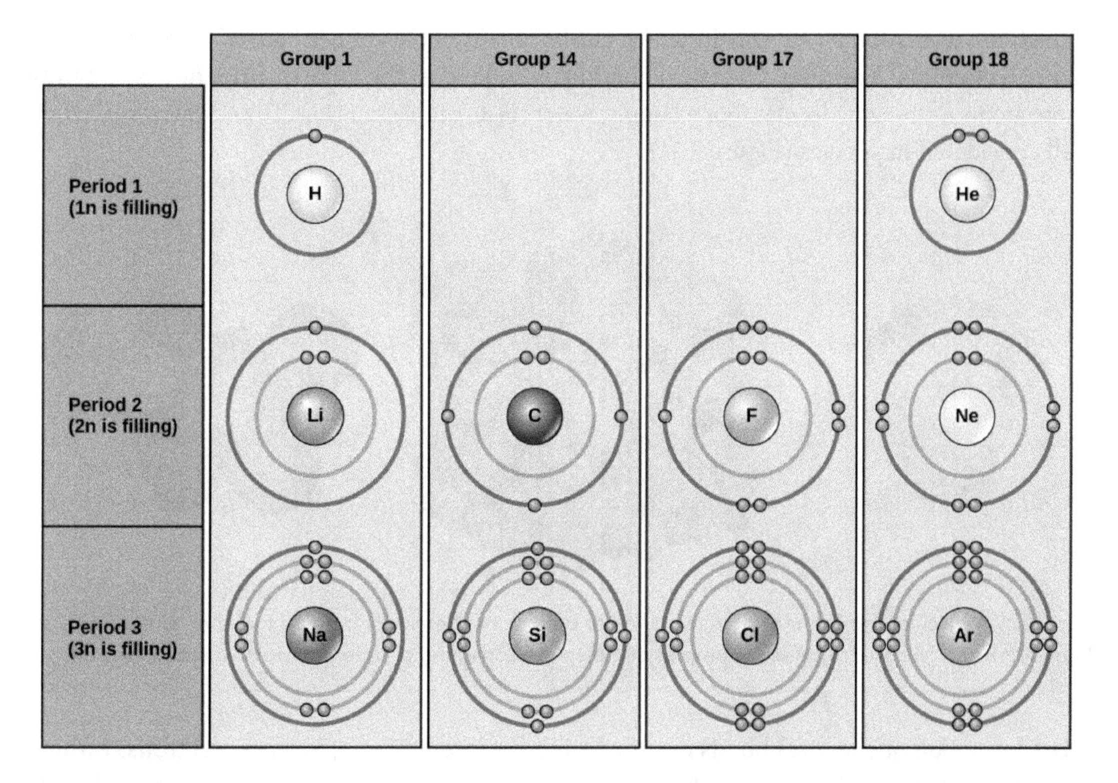

FIGURE 2.10 Electron configuration diagrams indicate the number of electrons found in the orbitals of an atom. The first orbital of every atom can only contain two electrons, while the outer orbitals can have up to eight electrons satisfying the octet rule, making the atom stable. Any partially filled outer orbital can gain or lose electrons to achieve the octet rule. The total number of orbitals an atom contains is represented by the period number found in the periodic table.

Noble gases are atoms that are extremely stable, because they have a maximum number of electrons in their outermost orbital, and they do not normally react with other atoms (often referred to as being *inert*). In order for an atom to reach the same electron configuration as that of the noble gas closest to it in the periodic table, the element will move either right or left by losing or gaining electrons during chemical reactions or ion formation. Atoms on the left side of the periodic table will move backward to the noble gas found in the previous row, while atoms on the right side of the table move to the next noble gas in the same row. Metals (found on the left side of the periodic table) will generally lose electrons to form positive ions in order to obtain noble gas configuration. Nonmetals (found on the right side of the periodic table) will typically gain electrons, forming negative ions, to obtain noble gas configuration. For example, the metal sodium (Na) will lose its one electron in its outermost orbital to become like the noble gas neon (Ne), and the nonmetal chlorine (Cl) will gain one electron to fill its outermost orbital to become like argon (Ar).

2-3. Chemical Bonding

Although ion formation is one way in which atoms obtain noble gas configuration, another means is through **chemical bonding**. Chemical bonding is defined as the force of attraction between two or more atoms forming molecules. For example, water (H_2O), table salt (NaCl), and glucose ($C_6H_{12}O_6$) are all considered molecules (Figure 2.11).

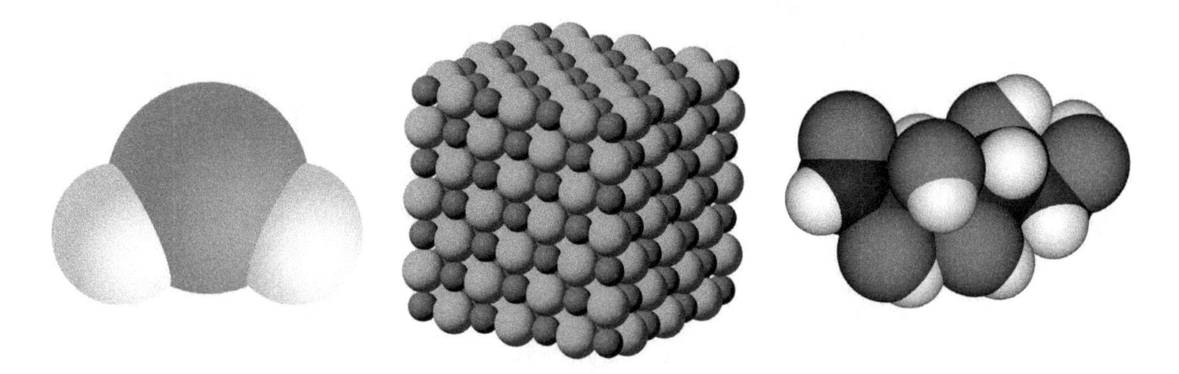

FIGURE 2.11 Molecules are formed when two or more atoms are chemically bonded together. Water (left), table salt (middle), and glucose (right) are examples of molecules consisting of different atoms bonded together.

The formation of chemical bonds occurs during the interaction of valence electrons. Although the number of valence electrons can be determined using the group number in the periodic table, the number available for bonding can only be learned using a **Lewis symbol**.

A Lewis symbol is a diagram created and introduced in 1916 by American chemist Gilbert N. Lewis in his journal article entitled *The Atom and the Molecule*. The concept involves using the elemental symbol of the atom and dots to correspond to the number of valence electrons found in the outermost orbital of the atom (Figure 2.12). The easiest approach to drawing out the Lewis symbol is to use the face of a clock to place the dots. One dot is placed at the 12 o'clock position, another placed at the three o'clock position, and so on. Dots are added to each side in turn until the number of dots placed in each position reaches the maximum total of valence electrons. If there are two dots in any position around the symbol, these electrons are not available for chemical bonding. However, any unpaired dot is available to form a chemical bond.

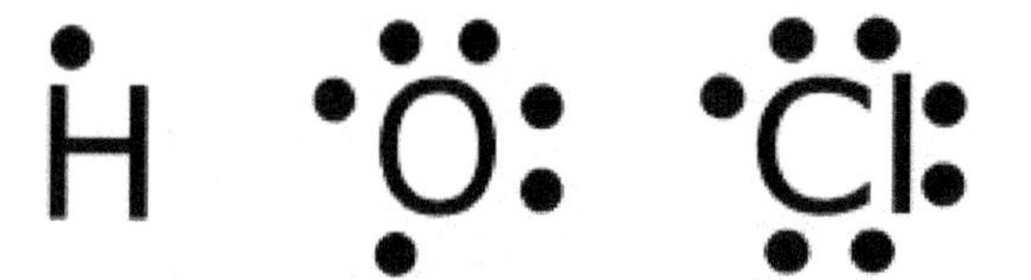

FIGURE 2.12 Lewis dot symbols were created to represent the total number of valence electrons that an atom has available for chemical bonding. Dots are used to represent the valence electrons and are placed around the chemical symbol in an order representative of a clock face. Any unpaired dots are the valence electrons available for chemical bonding. The total number of valence electrons is determined by atom's group number in the periodic table.

There are three main types of chemical bonds that exist in nature and are important to life.

Covalent bonds are bonds that form when valence electrons are shared between two or more atoms (Figure 2.13). Since these types of bonds are the strongest of all bonds that exist in nature, these bonds are important for holding together most of the important molecules in life. The valence electrons are shared between two or more nonmetal atoms, resulting in each atom obtaining noble gas configuration. Examples of covalent bonds occur between some of the simplest molecules, such as O_2, to more complex molecules like carbon dioxide (CO_2) (Figure 2.13).

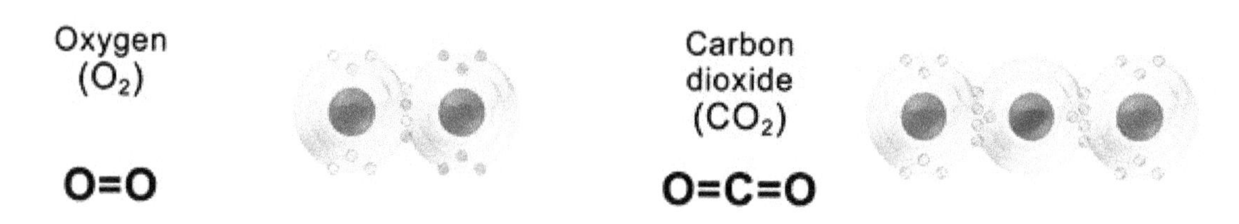

FIGURE 2.13 Covalent bonds are formed from the force of attraction between equally shared electrons.

When valence electrons are transferred from one atom to another, **ionic bonds** form between the atoms (Figure 2.14). These attractions occur because of the opposite charges that the atoms possess. These types of bonds occur between a metal and nonmetal, in which the metal atom loses an electron, and a nonmetal atom gains an electron. The most common example of an ionic bond occurs between the atoms of sodium (Na) and chlorine (Cl). According to the periodic table, sodium (Na) is located in group 1, which means that it has one valence electron, and chlorine (Cl) is located in group 7 and has seven valence electrons. When these two atoms interact, the one valence electron on the sodium (Na) atom is transferred to the chlorine (Cl) atom, thereby resulting in complete outermost orbitals and a noble gas configuration.

Hydrogen bonds are weak chemical bonds occurring between the slightly positive charge of a hydrogen atom on one molecule and the slightly negative charge of another atom, such as oxygen (O) or nitrogen (N), either on the same molecule or on a different molecule. The difference in charges between the atoms results from the unequal sharing of electrons in the covalent bond, or in other words, the negatively charged atoms, such as oxygen, has a stronger attraction to the

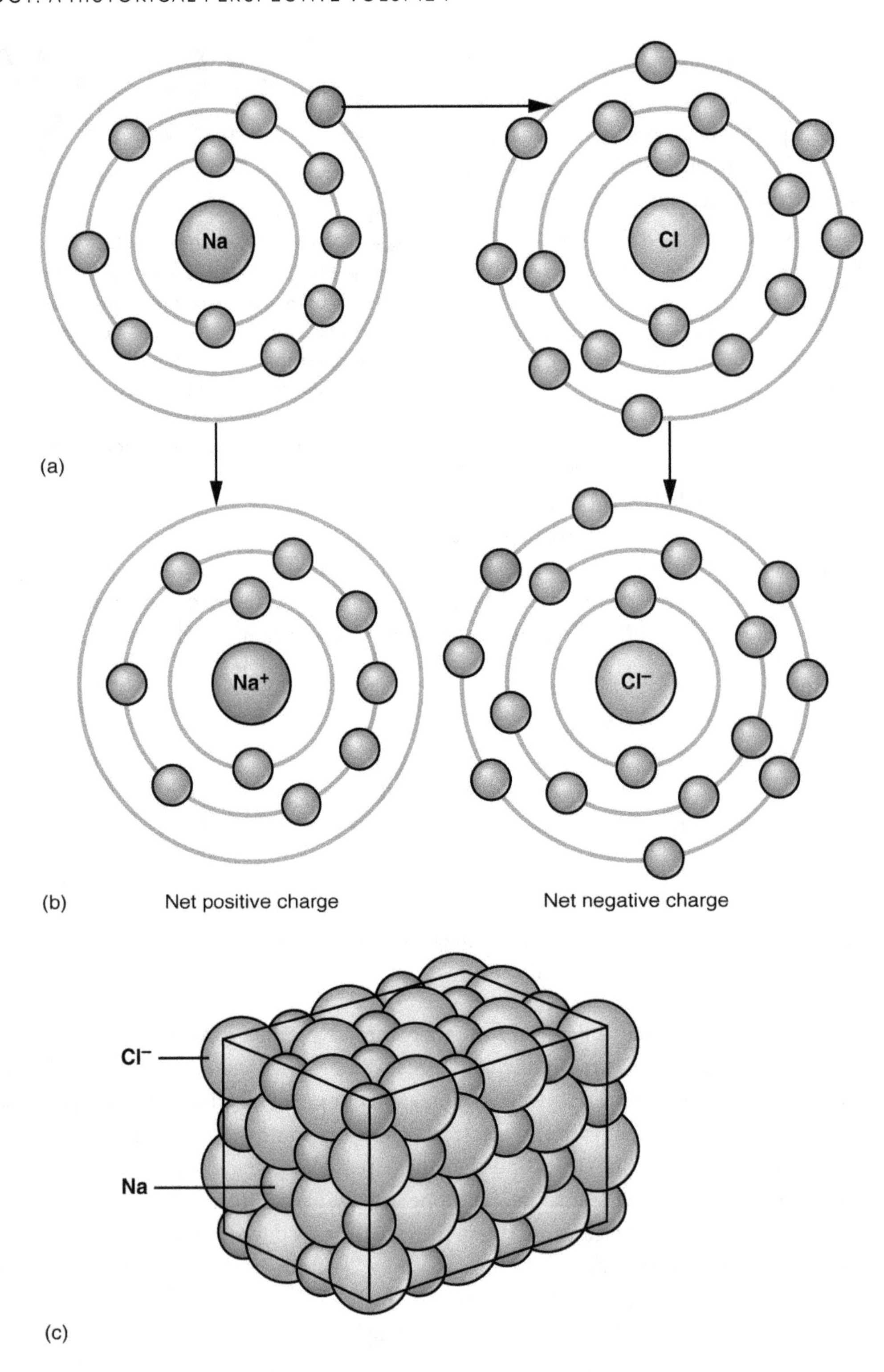

FIGURE 2.14 Ionic bonds are formed from the force of attraction between oppositely charged ions.

shared electrons in the molecule than the other atoms present. This phenomenon is known as *electronegativity*. Although hydrogen bonds are considered individually weak, multiple hydrogen bonds together are quite strong. Hydrogen bonds are found between complex biological molecules, such as DNA and proteins, and are responsible for helping maintain the structure and function of these important structures.

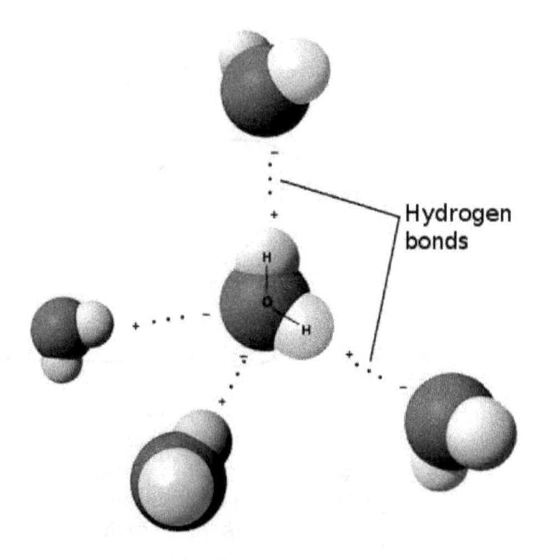

FIGURE 2.15 Hydrogen bonds are formed from the force of attraction between the slightly positive charge of a hydrogen atom on one molecule and the slightly negative charge of another atom.

2-4. Water and Its Importance to Life

Water could be considered the "mother" of all living organisms, as the first signs of primordial life in the form of living cells originated in water. Currently, all living organisms are composed of 70 to 95 percent water. Water is an interesting molecule with a V-shaped structure resulting from the unequal sharing of electrons between the oxygen and hydrogen atoms. Each water molecule is held together by hydrogen bonds, which are important in contributing to several properties that make it crucial to all life.

High Surface Tension

Hydrogen bonds exist between the oxygen atom of one water molecule and a hydrogen atom of another water molecule. Multiple water molecules held together by hydrogen bonds form a netlike property that is important in producing high surface tension. When pressure is applied to the surface of water, it is relatively strong and difficult to break. This is especially important for animals, such as predatory insects, who may need to walk quickly across the surface of water without falling through it (Figure 2.16).

FIGURE 2.16 The collective strength of hydrogen bonding between water molecules produces a net-like property that is strong and difficult to break. This enables some insects, including water striders to walk across the surface of water without falling through.

Cohesion and Adhesion

Cohesion of water is characterized by its ability to assist in helping water molecules stick together due to hydrogen bonds. This is an important feature in providing water with its high surface tension. **Adhesion** is another important feature of water and occurs when the hydrogen bonds of water molecules bond to other substances. Both of these features are especially important for plants (Figure 2.17), which rely on both to move water from their roots to their leaves. As roots absorb water from the soil, the water moves through water-conducting tubes toward the leaves. When water evaporates from the leaves, another water molecule replaces this lost molecule as a result of cohesion and the tension created by the hydrogen bonds. Adhesion attaches the water molecules to the cell walls within the conducting tubes, preventing the water from breaking apart as well as countering the effects of gravity.

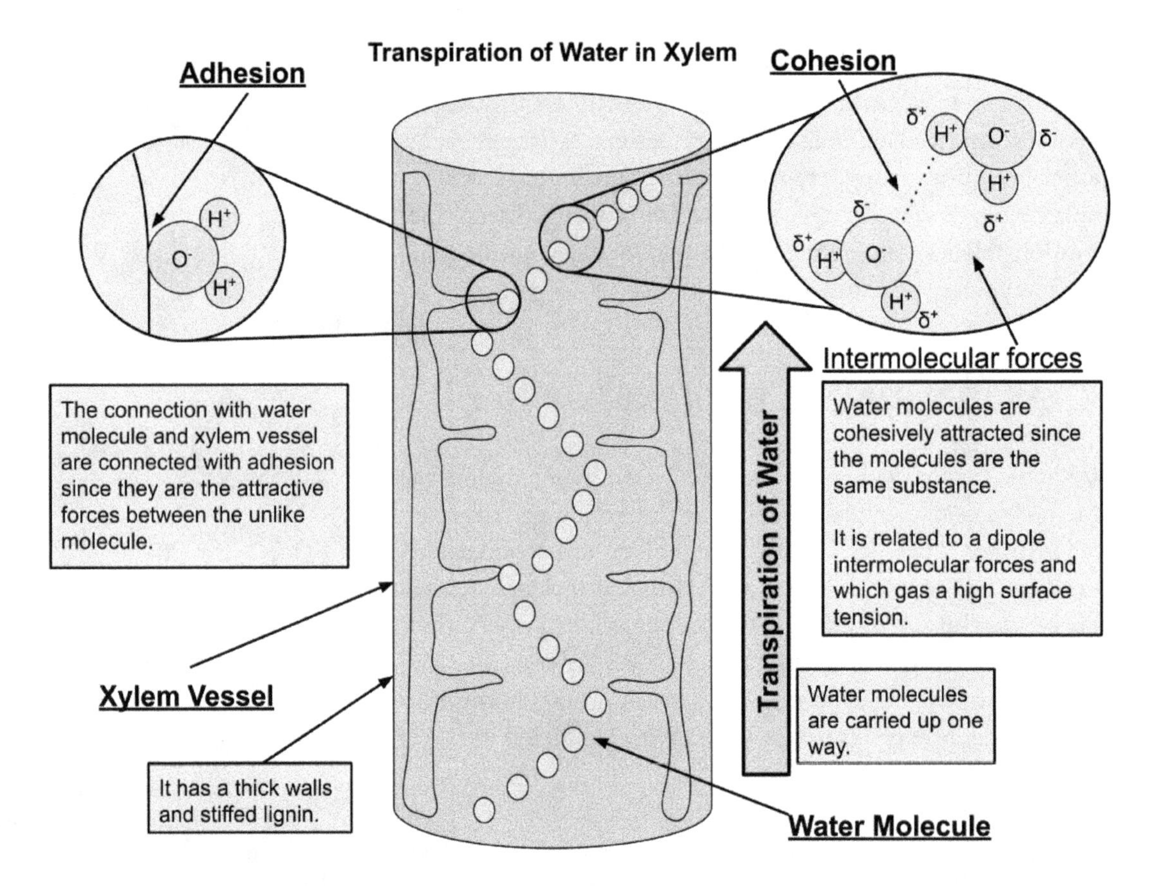

FIGURE 2.17 Cohesion and Adhesion of Water.

High Heat Capacity

Hydrogen bonds help keep water a liquid between 0°C and 100°C, which is important in helping maintain life. As water encounters high temperatures, hydrogen bonds are broken but are quickly reformed thereafter. As a result of the hydrogen bonds, the temperature of liquid water rises and falls more slowly than that of other types of liquids. This is important because it helps living organisms maintain their normal body temperature and be protected from quickly changing temperatures.

Decreased Density When Frozen

Water freezes at 0°C, and upon doing so, water molecules are arranged into a four-partner, hydrogen bonded, crystalline lattice, which distributes the water molecules farther apart (Figure 2.18a). This distribution results in ice having less density than that of liquid water, whose water molecules are moving more freely and are located much closer together. These qualities are an important feature of lakes and ponds, because as water freezes, the frozen water floats on the surface of the water, acting as an insulator to protect the organisms below during winter months (Figure 2.18b).

FIGURE 2.18 When water freezes, hydrogen bonds arrange water molecules farther apart, making ice less dense than liquid water (a), enabling ice to float on top of bodies of water (b) that act as an insulator for aquatic organisms.

Polar Solvent

Water is a **polar** molecule, consisting of a slightly positive hydrogen atom and slightly negative oxygen atom. As a result, water acts as an excellent solvent to most substances like itself. For example, table salt (NaCl) is an ionic molecule and considered a polar substance due to its positively charged sodium atom (Na^+) and its negatively charged chlorine atom (Cl^-). When placed in water, NaCl quickly dissolves, because the hydrogen (H) atom of water is attracted to the negatively charged chlorine (Cl) and the negatively charged oxygen of water is attracted to the positively charged sodium (Na^+), pulling the molecule apart (Figure 2.19). Nonpolar substances, such as oil, lack a charge and therefore are unable to dissolve in or be pulled apart by a water molecule. This is the reason oil separates from water and floats on top when both are placed in a container together.

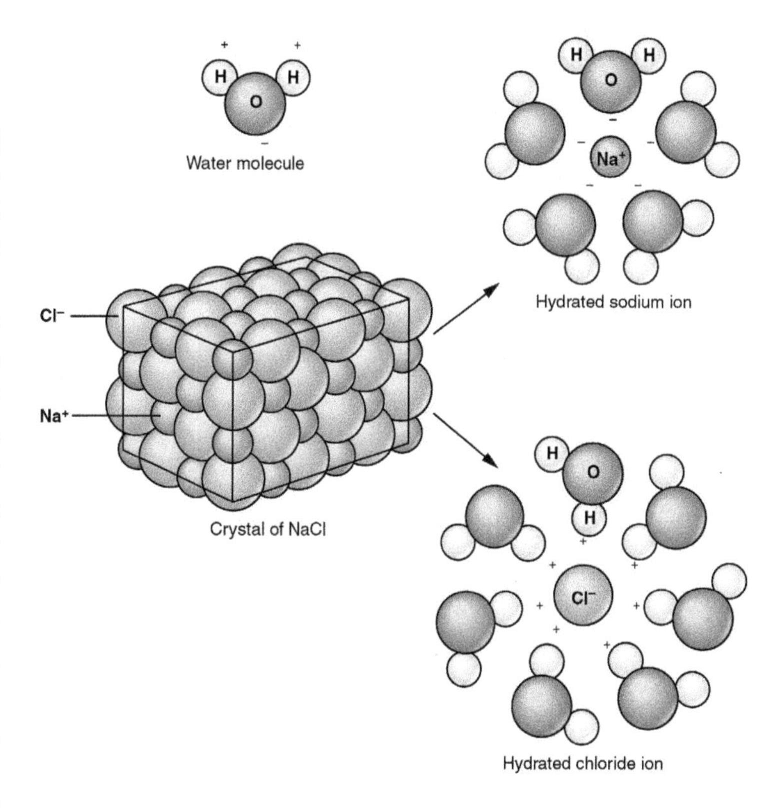

FIGURE 2.19 Due to its polar nature, water acts as an excellent solvent. When table salt (NaCl) is dissolved in water, Na^+ and Cl^- ions are produced and are completely surrounded by the oppositely charged atoms of the water molecules.

2-5. Acids and Bases

When pure water undergoes a process known as **ionization**, defined as a neutral molecule being converted to electrically charged ions, an equal amount of hydrogen protons (H^+) and hydroxide ions (OH^-) are produced, as seen below.

$$H_2O \leftrightarrow H^+ + OH^-$$

However, when certain substances are added to water, there occurs an imbalance in the concentration of hydrogen protons (H^+) and hydroxide ions (OH^-) produced. If a substance dissolved in water donates additional hydrogen protons (H^+) to the solution, this substance is an **acid** (Figure 2.20a). An example of an acid is hydrochloric acid (HCl), and when it is added to pure water, the amount of hydrogen protons (H^+) greatly increases, producing an acidic solution. Other examples of acids include lemon juice and vinegar. When a substance is dissolved in water and either releases hydroxide ions (OH^-) or takes up hydrogen protons (H^+), this substance is considered a **base** (Figure 2.20b). An example of a base that releases hydroxide ions (OH^-) into pure water is NaOH, which will completely dissolve, increasing the number of hydroxide ions in the water. An example of a base that takes up hydrogen protons (H^+) in pure water is ammonia, producing an ammonium ion (NH_4^+). In either case, the concentration of hydroxide ions (OH^-) is greater, producing a basic solution.

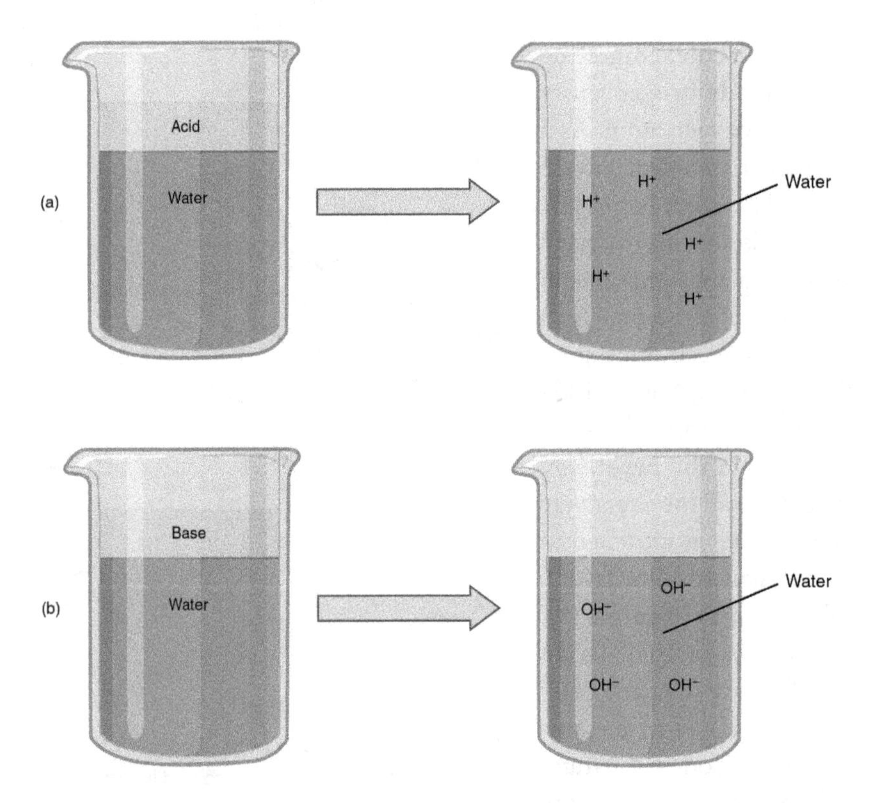

FIGURE 2.20 (a) Substances that dissolve in water and produce hydrogen protons (H^+) are called acids. (b) Bases, when dissolved in water, will typically release hydroxide ions (OH^-) ions; however, will also take up excess hydrogen protons (H^+).

In order to determine how acidic or basic a particular substance may be, a scientist will use a **pH scale**. The pH scale is based on the concentration of hydrogen protons (H^+) within a solution and ranges from a scale of 0 to 14. Pure water represents a neutral substance, with a pH of 7. Any substance that has a pH below 7 is considered an acid, while those substances above 7 are considered a base. Most fluids found in living organisms, including blood, will have a pH range of 6 to 8. However, there are some body fluids, like stomach juices, that are considerably acidic, with a pH of 2. Each unit on the pH scale represents a tenfold change in acidity. For example, a soda has a pH of 3, while coffee has a pH of 5; therefore, the soda is 10 x 10, or 100, times more acidic than coffee.

Buffers are chemicals responsible for maintaining pH at a steady level and within the normal range for living organisms. Buffers work to stabilize the pH of a solution by taking up additional hydrogen protons (H^+) or hydroxide ions (OH^-). A common example of work performed by a buffer in the body is a series of reactions responsible for maintaining the pH of the blood (Figure 2.22). Within the blood is a combination of carbonic acid (H_2CO_3) and bicarbonate ions (HCO_3^-). When an excess amount of hydrogen ions become present in the blood, bicarbonate ions (HCO_3^-) work to take up these excess hydrogen ions, producing carbonic acid, which is then converted to water and carbon dioxide (CO_2).

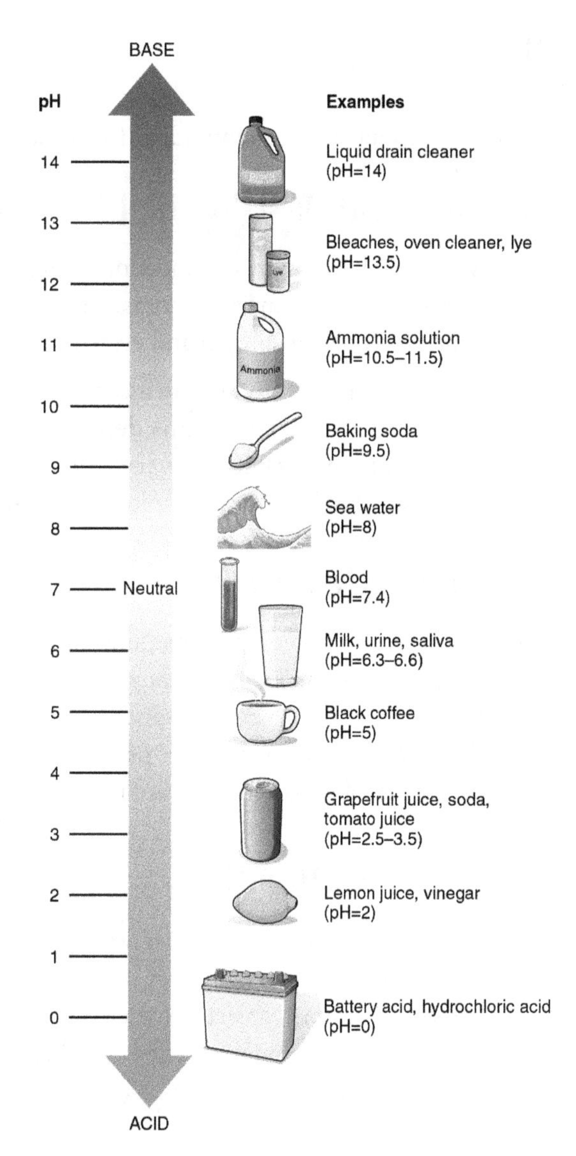

FIGURE 2.21 The pH scale represents the concentration of hydrogen protons (H^+) in a solution. The scale ranges from 0 to 14, with 0 being the most acidic and 14 being the most basic. Any substances with a pH of 7 is considered neutral.

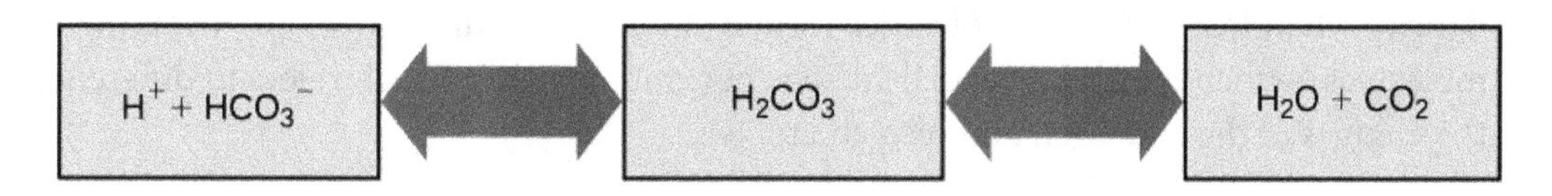

FIGURE 2.22 The human body system has an important buffering system that helps maintain the pH of the blood. The carbonic acid buffer system uses carbonic acid (H_2CO_3), the bicarbonate ion (HCO_3^-), and carbon dioxide (CO_2) to keep blood pH levels around 7.5 when hydrogen or hydroxide ions increase or decrease the pH of the blood.

Chapter Summary

Below is a summary of the main ideas from each section of the chapter.

2-1. Chemical Basis of the Living and Nonliving

- Matter is the primary component of all living and nonliving things and is composed of elements. Elements consist of atoms, which possess a nucleus with protons (positive charge) and neutrons (neutral charge) and electrons (negative charge) surrounding the nucleus.
- The number of protons in an atom defines its atomic number, while the number of protons plus the number of neutrons defines its atomic mass. Isotopes are atoms with the same number of protons but with varying numbers of neutrons. Ions, known as charged particles, are atoms that have gained or lost electrons.

2-2. The Periodic Table

- Atoms are arranged by their atomic number and chemical and physical properties in a periodic table. The periodic table arranges elements in a variety of ways, including groups (indicate the number of valence electrons); periods (indicate the number of electron orbitals); metals (located on the left side of the periodic table); and nonmetals (located on the right side of the periodic table). Noble gases are nonmetals that have satisfied the octet rule.

2-3. Chemical Bonding

- Chemical bonds form between valence electrons to make molecules. Lewis symbols are used to determine the number of valence electrons available to form chemical bonds. Lewis symbol diagrams consist of the element symbols and the arrangement of dots (valence electrons) around the symbol. There are three types of chemical bonds, including covalent bonds (shared electrons); ionic bonds (transferred electrons); and hydrogen bonds.

2-4. Water and Its Importance to Life

- Hydrogen bonds give water its unique properties, including high surface tension (water molecules held together by hydrogen bonds form a netlike property); cohesion (ability of water molecules to stick together); adhesion (ability of water molecules to stick to other substances); high heat capacity (ability of water to remain at a relatively stable temperature); decreased density when frozen (water molecules form a stable, crystalline lattice, distributing water molecules farther apart); and excellent solvent (water is polar, which refers to differences in the charges of the hydrogen and oxygen atoms).

2-5. Acids and Bases

- Acids are substances that release hydrogen protons (H^+) and bases are substances that release hydroxide ions (OH^-) or take up hydrogen protons (H^+). The pH scale indicates the acidity or basicity of a substance by determining the concentration of hydrogen protons (H^+) in a solution. Buffers are chemicals used to stabilize the pH of a solution by taking up hydrogen protons (H^+) or hydroxide ions (OH^-).

End-of-Chapter Activities and Questions

Directions: Please refer back to what you learned in this chapter to complete the following activities.

Define Each Term in Your Own Words

1. Atom
2. Periodic Table
3. Chemical Bonding
4. Polar
5. Buffer

Chapter Review

1. There are roughly 25 elements essential for life in the human body. Which four elements are the most abundant?
2. Describe what groups and periods indicate on the periodic table.
3. Chemical bonding is defined as the force of attraction between the valence electrons of two or more atoms. What type of diagram can be used in order to determine how many valence electrons are available to bond? Create your own diagram for the oxygen (O) atom to determine the number of valence electrons available for bonding.
4. Hydrogen bonding is important in giving water its unique properties. What are the five main properties of water? Given an example of each.
5. What are the major differences between acids and bases? Explain how buffers work.

Multiple Choice

1. Isotopes are atoms of the same element with the same number of ______, but a different number of ______.

   a. electrons, protons

   b. protons, neutrons

   c. neutrons, electrons

   d. protons, electrons

2. Which of the following elements was the first element synthetically discovered in a laboratory?

   a. technetium

   b. nihonium

   c. mendelevium

   d. curium

3. Oxygen has six valence electrons. How many of these valence electrons are available to bond with hydrogen atoms to make water? (Hint: Draw a Lewis diagram.)

   a. one

   b. two

   c. four

   d. six

4. Which of the following properties of water is (are) important in the transport of water from roots to leaves in plants?

   a. high heat capacity

   b. polar solvent

   c. adhesion and cohesion

   d. decreased density

5. An important pH buffering system in the human body is shown below:

   $$\_\_\_\_\_\_ + H^+ \leftrightarrow \_\_\_\_\_\_ \leftrightarrow CO_2 + H_2O$$

   Which molecules are missing from the equation?

   a. OH^-; HCO_3^-

   b. H_2CO_3; OH^-

   c. HCO_3^-; H_2CO_3

   d. none of the choices are correct

Image Credits

Fig. 2.1: Source: https://commons.wikimedia.org/wiki/File:DI-MendeleevCab.jpg.

Fig. 2.2: Copyright © by OpenStax (CC BY 3.0) at https://commons.wikimedia.org/wiki/File:201_Elements_of_the_Human_Body-01.jpg.

Fig. 2.3a: Copyright © by OpenStax (CC BY 4.0) at https://openstax.org/books/biology/pages/2-1-atoms-isotopes-ions-and-molecules-the-building-blocks.

Fig. 2.3b: Copyright © by OpenStax (CC BY 4.0) at https://openstax.org/books/anatomy-and-physiology/pages/2-1-elements-and-atoms-the-building-blocks-of-matter.

Fig. 2.4: Copyright © by OpenStax (CC BY 4.0) at https://openstax.org/books/chemistry-2e/pages/2-5-the-periodic-table.

Fig. 2.5: Copyright © 2021 Depositphotos/zizou07.

Fig. 2.6a: Source: https://commons.wikimedia.org/wiki/File:Mendeleev%27s_1869_periodic_table.png.

Fig. 2.6b: Source: https://commons.wikimedia.org/wiki/File:Periodic_Table_Mendeleev_1904.jpg.

Fig. 2.7: Source: https://commons.wikimedia.org/wiki/File:Periodic_Table_-_Atomic_Properties_of_the_Elements.png.

Fig. 2.8: Source: https://commons.wikimedia.org/wiki/File:Marie_Curie_in_her_Paris_Laboratory,_1912.jpg.

Fig. 2.9: Copyright © by OpenStax (CC BY 4.0) at https://openstax.org/books/biology/pages/2-1-atoms-isotopes-ions-and-molecules-the-building-blocks.

Fig. 2.10a: Source: https://commons.wikimedia.org/wiki/File:Water_molecule.svg.

Fig. 2.10b: Source: https://commons.wikimedia.org/wiki/File:-Sodium-chloride-3D-ionic.png.

Fig. 2.10c: Source: https://commons.wikimedia.org/wiki/File:D-glucose-chain-3D-vdW.png.

Fig. 2.11a: Source: https://commons.wikimedia.org/wiki/File:Lewis_dot_H.svg.

Fig. 2.11c: Source: https://commons.wikimedia.org/wiki/File:Lewis_dot_O.svg.

Fig. 2.11d: Source: https://commons.wikimedia.org/wiki/File:Elektronenformel_Punkte_HCl.svg.

Fig. 2.12a: Copyright © by BruceBlaus (CC BY-SA 4.0) at https://commons.wikimedia.org/wiki/File:Covalent_Bonds.png.

Fig. 2.13: Copyright © by OpenStax (CC BY 4.0) at https://openstax.org/books/anatomy-and-physiology/pages/2-2-chemical-bonds.

Fig. 2.14: Source: https://commons.wikimedia.org/wiki/File:3D_model_hydrogen_bonds_in_water.jpg.

Fig. 2.15: Copyright © by Schnobby (CC BY-SA 3.0) at https://commons.wikimedia.org/wiki/File:Water_strider_in_a_pond.jpg.

Fig. 2.16: Copyright © by FeltyRacketeer6 (CC BY-SA 4.0) at https://commons.wikimedia.org/wiki/File:Transpiration_of_Water_in_Xylem.svg.

Fig. 2.17: Copyright © by OpenStax (CC BY 4.0) at https://openstax.org/books/concepts-biology/pages/2-2-water.

Fig. 2.18: Copyright © by OpenStax (CC BY 4.0) at https://openstax.org/books/anatomy-and-physiology/pages/2-4-inorganic-compounds-essential-to-human-functioning.

Fig. 2.19: Copyright © by OpenStax (CC BY 4.0) at https://openstax.org/books/anatomy-and-physiology/pages/2-4-inorganic-compounds-essential-to-human-functioning.

Fig. 2.20: Copyright © by OpenStax (CC BY 4.0) at https://openstax.org/books/anatomy-and-physiology/pages/2-4-inorganic-compounds-essential-to-human-functioning.

Fig. 2.21: Copyright © by OpenStax (CC BY 4.0) at https://openstax.org/books/biology/pages/2-2-water.

CHAPTER

3

Macromolecules

PROFILES IN SCIENCE

FIGURE 3.1 Linus Pauling.

Linus Pauling was one of the greatest, most accomplished scientists in history for his work on chemical bonds and the structure of biological molecules. Born on February 28, 1901, in Portland, Oregon, Pauling died in California in 1994. His interest in chemistry began as a child and continued through high school, as he conducted experiments in his home from equipment and materials collected from an old, abandoned steel factory. He began attending Oregon State University in 1917, studying atomic structure, chemical bonding, and how each contributed to the physical and chemical properties of a molecule. He graduated with a chemical engineering degree in 1922. Upon graduation, Pauling attended Caltech, studying X-ray diffraction, and determining the structure of crystals, ultimately receiving his PhD in physical chemistry and mathematical physics in 1925. Pauling accepted an assistant professor position at Caltech in 1927, and he taught theoretical chemistry. By 1930, he became a full professor, teaching freshman chemistry. During the 1930s, Pauling supplemented his work on the nature of the chemical bond by introducing a variety of new concepts to the idea, including electronegativity, hybridization, and resonance, which culminated in his most famous publication, *The Nature of the Chemical Bond and the Structure of Molecules and Crystals*, in 1939. Pauling refocused his research on the structure of biological molecules during the mid-1930s, studying the structure of hemoglobin, the binding of antibodies to antigens, enzymatic reactions, and the structure and replication of DNA. By 1951, with the use of X-ray diffraction, model building, and his knowledge of the structure of hemoglobin, he discovered the α-helix and β-pleated sheet structures of proteins, for which he was given the title "father of molecular biology." After winning the Nobel Prize in Chemistry in 1954, he spent the 1960s and 1970s making significant contributions to medicine, wartime activism, and the nuclear test ban treaty of 1963, for which he received the 1962 Nobel Peace Prize. Pauling continued teaching alongside his scientific research at the University of California–San Diego (1967–1969) and Stanford University (1969–1973). In 1973, Pauling established the Linus Pauling Institute of Science and Medicine, whose focus is to conduct research, develop diagnostic tools, and provide education in orthomolecular medicine.

Introduction to the Chapter

As we found out in chapter 2, chemistry plays a central role in a variety of disciplines, especially biology. Recall that atoms are extremely important because they provide the primary foundation for molecules, which are responsible for helping support and maintain living organisms. One of the most important and essential atoms from which all molecules are built is the carbon atom. Unique in its size and chemistry, the carbon atom has the ability to bond to a variety of other atoms, as well as itself, to produce an array of diverse organic molecules. Macromolecules, such as carbohydrates, lipids, proteins, and nucleic acids, are examples of organic molecules, which are essential in providing all living organisms with the required materials they need in order to grow, develop, and sustain life. Within this chapter, the chemistry of the carbon atom will be discussed, along with how the atom plays a central role in the formation of important biological organic molecules, such as macromolecules. In addition, the chapter will provide a detailed discussion on each of the four classes of macromolecules, with emphasis on structure, function, and their importance to living organisms.

Chapter Objectives

In this chapter, students will learn the following:

3-1. Carbon is an essential atom in the formation of organic molecules such as macromolecules, which are important molecules found and used in all living organisms.

3-2. Carbohydrates are complex macromolecules composed of carbon, hydrogen, and oxygen in a 1:2:1 ratio that have a function in providing energy.

3-3. Lipids are complex macromolecules consisting of carbon, hydrogen, and oxygen that are responsible for a variety of biological functions vital to life.

3-4. Proteins are complex macromolecules composed of amino acids that sustain life through various different biological functions.

3-5. Nucleic acids are complex macromolecules composed of nucleotides that have a function in storing and translating genetic information.

3-1. The Carbon Atom and Macromolecules

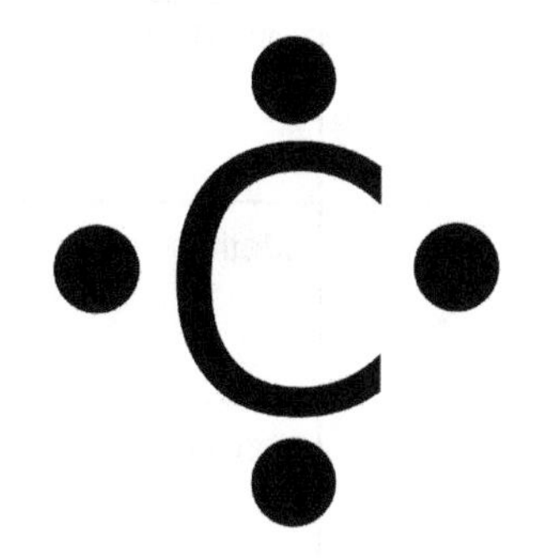

FIGURE 3.2 Carbon has four valence electrons, which can covalently bond to other types of atoms, including itself.

The carbon atom is a key component to the diversity of molecules found within living organisms. Living organisms use the carbon atom to construct important biological molecules, which are used as the starting materials for cellular structures. The chemistry of the carbon atom is unique because it is relatively small and contains four valence electrons (Figure 3.2). The available valence electrons enable the carbon atom to bond to four other atoms, such as hydrogen, oxygen, nitrogen, sulfur, and phosphorus, stabilizing the atom.

The carbon atom also has the ability to bond to itself, forming what are known as *carbon skeletons*, providing the backbone from which biological molecules are built. The way in which carbon bonds to other atoms and itself results in a variety of carbon skeleton shapes and structures, including long chains, rings, and branches (Figure 3.3). This characteristic of carbon provides living organisms an infinite number of ways to construct the necessary molecules needed for a variety of different cellular structures.

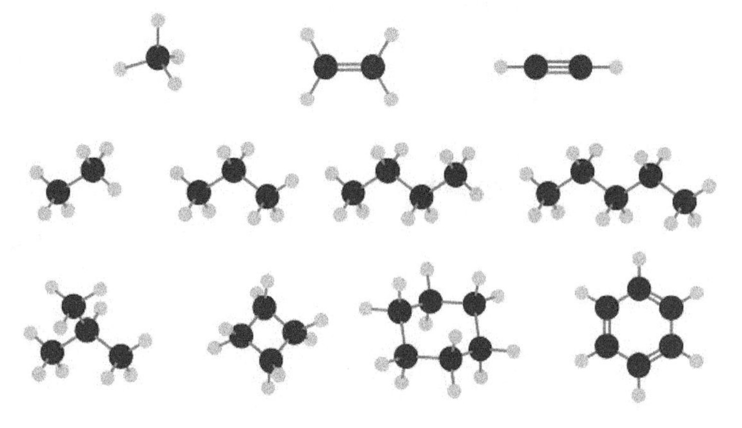

FIGURE 3.3 Carbon can bond to other atoms and itself in a variety of ways to form chains, branches, and rings.

Molecules composed of a carbon skeleton and attached hydrogen atoms are called *organic molecules*. Organic molecules have unique chemical properties, which are determined by groups of atoms bonded to the carbon skeleton, known as *functional groups* (Figure 3.4). Functional groups contain a mixture of other atoms, including oxygen, sulfur, and nitrogen, that contribute to the overall structure of the molecule. In addition, functional groups dictate how the molecule will function in the presence of other molecules and are responsible for a variety of chemical reactions that occur within the cell. The combination of atoms within functional groups is what ultimately accounts for the diversity of biological molecules, such as macromolecules. Examples of functional groups include a hydroxyl group, carboxyl group, and an amino group.

Macromolecules are large, chain-like molecules called *polymers*. Polymers are composed of numerous monomers (molecules existing as small, repeating subunits) linked together by covalent bonds. As an individual molecule, a monomer has its own distinct properties. However, when monomers are linked together, the resulting polymer possesses its own unique properties. Macromolecules are an important group of diverse organic molecules essential in sustaining and maintaining life. All living organisms use macromolecules, either directly or by rearranging them to construct necessary cellular structures. There are four main classes of biologically important macromolecules: carbohydrates, lipids, proteins, and nucleic acids.

Functional Group	Structure	Properties
Hydroxyl	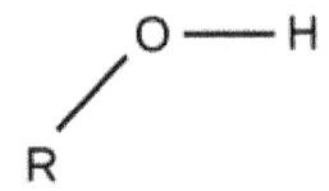 R—O—H	Polar
Methyl	R—CH_3	Nonpolar
Carbonyl	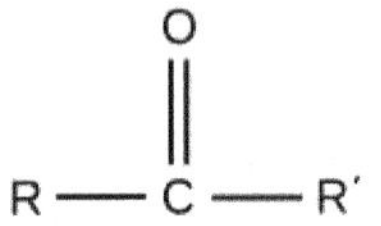 R—C(=O)—R′	Polar
Carboxyl	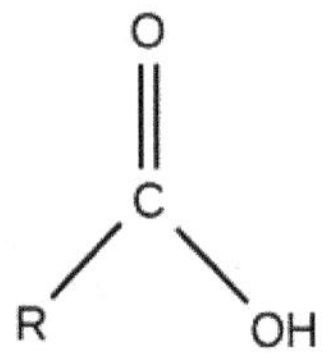 R—C(=O)—OH	Charged, ionizes to release H^+. Since carboxyl groups can release H^+ ions into solution, they are considered acidic.
Amino	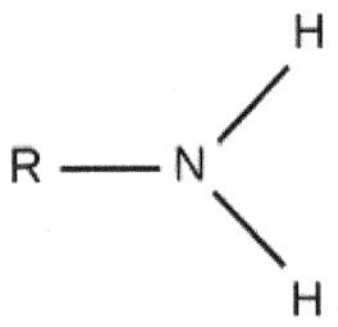 R—N(—H)—H	Charged, accepts H^+ to form NH_3^+. Since amino groups can remove H^+ from solution, they are considered basic.
Phosphate	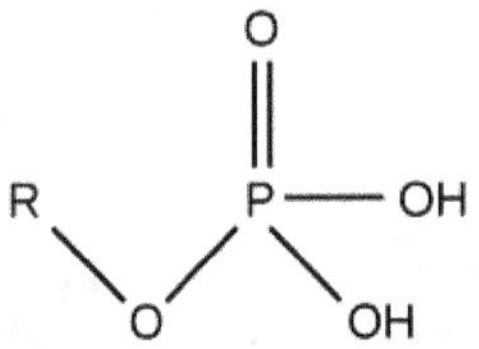 R—O—P(=O)(—OH)—OH	Charged, ionizes to release H^+. Since phosphate groups can release H^+ ions into solution, they are considered acidic.
Sulfhydryl	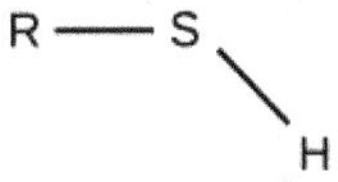 R—S—H	Polar

FIGURE 3.4 Functional groups are an important group of atoms that provide organic molecules with their unique chemical and functional properties.

3-2. Carbohydrates

Carbohydrates are complex molecules composed of carbon, hydrogen, and oxygen existing in a 1:2:1 ratio. In other words, there are twice as many hydrogen atoms as both carbon and oxygen, with a general formula of $(CH_2O)_n$. Carbohydrates generally exist as either single sugar molecules or as long chains consisting of several to hundreds to thousands of sugar molecules bonded together. Carbohydrates are synthesized by plants through the process of photosynthesis, and common examples include grains, rice, and sugar cane. Carbohydrates play important structural roles in the cells and tissues of different types of organisms, such as plants, fungi, and bacteria. However, their most important purpose is that of a major energy source to all living organisms, producing roughly four kilocalories of energy per gram when broken down.

Carbohydrates can exist as either simple or complex carbohydrates. Simple carbohydrates include **monosaccharides** and **disaccharides**. Monosaccharides are also known as simple sugars because they consist of only one sugar, or saccharide, molecule (Figure 3.5). These types of sugars contain between three and seven carbon atoms, with most containing six carbons. The most common example of a monosaccharide is glucose, which has the molecular formula $C_6H_{12}O_6$. Glucose is the carbohydrate that is important in providing energy to all living organisms. Other examples of monosaccharides include fructose (sugar found in fruits, vegetables, and honey), galactose (sugar found in milk), and ribose and deoxyribose (five carbon sugars found in nucleic acids).

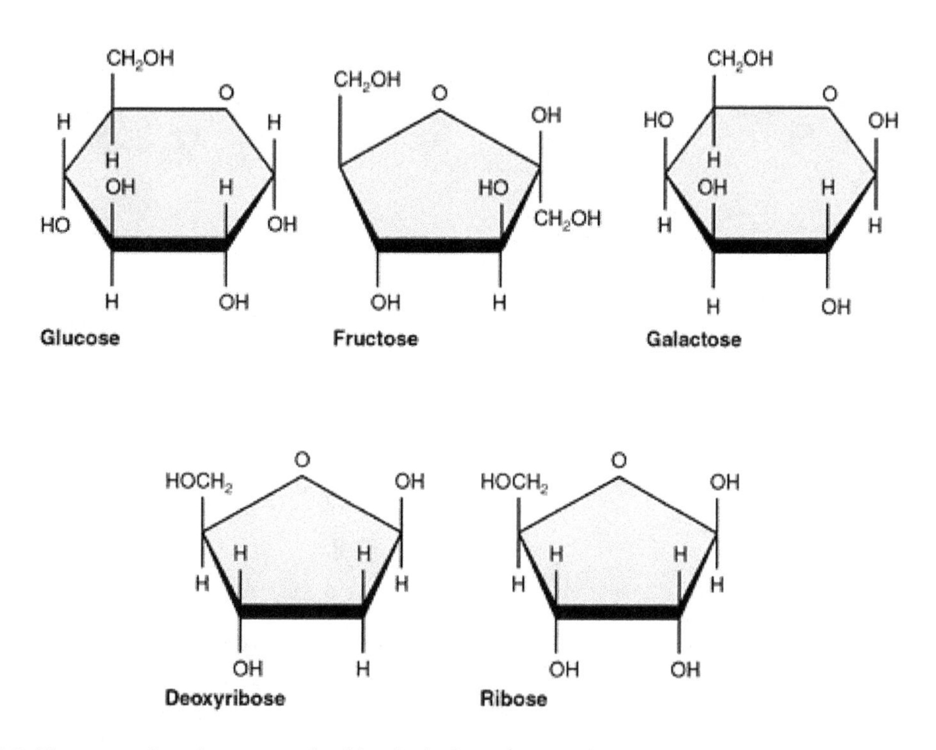

FIGURE 3.5 Five examples of monosaccharides, including glucose, the most important type of monosaccharide, responsible for the generation of energy.

Glucose is an important carbohydrate found in many of the different types of foods we consume, including fruits. Carbohydrates found in other types of starch-based foods and food additives, such as sucrose, are converted to glucose, then circulated in the bloodstream to areas in the body where it is needed. Once in the bloodstream, glucose will be used in various ways depending on the needs of the body cells. The primary use of glucose in body cells is the generation of energy in the form of ATP, during a process known as cellular respiration. ATP is used by the body for cellular activity and physiological processes such as muscle contraction. Excess glucose not used to produce ATP is stored as glycogen in a process known as glycogenesis. This process is performed by liver cells and muscle fibers, and once glycogen is formed, about 500 grams are stored in the body, with roughly 125 grams stored in the liver and the remaining 375 grams stored in skeletal muscle. Finally, once the storage capacity for glycogen has been reached in the liver and skeletal muscles, excess glucose is transformed into triglycerides in the process of lipogenesis. The triglycerides are stored in adipose (fat) tissue, in which there is an unlimited storage capacity for this molecule.

A disaccharide is a carbohydrate formed when two monosaccharides are joined together during a dehydration synthesis reaction. A dehydration synthesis reaction employs the use of enzymes that remove a hydrogen atom (-H) and a hydroxyl group (-OH) from the monosaccharides to produce a water molecule. The result is the formation of a covalent bond between the monosaccharides known as a glycosidic bond (Figure 3.6). Common examples of disaccharides include sucrose (glucose + fructose), maltose (glucose + glucose), and lactose (glucose + galactose) (Figure 3.6). Sucrose and maltose are important disaccharides produced in abundance and used by plants, while lactose is found in milk.

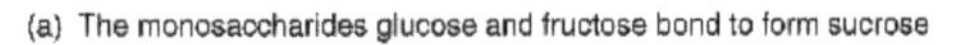

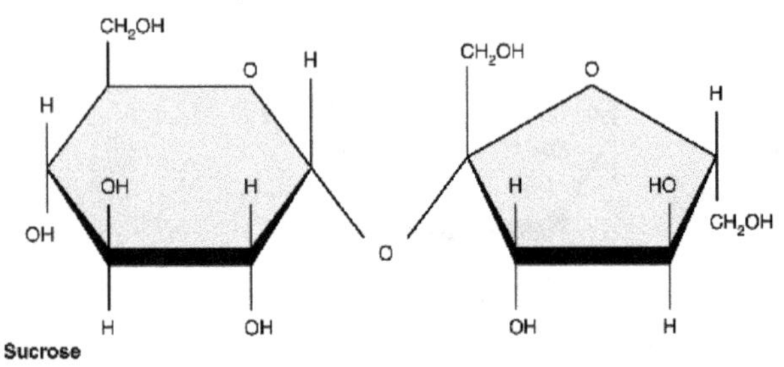

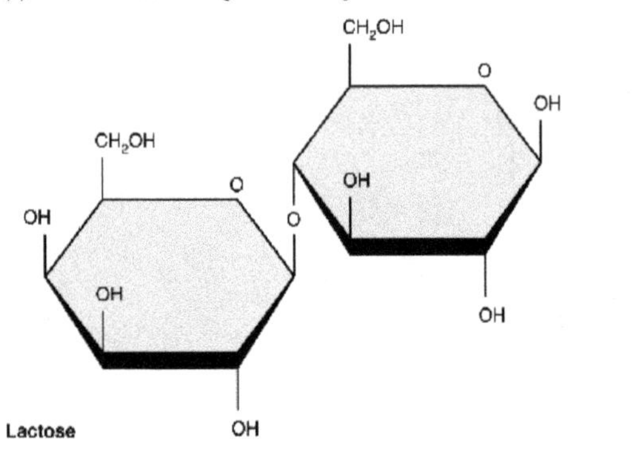

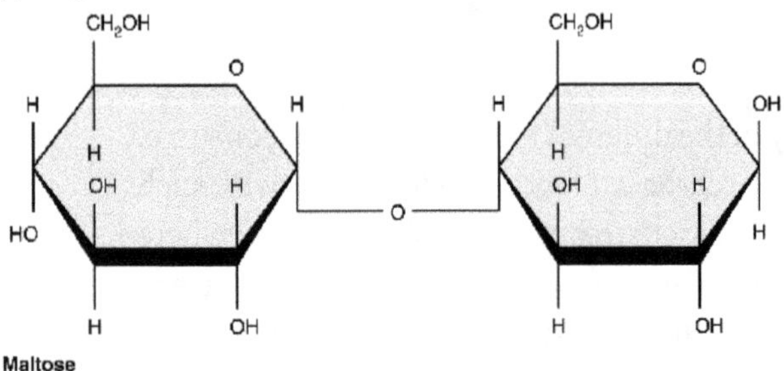

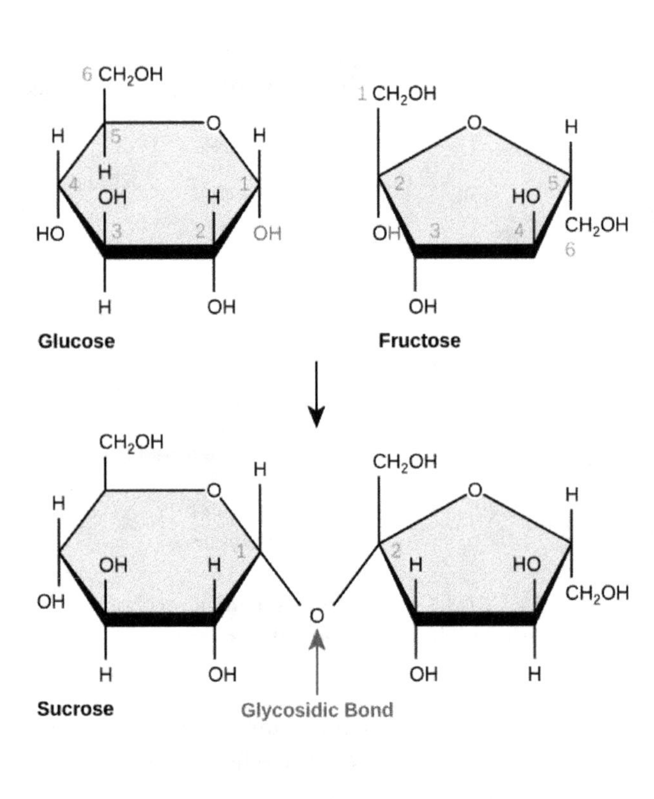

FIGURE 3.6 Disaccharides (left) consist of two monosaccharides bonded together. Common examples include (a) sucrose, (b) lactose, and (c) maltose. Glycosidic bonds (right) are covalent bonds that form when two or more monosaccharides are joined together when a dehydration synthesis reaction initiates the removal of a water molecule.

Polysaccharides are complex carbohydrates consisting of hundreds of thousands of sugar molecules bonded together by glycosidic bonds (Figure 3.7). Polysaccharides act as energy storage molecules, in which energy stored within the sugar molecules is released when the carbohydrate is broken down by the organism. A common polysaccharide is starch; plants store energy in starch and use it for growth and development. There are two main types of starch, including amylose, which is the simplest form and consists of unbranched glucose molecules, while amylopectin is much more complex and contains branches of glucose molecules. Glycogen is another type of energy-storage polysaccharide found in animals. Animals store glycogen within muscle tissue and the liver, and this glycogen is used when needed. Polysaccharides also play an important structural role in living organisms. For example, cellulose, the most abundant carbohydrate and common organic molecule on Earth, is the primary sugar found in wood and in the cell walls of plants. Cellulose is unique in that it is similar to starch in composition; however, a small structural difference between the molecules makes cellulose indigestible in animals. Another example of a structural polysaccharide is chitin, found in fungal cell walls and the exoskeletons of insects and crustaceans. Chitin, like cellulose, is also an indigestible polysaccharide.

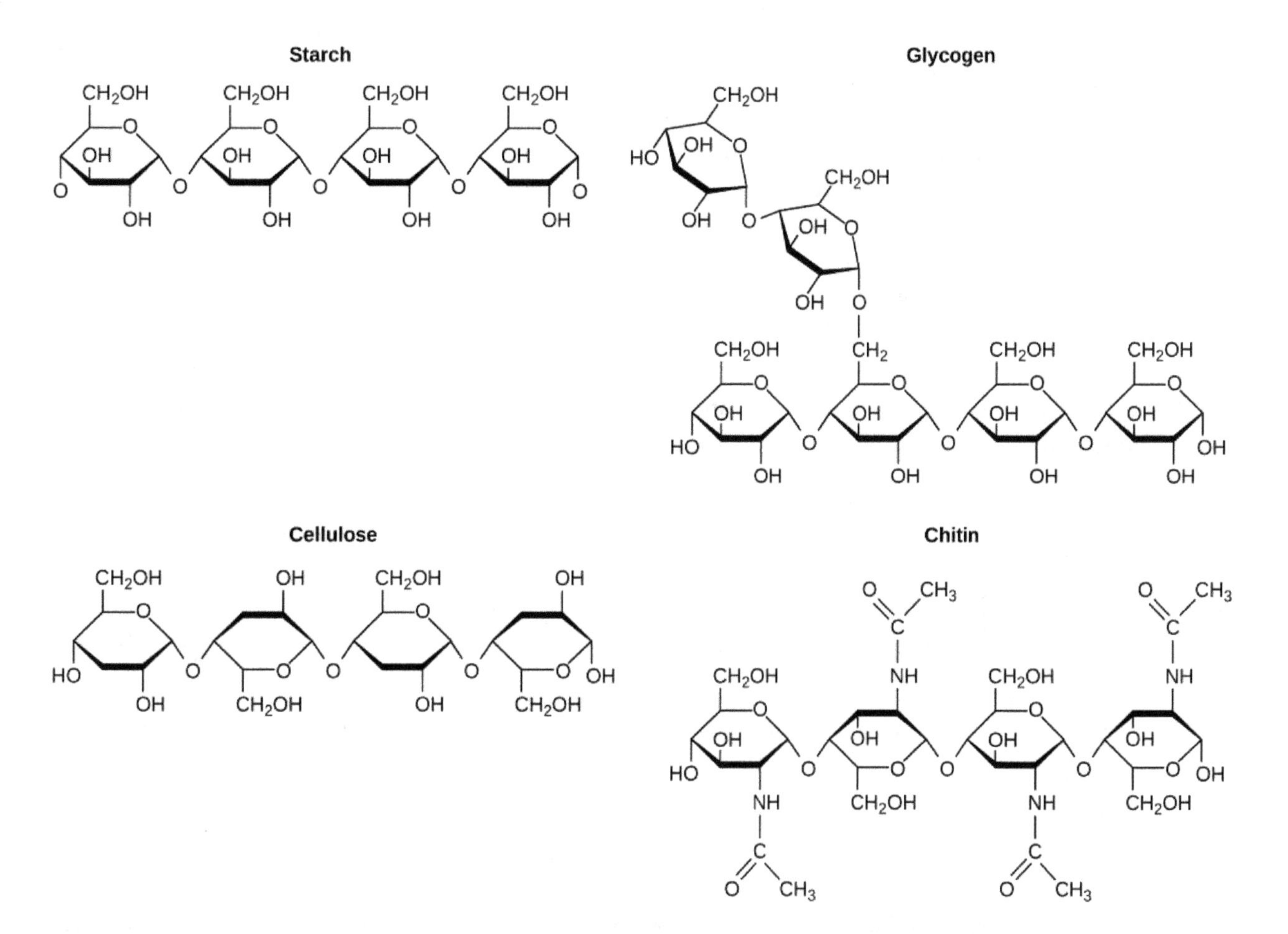

FIGURE 3.7 Polysaccharides are complex carbohydrates that act as either storage molecules, such as starch and glycogen or as structural molecules, like cellulose and chitin.

3-3. Lipids

Lipids are a group of macromolecules similar in composition to the structure of carbohydrates, because they consist of carbon, hydrogen, and oxygen. However, the overall composition of lipids lacks the defined 1:2:1 ratio found in carbohydrates. One of the main defining structural characteristics of lipids is the long carbon-hydrogen chains of which they are composed. As a result, lipids lack the ability to dissolve in water, thereby making them hydrophobic (water-fearing) molecules.

Lipids have many biological functions, making them vital to life. Lipids are a major energy source, with each gram of fat releasing about nine kilocalories of energy, which is twice as much as carbohydrates. In addition, lipids store energy in the form of fat cells called adipocytes. Lipids, such as phosphoglycerides, sphingolipids, and steroids, are important structural components of cell membranes, the structure responsible for controlling the flow of substances into and out of a cell. Steroids are not only an important structural component of cell membranes but also an important component of steroid hormones, such as estrogen and testosterone, which are chemical messengers that allow body tissues to efficiently communicate. Lipids also play an important role in vitamins, such as vitamins A, E, D, and K. These vitamins are lipid-soluble and require dietary fats for transportation throughout the body and small intestine. Generally, a diet low in dietary fat can result in lipid-soluble vitamin deficiency. Lipid-soluble vitamins are important in critical biological processes, including blood clotting and vision. Lipids also provide protection, as they serve as shock absorbers around vital organs such as the heart and play a role in insulation, as lipids are stored beneath the skin, helping to insulate the bodies of organisms from cold temperatures.

There are four main types of lipids that exist in nature.

Triglycerides, also known as fats, are large molecules consisting of two main components: glycerol and three fatty acid chains (Figure 3.8). Glycerol makes up the "head" region and is a water-soluble molecule consisting of three carbon atoms attached to three hydroxyl groups (-OH). Fatty acids make up the "tail" region and are composed of long hydrocarbon chains containing up to 18 carbon atoms, with a carboxyl (acid) group at each end.

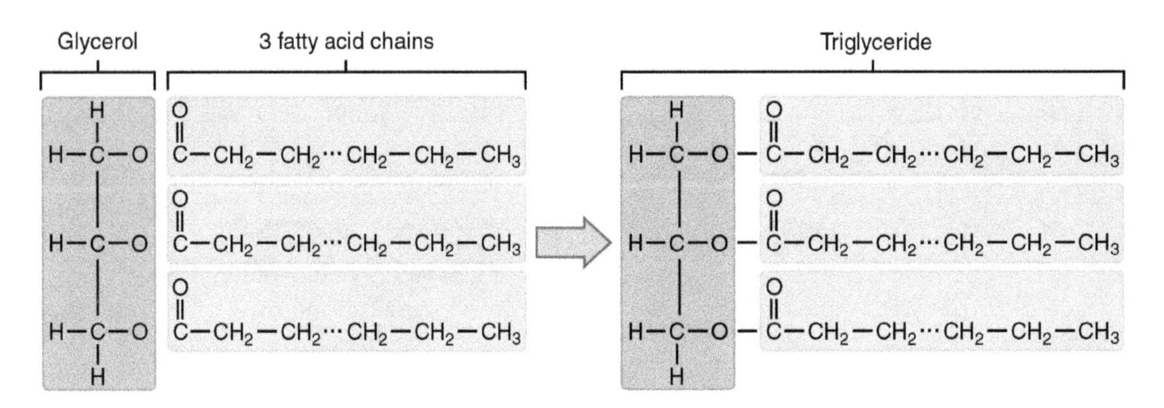

FIGURE 3.8 Triglycerides consists of two main components, a glycerol "head" attached to three fatty acid "tails" or chains.

Based on the structural composition of the hydrocarbon chains, fatty acids can be either saturated or unsaturated. Saturated fatty acids lack double bonds, producing straight hydrocarbon chains (Figure 3.9). This produces what are known as saturated fats. Due to the structural composition of the hydrocarbon chains, saturated fatty acid tails can tightly pack together, resulting in saturated fats being solids at room temperature, such as butter and lard.

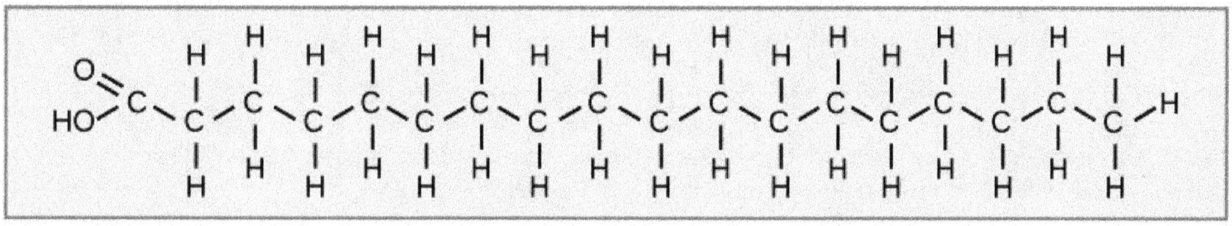

FIGURE 3.9 Saturated fats are characterized by having straight fatty acid tails, which allows them to tightly pack together, forming solids at room temperature.

Unsaturated fatty acids contain double bonds within their hydrocarbon chains, producing kinks in the tails of the fatty acid (Figure 3.10). This results in a lower number of hydrogens per carbon atom, making the hydrocarbon chains less likely to pack tightly together. This configuration produces unsaturated fats, which are also known as oils and exist as liquids at room temperature, such as corn, peanut, and olive oils.

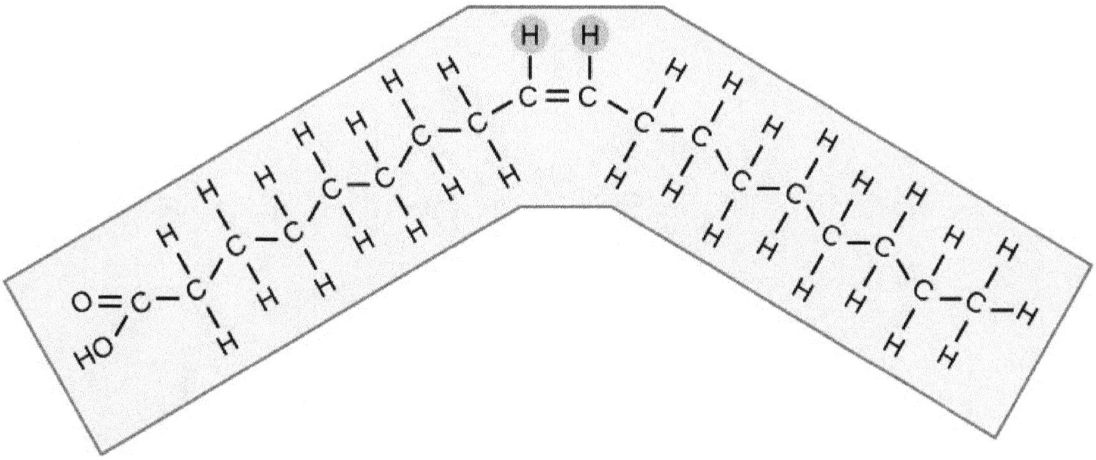

FIGURE 3.10 Unsaturated fats consist of double bonds that result in the fatty acid tails having kinks. An unsaturated fat cannot tightly pack together, therefore, exists as liquids at room temperature.

Another example of fats are trans-fats, which are the result of a process known as *hydrogenation*. This process involves the addition of hydrogen atoms to unsaturated fats, resulting in a structural change to the molecule that resembles a saturated fat (Figure 3.11). Trans-fats are commonly added to processed food products, such as margarine and crackers, to provide a more desirable taste and texture.

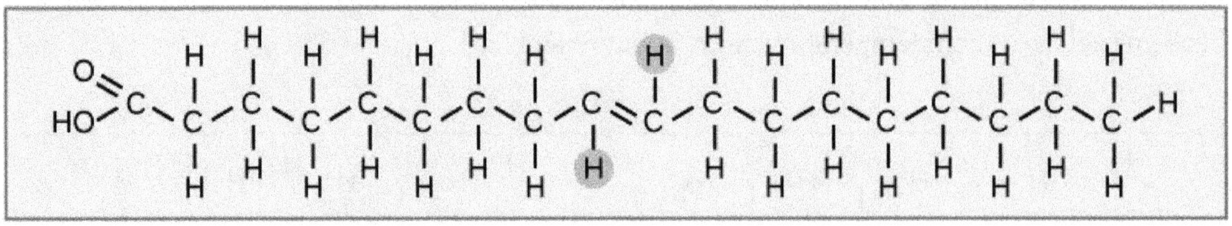

FIGURE 3.11 Trans-fats are produced during a process known as hydrogenation. The additional hydrogens added to unsaturated fats, make them resemble the structure of a saturated fat.

Sterols are a unique group of lipids consisting of four fused carbon skeletons with various attached functional groups. Functional groups are chemical structures that determine the overall biological function of the sterol type. Unlike other types of lipids, sterols function in regulating growth and development in an organism. For example, cholesterol (Figure 3.12) is a sterol produced by the liver and becomes an integral component of cell membranes, helping maintain their fluidity and stability. In addition, cholesterol is a precursor to other types of sterol-based lipids, including the steroid hormones estrogen and testosterone (Figure 3.12), which both regulate sexual development, maturation, sex cell production, as well as influence memory and mood (estrogen) and stimulate muscle growth (testosterone).

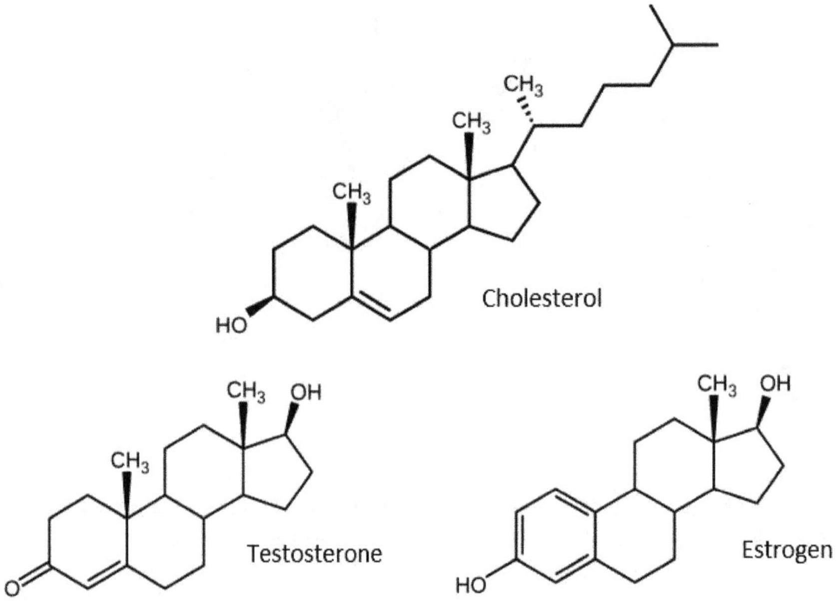

FIGURE 3.12 Sterols consist of four fused rings with different functional groups attached. The most common type of sterol is cholesterol and is a precursor to different types of steroid hormones, including testosterone and estrogen.

Phospholipids are another group of lipids and a fundamental component of cell membranes. As the name suggests, phospholipids contain a phosphate group that has replaced one of the fatty acid chains in triglycerides, leaving two fatty acid chains attached to glycerol. (Figure 3.13). These two different ends exhibit a unique behavior in that the phosphate group end is attracted to water; therefore, it exists as a hydrophilic head. The fatty acid tails are repelled by water and are hydrophobic in nature.

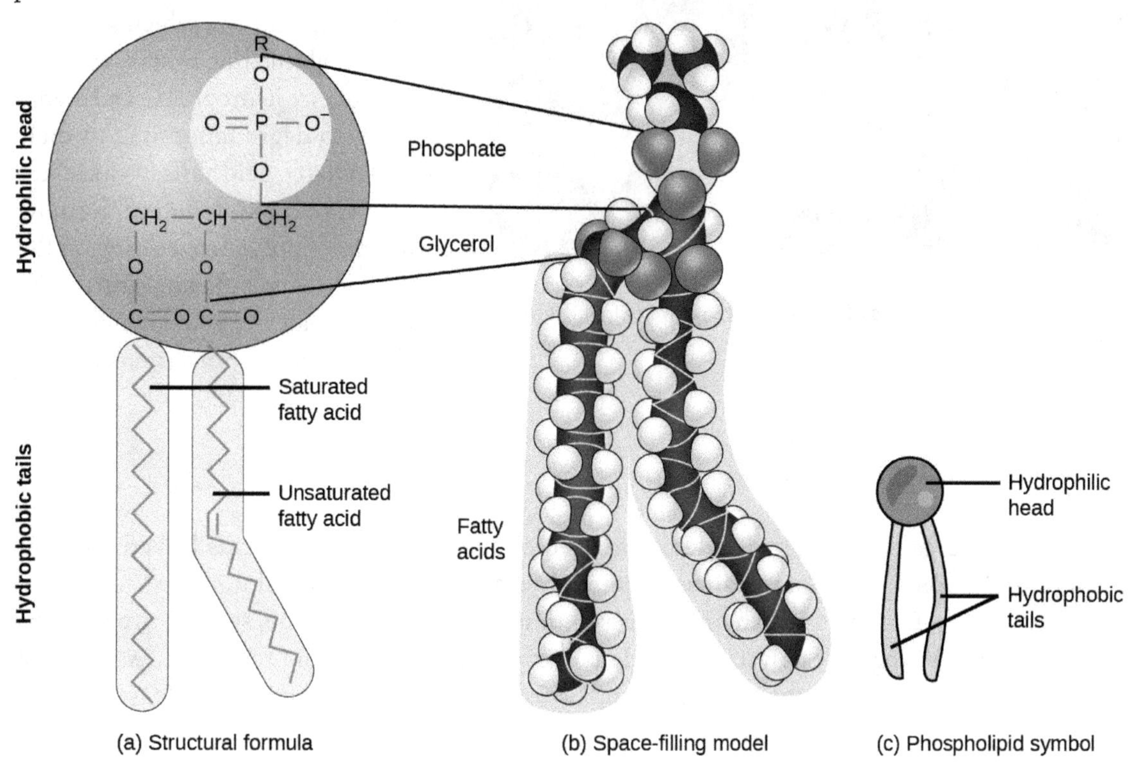

FIGURE 3.13 Phospholipids consists of phosphated glycerol head and two fatty acid tails.

As a result, phospholipids arrange themselves in a bilayer structure, with the hydrophilic heads facing outward, while the hydrophobic tails are directed inward (Figure 3.14). This design enables the phospholipids to separate themselves from both interior and exterior solutions associated with the cell. The cell membrane consists of a phospholipid bilayer and is essential in regulating the movement of substances into and out of the cell. This structure will be discussed in further detail in chapter 4.

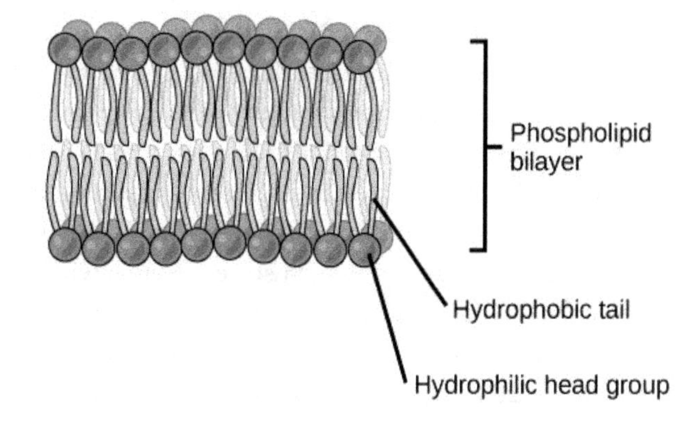

FIGURE 3.14 The phosphated glycerol head is attracted to water (hydrophilic), while the fatty acid tails are not (hydrophobic). This results in phospholipids creating a bilayer structure, which is an important component of cell membranes.

FIGURE 3.15 Some leaves of plants contain a waxy layer that helps prevent the plant from losing water.

Waxes are another type of lipid similar to triglycerides; however, they consist of only one fatty acid chain bonded to a glycerol head and a long-chain alcohol. Due to their high saturation, waxes exist as solids at normal temperatures, which accounts for their high melting points. In addition, waxes are hydrophobic and highly resistant to degradation due to their nonpolar fatty acid tails. Waxes are an important lipid in nature, forming a waterproof protective coating on the leaves and stems of plants (Figure 3.15), preventing water loss, and in maintaining the skin and fur of animals.

3-4. Proteins

Proteins are large, complex molecules composed of one or more amino acids linked together by **peptide bonds** to form long chains. Proteins have a variety of biological functions important in sustaining life. Enzymes, which are discussed in more detail later in this section, are proteins that function as biological catalysts to speed up chemical reactions within the body. Examples of enzymes are two digestive enzymes, pepsin and trypsin. Proteins also function as defense proteins in the form of antibodies, specialized cells produced by the body's immune system that respond to the presence of foreign antigens or molecules. Hemoglobin is an important transport protein that serves to carry oxygen in the bloodstream from the lungs to the body tissues. Cellular function, including metabolism and reproduction, along with maintaining the body within a homeostatic state, is performed by regulatory proteins. For example, glucose levels within the blood are regulated by two proteinaceous hormones produced and secreted by the pancreas: glucagon and insulin. Proteins also provide structural and mechanical support. Keratin is a structural protein found in fingernails and hair, while collagen is a mechanical protein found in bones, tendons, ligaments, and skin. Muscle contraction occurs as a result of movement proteins found in muscle fibers, including actin and myosin. Myoglobin, a protein similar in structure to that of hemoglobin, is an example of a storage protein, which stores oxygen within muscle cells. Other examples of storage proteins include albumin and casein, found in milk.

Amino acids are a group of molecules that, when linked together, form proteins. Organisms utilize 20 different amino acid molecules in order to build proteins, and the combination of these amino acids results in a variety of unique protein structures and functions. All amino acids have the same basic structure, starting with a central carbon, also known as an alpha carbon (α-carbon), on which are bonded four different functional groups: a hydrogen atom, an amino group ($-NH_2$), a carboxyl group (-COOH), and an R-group (Figure 3.16).

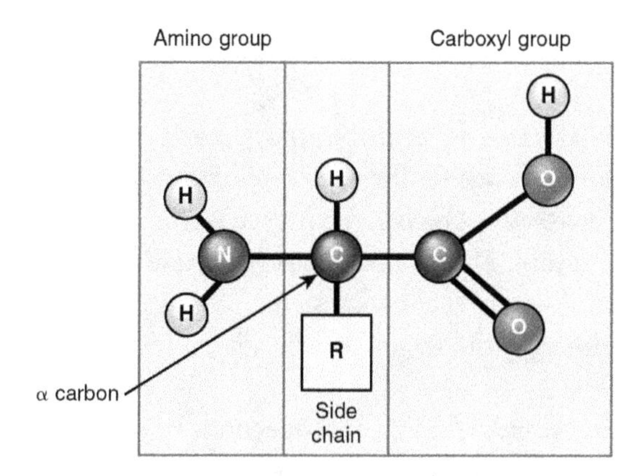

FIGURE 3.16 The structure of an amino acid consists of a central (or alpha) carbon to which an amino group, carboxyl group, and an R-group are attached.

FIGURE 3.17 Peptide bonds are covalent bonds that form between the amino group of one amino acid and the carboxyl group of another during a dehydration synthesis reaction.

During the formation of a protein, a dehydration synthesis reaction occurs between the amino group of one amino acid and the carboxyl group of another to form a covalent peptide bond that links the amino acids together (Figure 3.17).

The R-group is what distinguishes the 20 different amino acids in size, solubility, and electrical charge, but it is ultimately responsible for causing proteins to fold into a unique three-dimensional configuration (Figure 3.18).

When proteins fold into their unique three-dimensional structure, this structure determines the function of the protein. There are four different structures in which a protein can form based on the composition of amino acids.

The **primary structure** is a simple linear sequence of amino acids covalently bonded together to make up the protein (Figure 3.19a). This sequence is determined by the genetic information contained within a DNA molecule. The overall sequence of amino acids

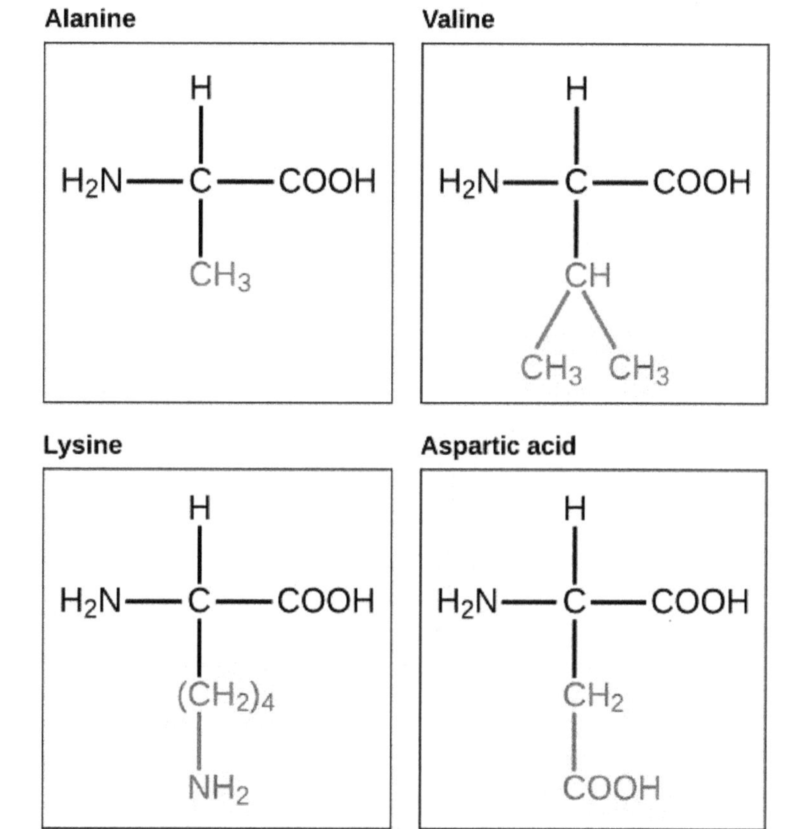

FIGURE 3.18 All amino acids have the same general structure but are distinguished based upon their unique R-group.

helps determine where the R-groups will be located and how the protein will fold into its final three-dimensional shape.

As the amino acid chains begin to fold or coil, they form a **secondary structure**, a structure maintained by hydrogen bonds (Figure 3.19b). The folds and coils for the amino acid chain produce two different types of secondary structures: The first is *α-helix*, which is the most common type of secondary structure and exhibits a coiled or helical structure. The most common example of this is keratin, the protein found in hair and nails. The second type of secondary structure is a *β-pleated sheet*, characterized by its pleated folding. Silk, the protein found in spider webs, is an example of β-pleated secondary structure.

The **tertiary structure** of a protein forms from the continual folding of the secondary structure because of the interactions between R-groups and their exposure to water. These interactions are responsible for the final protein shape (Figure 3.19c).

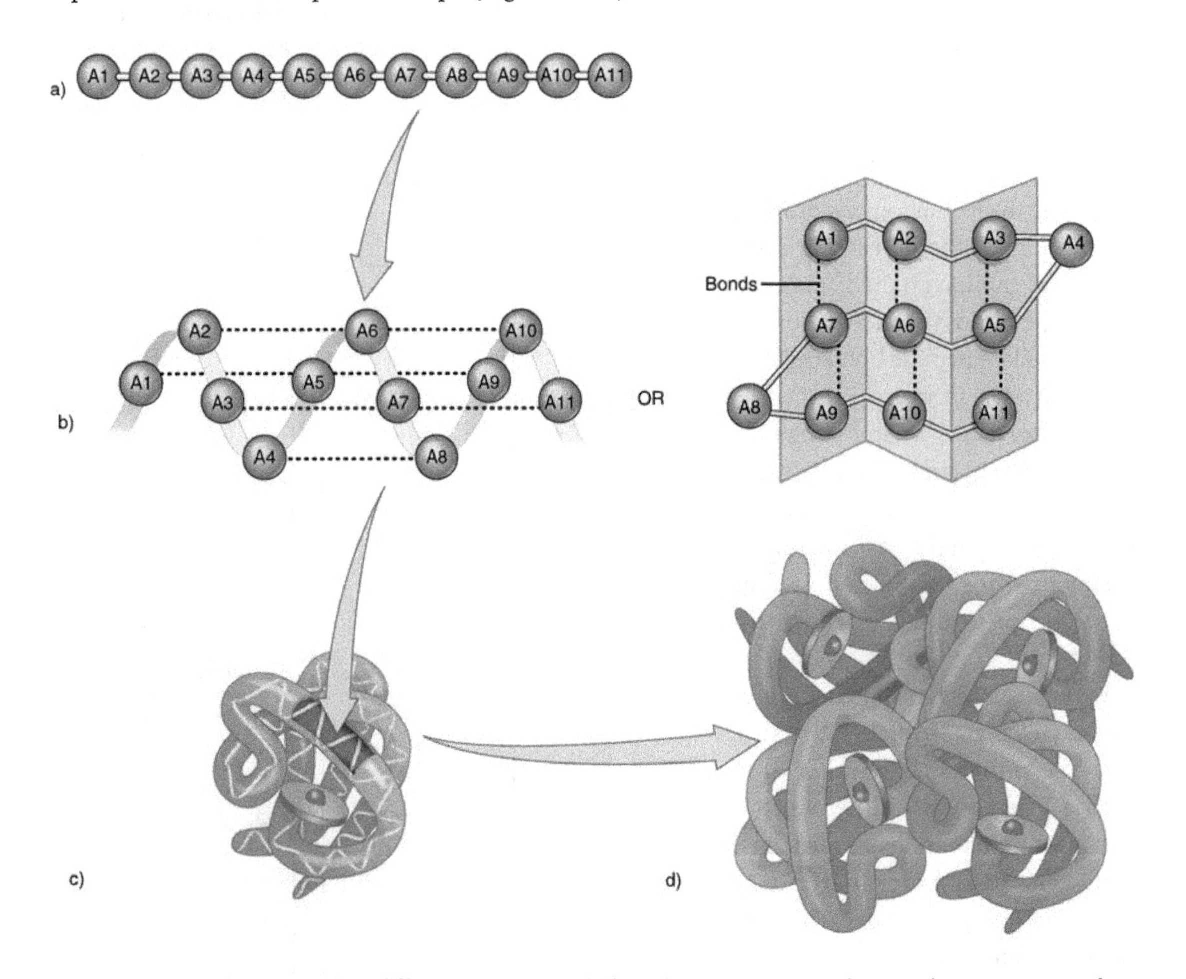

FIGURE 3.19 Proteins exist in four different structures. a) The primary structure exists as a long sequence of amino acids bonded together by peptide bonds. b) Folding of the primary structure produces a secondary structure maintained by hydrogen bonds, which can take on the form of either an alpha-helix (left) or beta-pleated sheet (right). c) The tertiary structure is produced because of continual folding of the secondary structure producing a three-dimensional form maintained by hydrogen bonding. d) A quaternary structure consists of the interaction between two or more tertiary structures held together by hydrogen bonding.

Some proteins form a **quaternary structure**, which occurs when a protein contains two or more small tertiary proteins held together by hydrogen or ionic bonds (Figure 3.19d). Hemoglobin, the protein responsible for transporting oxygen throughout the body, is a common example of a quaternary structure, consisting of four small globular proteins, which exhibit primary, secondary, and tertiary structures held together by hydrogen bonds.

As a protein folds into its final secondary or tertiary structure, this structure determines its function. However, if any disruption within the interactions responsible for maintaining this structure occurs, then the protein is said to be **denatured**. When a protein becomes denatured, it begins to unfold, resulting in a nonfunctional protein. There are several different environmental factors that can result in denaturation, including elevated temperatures, changes in pH, alcohols, and detergents.

Enzymes are catalytic proteins that increase the rate at which chemical reactions occur. Enzymes work by lowering activation energy, the energy needed for a chemical reaction to begin (Figure 3.20).

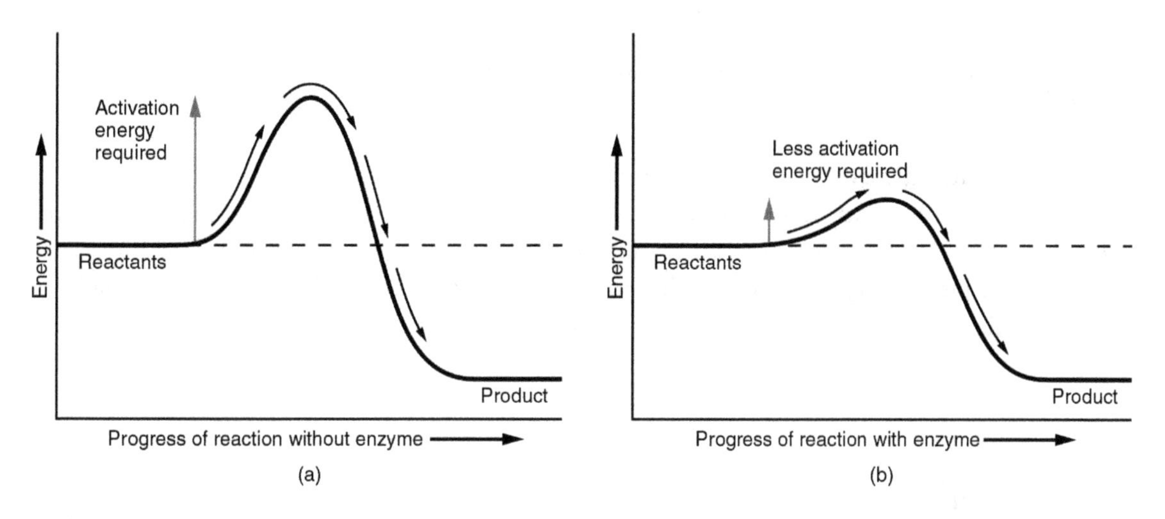

FIGURE 3.20 Enzymes help speed up chemical reactions by lowering activation energy, the energy needed to initiate a chemical reaction. Without an enzyme, energy input is considerably higher than with an enzyme.

All chemical reactions require a set amount of energy to begin; however, enzymes lower this energy to speed up the chemical reaction. Enzymatic activity is a four-step process and begins with the formation of an **enzyme-substrate complex** (Figure 3.21). Each enzyme contains an **active site**, a groove or pocket specific to a particular **substrate** or reactant, to which the substrate binds. The active site is the most important part of the enzyme, and its characteristics are important for optimal enzymatic activity. The active site is found on the surface of the enzyme, and its shape is complementarily shaped to that of the substrate. For example, if the active site is shaped like a triangle, the shape of the substrate is also that of a triangle. Once the substrate attaches to the active site, it is held in place by noncovalent interactions, which is important in allowing the final product to be easily released. This final attachment of the substrate to the active site forms the enzyme-substrate complex.

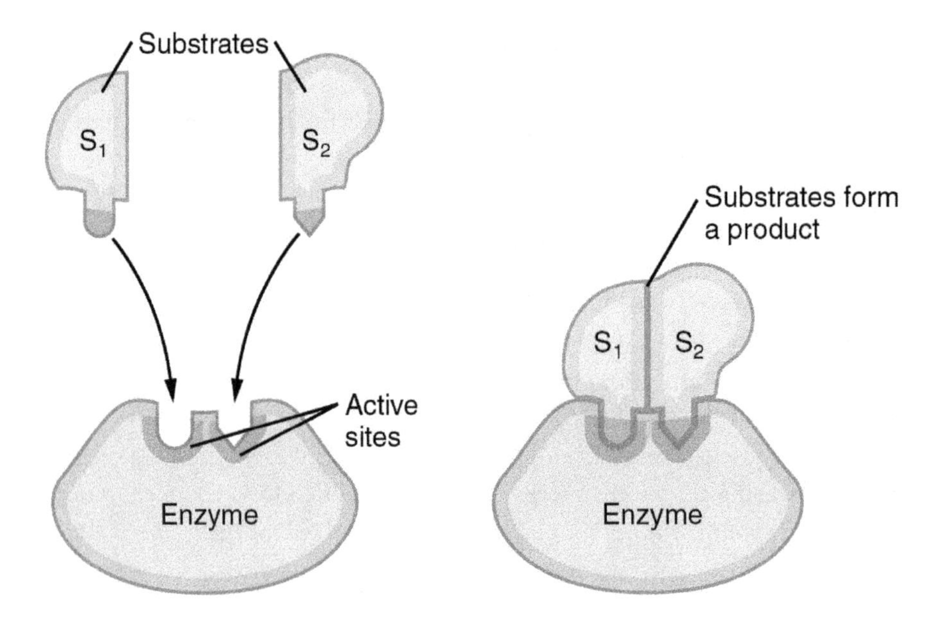

FIGURE 3.21 An enzyme-substrate complex forms when a substrate fits into an enzyme's active site. The shape of the substrate and active site complement one another for a tight fit.

There are currently two models that attempt to explain the interaction of the substrate with the active site to form the enzyme-substrate complex. The first model is known as the *lock-and-key model* (Figure 3.22), which describes the active site as rigid, with the substrate snapping into place, like that of two pieces of a jigsaw puzzle being put together. In this model, the active site is considered the lock, while the substrate is the key, with the substrate fitting perfectly within the active site.

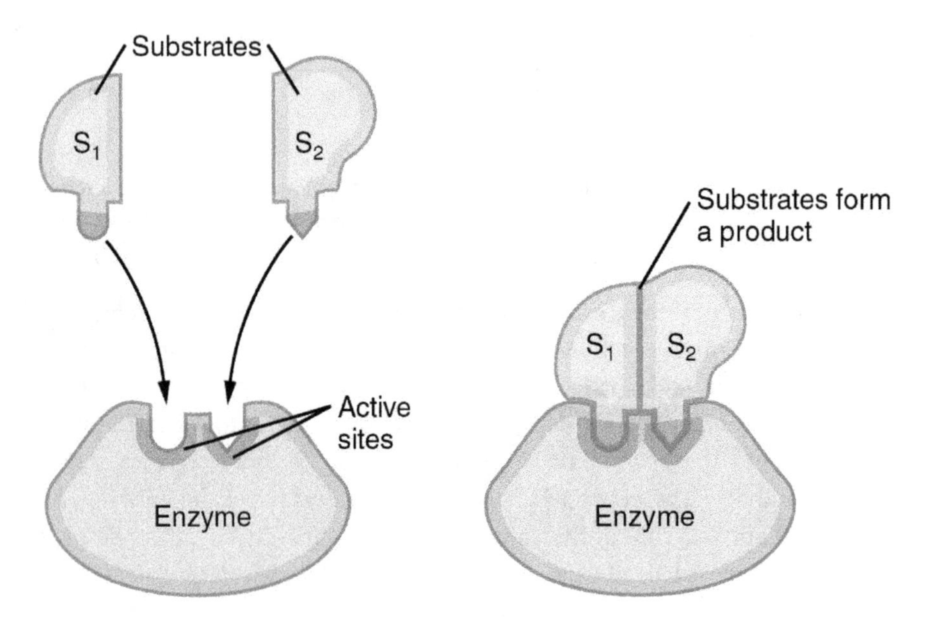

FIGURE 3.22 The lock-and-key model suggests that the substrate fits into the active site much like a jigsaw puzzle piece.

However, what this model fails to take into consideration are conformational changes a protein may undergo to accommodate a substrate. Therefore, a second model has been suggested that describes the active site as more flexible, not as rigid. This model is known as the *induced-fit model* (Figure 3.23). This model looks at the active site as a pocket that will "mold" itself around the substrate as it enters the active site, ultimately producing a perfect fit. Both models are generally accepted by scientists.

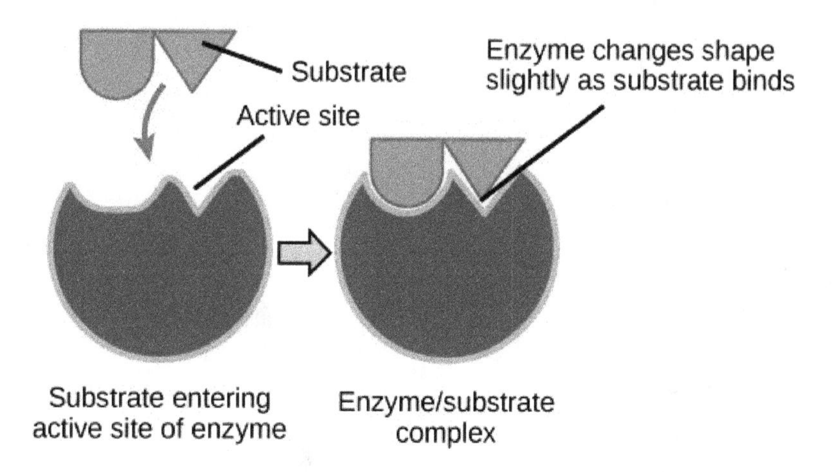

FIGURE 3.23 The inducted fit model takes into consideration conformational changes of the protein, therefore, suggests that the active site "molds" itself around the substrate.

Regardless of whether the enzymatic activity involves synthesizing a new molecule or breaking down an existing molecule, the pathway begins upon the formation of the enzyme-substrate complex. This step is proceeded by a series of molecular changes by which the substrate is converted to its final product(s) (Figure 3.24). The transition state is the step in which the shape of the substrate exhibits an intermediate form between the substrate and the final product. Product formation begins next, as the substrate is converted into the final product. The last step is when the product is released from the active site, leaving the enzyme unaltered and available to pick up more substrate.

Enzymatic activity occurs under specific chemical and physical factors. Any alteration in these factors can greatly affect how an enzyme works and the speed at which an enzyme catalyzes a reaction. These factors include temperature, pH, and enzyme/substrate concentrations.

Enzymatic activity increases as temperature increases; however, at extremely high temperatures, enzymes can become denatured, resulting in a decrease in reaction rates (Figure 3.25). Generally, enzymes work best at an optimum temperature for that enzyme, which for body enzymes is around 98.6°F (37°C).

Reaction rates increase as pH increases to an optimum level for that enzyme, but a pH either above or below that pH can greatly affect reaction rates (Figure 3.25). A pH of 7 is usually an ideal pH for most enzymes; however, some enzymes work best at a lower pH, such as pepsin, with an optimal pH of 2.

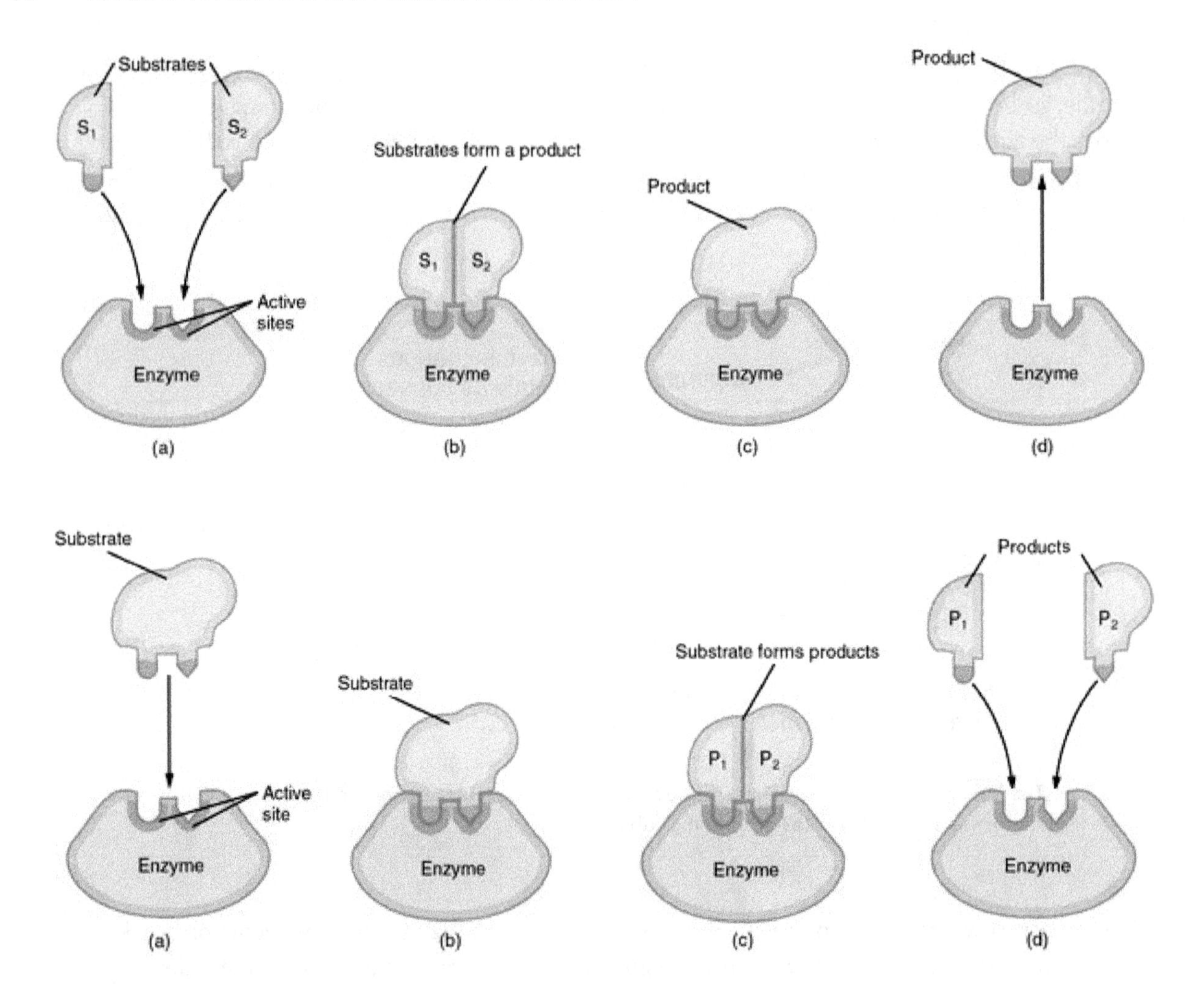

FIGURE 3.24 Once the enzyme-substrate complex forms (a & b), enzymatic activity proceeds with the transition state, in which the substrate exists in an intermediate state to the final product (b). During product formation, the substrate is converted to the final product (c) and product release (d) occurs when the final product(s) is/are released from the enzyme, making the enzyme available for another substrate molecule.

Finally, as the concentration of enzymes and substrates increases, reaction rates also increase (Figure 3.25). However, these rates only increase to a certain point, because once all the enzymes are bound with substrate, this leaves no enzymes available for catalyzing. As a result, additional increases in either enzymes or substrates do not increase reaction rates.

Activators and inhibitors (Figure 3.26) are groups of molecules that can influence the reaction rates of enzymes. As the names suggest, activators increase reaction rates, while inhibitors decrease reaction rates. There are two different types of inhibitors—competitive, which bind directly to the active site, blocking the ability of the substrate to attach, and noncompetitive, which bind to an area of the enzyme other than the active site, altering the shape of the active site and thereby blocking the ability of the substrate to attach.

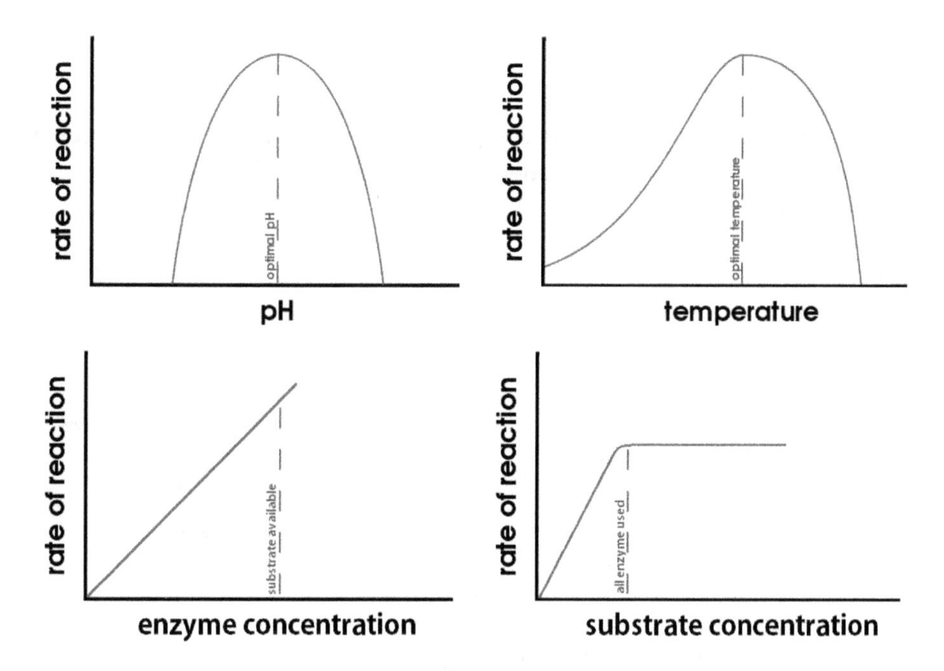

FIGURE 3.25 As temperature increases, reaction rates increase, but at high temperatures, reaction rates decrease due to the enzymes becoming denatured. Reaction rates increase as pH increases, but below or above the optimum pH of 7, the enzyme structure denatures, also influencing reaction rates. An increase in enzyme/substrate concentrations stimulates an increase in reaction rates, but only until all enzymes are bound with substrate.

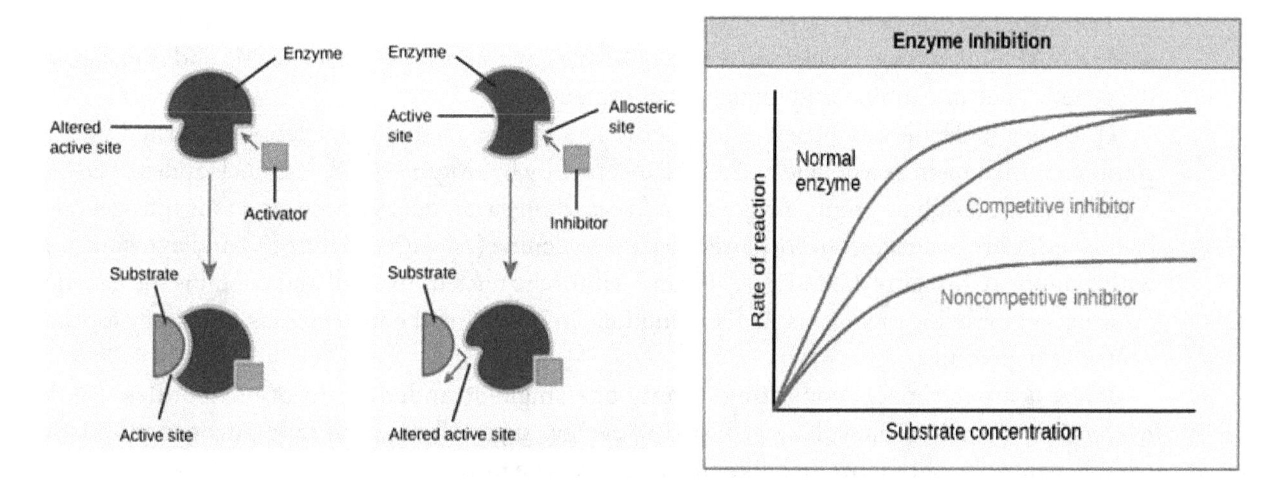

FIGURE 3.26 Activators (left) are molecules that can increase the reaction rate of enzymes and inhibitors (middle) are molecules that can decrease the reaction rate of enzymes. Inhibitors can exist as competitive inhibitors, which affect initial reaction rates because they attach directly to an enzyme's active site, while noncompetitive inhibitors affect maximal reaction rate by binding to an area other than the active site, thereby altering its overall shape (right).

3-5. Nucleic Acids

Nucleic acids are composed of a group of molecules linked together in a long chain known as **nucleotides**. Nucleotides have three main components, including a phosphate group, five-carbon sugar, and nitrogenous base (Figure 3.27).

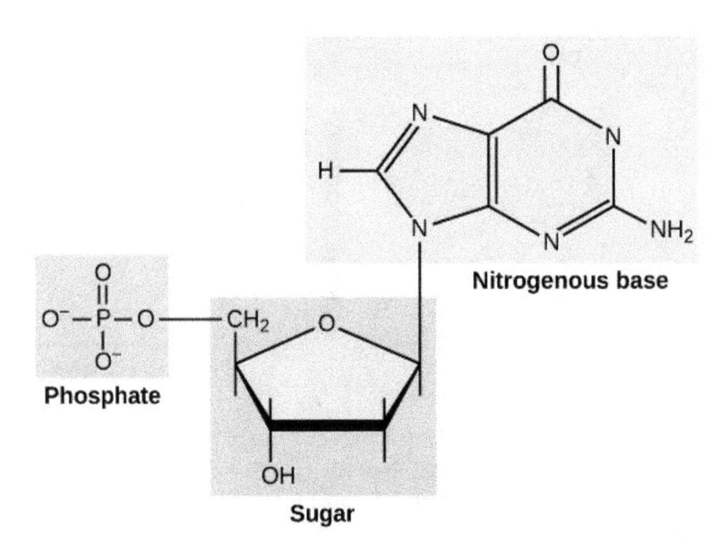

FIGURE 3.27 The basic structure of a nucleotide includes a phosphate group, a five-carbon sugar, and a nitrogenous base.

There are two different types of nucleic acids, **deoxyribonucleic acid (DNA)** and **ribonucleic acid (RNA)**. Both of these nucleic acid molecules play an important role in storing and translating the genetic code into amino acid sequences of proteins.

DNA is a molecule composed of long, double-stranded chains of nucleotides twisted into a spiral pattern, known as a double helix, held together by hydrogen bonds. The nucleotides of DNA consist of the phosphate group, along with a five-carbon sugar, deoxyribose, and four nitrogenous bases, which are complementarily paired together, adenine (A) with thymine (T) and cytosine (C) with guanine (G) (Figure 3.28). DNA is found within the nucleus of a cell and contains the genetic information for living organisms. It also functions in providing the instructions necessary for the synthesis of proteins.

RNA is a nucleic acid consisting of only one single-stranded chain of nucleotides. RNA nucleotides contain a phosphate group, five-carbon sugar ribose, and four nitrogenous bases complementarily paired together, like that of DNA. The nitrogenous bases of RNA are similar to DNA, with one exception: RNA contains uracil (U), not thymine (T). Nitrogenous base pairing is, however, the same, with adenine (A) pairing with uracil (U) and cytosine (C) with guanine (G) (Figure 3.29). RNA is located outside of the cell and is important in translating the DNA code to build proteins.

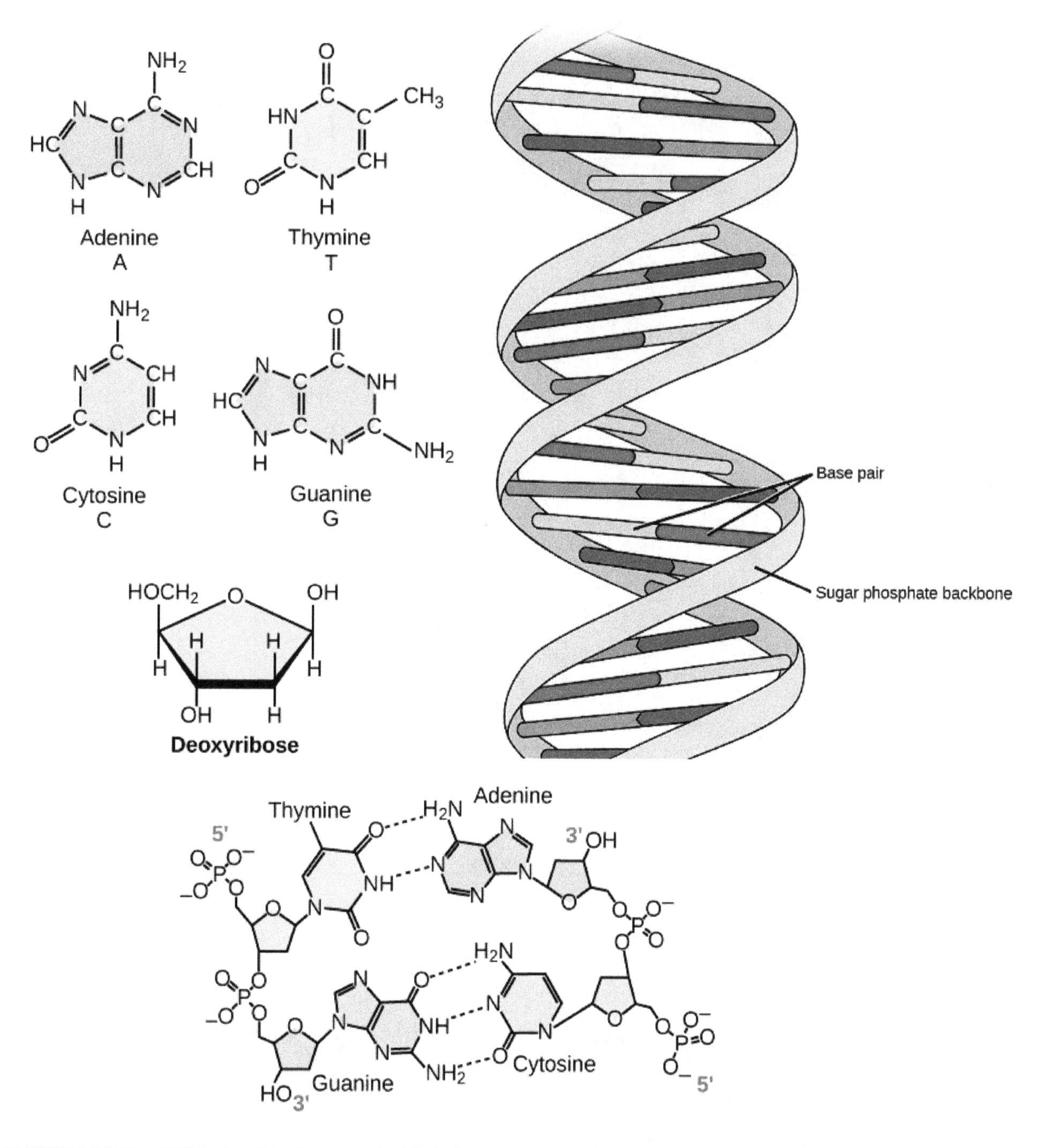

FIGURE 3.28 The DNA molecule exists as a double helix structure consisting of two sugar-phosphate backbones and nitrogenous bases that complementarily pair as follows: adenine (A) with thymine (T) and cytosine (C) with guanine (G). The nitrogenous base pairs are held together with hydrogen bonds.

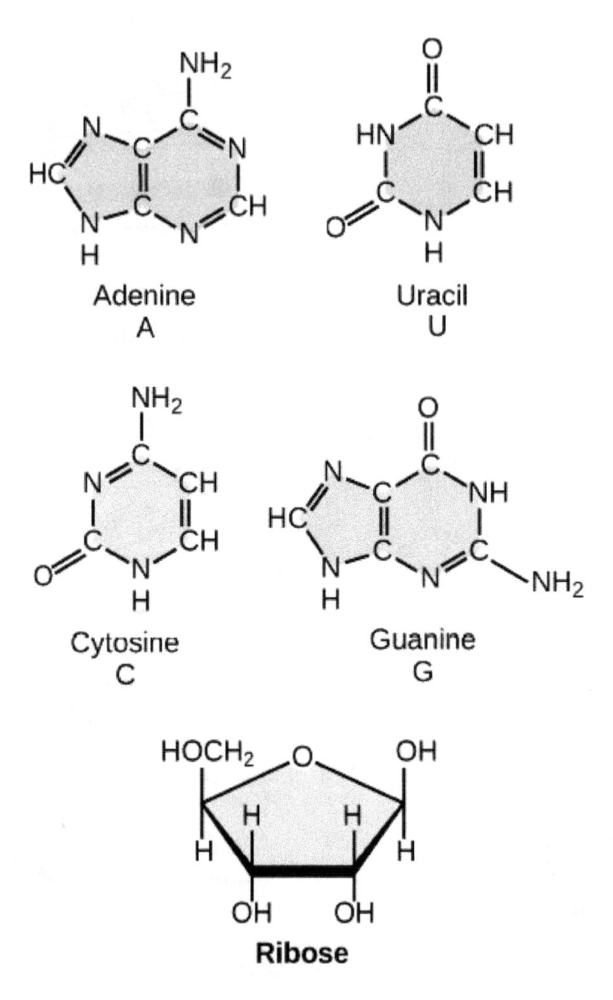

FIGURE 3.29 RNA molecules consist of a five-carbon sugar: ribose and four different nitrogenous bases: adenine (A), uracil (U), cytosine (C), and guanine (G), which also complementarily pair, with uracil (U) replacing thymine (T).

Chapter Summary

Below is a summary of the main ideas from each section of the chapter.

3-1. The Carbon Atom and Macromolecules

- Carbon is an important atom used to construct a variety of biological molecules required by all living organisms. Carbon has four valence electrons, which allows the atom to bond to various atoms, including hydrogen, oxygen, and sulfur. When carbon bonds to itself, it produces a carbon skeleton, providing the framework for building a biological molecule. The way carbon bonds to other atoms, including itself, results in a variety of shapes and structures, such as long chains, rings, and branches, thereby providing living organisms multiple ways to build biological molecules.

- Molecules composed of carbon and hydrogen atoms are called organic molecules. The chemical properties, overall structure, and function of an organic molecule are determined by a group of atoms attached to the carbon skeleton, known as a functional group.
- Macromolecules are large chain polymers composed of numerous small, repeating subunits called monomers covalently bonded together. Macromolecules are important organic molecules because they provide living organisms with the starting material required to build cellular structures. The four main classes of macromolecules are carbohydrates, lipids, proteins, and nucleic acids.

3-2. Carbohydrates

- Carbohydrates are macromolecules composed of carbon, hydrogen, and oxygen existing in a 1:2:1 ratio. Carbohydrates provide a structural role for a variety of organisms and play an important role as a major energy source for sustaining life. Carbohydrates can exist as simple or complex carbohydrates.
- Monosaccharides are simple carbohydrates consisting of one saccharide molecule. The most common example of a monosaccharide is glucose, which is an important carbohydrate that provides energy to all living organisms. Excess glucose is either stored as glycogen in muscle or liver cells or converted to triglyceride and stored in adipose tissue.
- Disaccharides are simple carbohydrates composed of two saccharide molecules bonded together by a glycosidic bond. A common example of a disaccharide is lactose (glucose + galactose), an important carbohydrate found in milk.
- Polysaccharides are complex carbohydrates composed of multiple saccharide molecules bonded together by glycosidic bonds. Polysaccharides function as energy storage molecules. Common examples of polysaccharides include starch, glycogen, cellulose, and chitin.

3-3. Lipids

- Lipids are macromolecules composed of carbon, hydrogen, and oxygen, but they lack the defined composition ratio found in carbohydrates. Due to the structural characteristics of lipids, they lack the ability to dissolve in water, making them hydrophobic, or water-fearing, molecules. Lipids provide a variety of biological functions, including energy storage, major energy source, structural, vitamin composition and absorption, protection, and insulation. There are four main types of lipids: triglycerides, sterols, phospholipids, and waxes.
- Triglycerides are fats composed of a glycerol head and three fatty acid chains or tails. Triglycerides can exist as saturated, unsaturated, or trans-fats. Saturated fats are characterized by their straight fatty acid tails, which enable them to tightly pack together and remain solids at room temperature. Unsaturated fats are characterized by the presence of kinks in their fatty acid tails, which make them less likely to tightly pack together and more likely to remain liquids at room temperature. Trans-fats result from a process known as hydrogenation, which changes the structural confirmation of unsaturated fats, resulting in a fat that more closely resembles saturated fats.

- Sterols are fats composed of four fused carbon skeletons with unique functional groups that define their biological function. Sterols are responsible for growth and development. Cholesterol is a common example of a sterol and plays a variety of important biological roles. Cholesterol is an integral component of cell membranes and is the structural precursor to steroid hormones such as estrogen and testosterone.
- Phospholipids are fats composed of a glycerol head, two fatty acid chains, and a phosphate group. Phospholipids are the main structural component of cell membranes. Phospholipids arrange themselves in a bilayer structure based on their reaction to water. The glycerol head, with the bonded phosphate group, is hydrophilic (water-loving) and faces outward, while the fatty acid tails are hydrophobic (water-fearing) and are oriented inward. The bilayer nature of the cell membrane enables the cell to regulate the movement of substances into or out of the cell.
- Waxes are fats consisting of one fatty acid chain, a glycerol head, and a long-chain alcohol. Waxes have a variety of characteristics, including high melting points, existing as solids at room temperature, and being hydrophobic. Waxes are found in the protective coating of plant leaves and stems and are responsible for maintaining animal skin and fur.

3-4. Proteins

- Proteins are macromolecules composed of amino acids bonded together by peptide bonds. Proteins have many different types of biological functions, including enzymes, defense, transport, regulation, structural, mechanical, movement, and storage.
- Amino acids are molecules that form proteins. There are 20 different types of amino acids, and the combination of these amino acids provides the function and structure of proteins. All amino acids contain the same basic molecules, including an alpha carbon, to which is bonded a hydrogen atom, an amino group, a carboxylic acid group, and an R-group. The R-group distinguishes the different types of amino acids and is responsible for how a protein folds into its unique three-dimensional configuration. The three-dimensional configuration determines the function of a protein.
- Proteins exist in one of four different structures, based on the amino acid composition of the protein. Primary protein structure consists of a simple, linear sequence of amino acids. As chains of amino acids begin to fold based on the location of the R-groups, they produce a secondary protein structure, which is maintained by hydrogen bonds. There are two different types of secondary protein structures produced by how the amino acids fold: α-helix or β-pleated sheet. A tertiary protein structure is formed as amino acids in a secondary protein structure continue folding because of how the R-groups of amino acids interact with water. Quaternary protein structures are formed when two or more small, tertiary protein structures are bonded together.
- A denatured protein is a protein whose structural interactions have been disrupted by means of various environmental factors (temperature, pH), causing it to unfold. The unfolding of a protein can result in overall loss of function.
- Enzymes are proteins that are responsible for speeding up chemical reactions by lowering activation energy (energy required to begin a chemical reaction). Enzymatic activity is a

four-step process, beginning with the formation of an enzyme-substrate complex (substrate fits into the active site of an enzyme), which can be explained using either the lock-and-key model (substrate snaps into the active site) or the induced-fit model (active site molds itself around the substrate). This step is followed by a transition state (substrates exists as an intermediate form), product formation (substrate is converted to a final product), and product release (product is released from the active site). Enzymatic activity can be affected by a variety of chemical and physical factors, including enzyme-substrate concentration, temperature, pH, and activators or inhibitors.

3-5. Nucleic Acids

- Nucleic acids are composed of molecules called nucleotides. Nucleotides have three main components: phosphate group, five-carbon sugar, and nitrogenous base. There are two different types of nucleic acids: deoxyribonucleic acid (DNA) and ribonucleic acid (RNA).
- Deoxyribonucleic acid (DNA) is composed of a double strand of nucleotides twisted into a helix. The nucleotides of DNA consist of a phosphate group, a five-carbon sugar (deoxyribose), and four nitrogenous bases (adenine, guanine, cytosine, and thymine). DNA is found within the nucleus of a cell and outlines the genetic makeup of an organism and provides a cell with the necessary instructions for building a protein.
- Ribonucleic acid is composed of a single strand of nucleotides. The nucleotides of RNA consist of a phosphate group, a five-carbon sugar (ribose), and four nitrogenous bases (adenine, guanine, cytosine, and uracil). RNA is found outside the cell and is responsible for translating the DNA genetic code to make proteins.

End-of-Chapter Activities and Questions

Directions: Please refer back to what you learned in this chapter to complete the following activities.

Define Each Term in Your Own Words

1. Macromolecule
2. 1:2:1 Ratio
3. Hydrophobic
4. Peptide Bond
5. Nucleotide

Chapter Review

1. A functional group is an important group of atoms that provides organic molecules with their unique function. Perform an online search for an example of an organic molecule and define the functional group(s) responsible for the molecule's function.
2. Define monosaccharide, disaccharide, and polysaccharide using your own words, and give an example of each.

3. How are saturated and unsaturated fatty acids structurally different from one another?
4. Enzymatic activity is greatly affected by both chemical and physical factors. List the different types of chemical and physical factors and explain how each affects enzymatic activity.
5. Outline the main differences between DNA and RNA.

Multiple Choice

1. The chemistry of the carbon atom is unique in that it ______________.
   - **a.** can form ionic bonds
   - **b.** has four valence electrons
   - **c.** is a relatively large atom
   - **d.** cannot bond to itself

2. What type of bond is found between two monosaccharides to form a disaccharide?
   - **a.** glycosidic bond
   - **b.** peptide bond
   - **c.** hydrogen bond
   - **d.** ionic bond

3. Which of the following best describes a triglyceride?
   - **a.** four fused carbon skeletons with a bonded functional group
   - **b.** glycerol head bonded to three fatty acid tails
   - **c.** fatty acid tail bonded to a glycerol head and long-chain alcohol
   - **d.** glycerol head bonded to two fatty acid tails and a phosphate group

4. An amino acid consists of all the following functional groups except:
   - **a.** carboxylic acid
   - **b.** R-group
   - **c.** phosphate group
   - **d.** amino group

5. Which of the following nitrogenous bases pertains to an RNA molecule and not to a DNA molecule?
   - **a.** adenine
   - **b.** thymine
   - **c.** uracil
   - **d.** guanine

Image Credits

Fig. 3.1: Source: https://commons.wikimedia.org/wiki/File:-Linus_Pauling.jpg.

Fig. 3.2: Source: https://commons.wikimedia.org/wiki/File:Lewis_dot_C.svg.

Fig. 3.3: Copyright © 2016 by istockphoto/jack0m.

Fig. 3.4: Copyright © by OpenStax (CC BY 4.0) at https://openstax.org/books/biology/pages/2-3-carbon.

Fig. 3.5: Copyright © by OpenStax (CC BY 4.0) at https://openstax.org/books/anatomy-and-physiology/pages/2-5-organic-compounds-essential-to-human-functioning.

Fig. 3.6a: Copyright © by OpenStax (CC BY 4.0) at https://openstax.org/books/anatomy-and-physiology/pages/2-5-organic-compounds-essential-to-human-functioning.

Fig. 3.6b: Copyright © by OpenStax (CC BY 4.0) at https://openstax.org/books/biology/pages/3-2-carbohydrates.

Fig. 3.7: Copyright © by OpenStax (CC BY 4.0) at https://openstax.org/books/concepts-biology/pages/2-3-biological-molecules.

Fig. 3.8: Copyright © by OpenStax (CC BY 4.0) at https://openstax.org/books/anatomy-and-physiology/pages/2-5-organic-compounds-essential-to-human-functioning.

Fig. 3.9: Copyright © by OpenStax (CC BY 4.0) at https://openstax.org/books/biology/pages/3-3-lipids.

Fig. 3.10: Copyright © by OpenStax (CC BY 4.0) at https://openstax.org/books/biology/pages/3-3-lipids.

Fig. 3.11: Copyright © by OpenStax (CC BY 4.0) at https://openstax.org/books/biology/pages/3-3-lipids.

Fig. 3.12: Copyright © by OpenStax (CC BY 4.0) at https://commons.wikimedia.org/wiki/File:Figure_37_01_01abc.jpg.

Fig. 3.13: Copyright © by OpenStax (CC BY 4.0) at https://openstax.org/books/biology-2e/pages/5-1-components-and-structure#fig-ch05_01_02.

Fig. 3.14: Copyright © by OpenStax (CC BY 4.0) at https://openstax.org/books/biology/pages/3-3-lipids.

Fig. 3.15: Copyright © by Roger Griffith / OpenStax (CC BY 4.0) at https://openstax.org/books/biology/pages/3-3-lipids.

Fig. 3.16: Copyright © by OpenStax (CC BY 4.0) at https://openstax.org/books/anatomy-and-physiology/pages/2-5-organic-compounds-essential-to-human-functioning.

Fig. 3.17: Copyright © by OpenStax (CC BY 4.0) at https://openstax.org/books/biology/pages/3-4-proteins.

Fig. 3.18: Copyright © by OpenStax (CC BY 4.0) at https://openstax.org/books/concepts-biology/pages/2-3-biological-molecules.

Fig. 3.19: Copyright © by OpenStax (CC BY 4.0) at https://openstax.org/books/anatomy-and-physiology/pages/2-5-organic-compounds-essential-to-human-functioning.

Fig. 3.20: Copyright © by OpenStax (CC BY 3.0) at https://commons.wikimedia.org/wiki/File:212_Enzymes-01.jpg.

Fig. 3.21: Copyright © by OpenStax (CC BY 4.0) at https://openstax.org/books/anatomy-and-physiology/pages/2-5-organic-compounds-essential-to-human-functioning.

Fig. 3.22: Copyright © by OpenStax (CC BY 4.0) at https://openstax.org/books/anatomy-and-physiology/pages/2-5-organic-compounds-essential-to-human-functioning.

Fig. 3.23: Copyright © by OpenStax (CC BY 4.0) at https://openstax.org/books/biology/pages/6-5-enzymes.

Fig. 3.24: Copyright © by OpenStax (CC BY 4.0) at https://openstax.org/books/anatomy-and-physiology/pages/2-5-organic-compounds-essential-to-human-functioning.

Fig. 3.26: Copyright © by domdomegg (CC BY 4.0) at https://commons.wikimedia.org/wiki/File:Effect_of_temperature_on_enzymes.svg.

Fig. 3.27: Copyright © by domdomegg (CC BY 4.0) at https://commons.wikimedia.org/wiki/File:Effect_of_pH_on_enzymes.svg.

Fig. 3.25c: Copyright © by domdomegg (CC BY 4.0) at https://commons.wikimedia.org/wiki/File:Effect_of_temperature_on_enzymes.svg.

Fig. 3.25d: Copyright © by domdomegg (CC BY 4.0) at https://commons.wikimedia.org/wiki/File:Effect_of_pH_on_enzymes.svg.

Fig. 3.26a: Copyright © by OpenStax (CC BY 4.0) at https://openstax.org/books/biology/pages/6-5-enzymes#fig-ch06_05_03.

Fig. 3.26b: Copyright © by OpenStax (CC BY 4.0) at https://openstax.org/books/biology/pages/6-5-enzymes#fig-ch06_05_04.

Fig. 3.29: Copyright © by OpenStax (CC BY 4.0) at https://openstax.org/books/concepts-biology/pages/2-3-biological-molecules.

Fig. 3.30a: Copyright © by OpenStax (CC BY 4.0) at https://openstax.org/books/anatomy-and-physiology/pages/2-5-organic-compounds-essential-to-human-functioning.

Fig. 3.30b: Copyright © by OpenStax (CC BY 4.0) at https://openstax.org/books/biology/pages/3-5-nucleic-acids.

Fig. 3.30c: Copyright © by OpenStax (CC BY 4.0) at https://openstax.org/books/anatomy-and-physiology/pages/2-5-organic-compounds-essential-to-human-functioning.

Fig. 3.31: Copyright © by OpenStax (CC BY 4.0) at https://openstax.org/books/anatomy-and-physiology/pages/2-5-organic-compounds-essential-to-human-functioning.

CHAPTER

4

Cell Structure and Function

PROFILES IN SCIENCE

Robert Hooke was born July 18, 1635, in Freshwater, a village located on the western end of England's Isle of Wight, and died in London, England, in 1703. Hooke's father was an English priest who largely presided over his early education, which may have resulted from Hooke's physical and emotional state after recovering from smallpox. Despite the family's expectation for him to join the church, Hooke was more interested in mechanics and art, which he began studying in 1648 at London's Westminster School, in addition to Greek, Latin, and mathematics. In 1653, Hooke enrolled at the University of Oxford's Christ Church College, where he studied experimental science and worked alongside famed chemist Robert Boyle (Boyle's Law). One of Hooke's first notable contributions to science was his law of elasticity, which stated that the force of tension in an elastic spring was directly proportional to its length. Hooke became a member of the Royal Society in 1663, holding the position of curator of experiments, where he demonstrated new experiments weekly and consulted and confirmed the early microscopy work of Antonie van Leeuwenhoek. In 1665, Hooke accepted a position as a professor of geometry at Gresham College in London, where he carried out many notable astronomical observations, including becoming the first to observe the rings of Saturn and the planetary rotations of both Mars and Jupiter. Many of Hooke's astronomical observations and views led to various disputes with fellow scientists, most notably Sir Isaac Newton. Also in 1665, Hooke published his most famous work, *Micrographia*, which provided detailed descriptions and illustrations of observations he made with microscopes and telescopes and included the

FIGURE 4.1 Robert Hooke.

description of the cork cells, from which the term *cells* originated. Hooke is also known for inventions that led to improvements in the science of timekeeping, his observations of fossils, and his work as one of the lead architects responsible for designing and rebuilding London after the great fire of 1666. As a result of his many contributions to science, Hooke is often considered one of the most brilliant English scientists of the 17th century, hence given the title by some as "the English Leonardo."

Introduction to the Chapter

According to the hierarchical organization of life, cells exhibit both structure and function, two characteristics scientists take into consideration when asserting that cells are the smallest form of life. Each individual cell is a microscopic structure surrounded by a complex and dynamic layer of lipids and proteins that provide the cell with a unique set of biological functions. Some cells contain important structures called organelles, which house the necessary biological machinery required for initiating chemical reactions and other cellular functions, such as making proteins, producing energy, and metabolism. All organisms are made of cells, but the number of cells composing an organism varies. Some life forms exist as single-celled organisms, like bacteria, while other organisms are multicellular, consisting of hundreds of thousands of role-specific cells, such as plants and animals. Although each individual cell has a specialized role, every cell within a living organism shares common features, and all ultimately work together as a cohesive whole in helping sustain life. In this chapter, we will introduce the scientists who were instrumental in building our early understanding of cells, compare and contrast the two main types of cells, and describe the various different organelles, how they communicate, and the two theories suggesting their origins.

Chapter Objectives

In this chapter, students will learn the following:

4-1. The observation of cells began with the invention of the simple microscope, which helped paved the way to the development of the field of cytology and the cell theory.

4-2. Prokaryotic cells are small, simple cells characterized by their lack of membrane-bound organelles and presence of unique cellular structures and features.

4-3. Eukaryotic cells are large, complex cells characterized by the presence of membrane-bound organelles, including a nucleus, and are classified into one taxonomic domain, Eukarya, which consists of algae, protozoa, fungi, plants, and animals.

4-4. Eukaryotic cells may contain a variety of different types of nonmembranous structures with important and specified functions, such as glycocalyx, cell wall, plasma membrane, cytoplasm, and cytoskeleton.

4-5. Eukaryotic cells contain many different types of membranous structures called organelles, which have important, specialized functions, including the nucleus, endoplasmic reticulum, Golgi apparatus, lysosomes, vacuoles, mitochondria, and chloroplasts.

4-6. Membrane contact sites are interactions that exist between organelles and are responsible for helping maintain a cell's normal, healthy physiology and homeostasis.

4-7. The endosymbiotic theory suggests a bacterial origin of mitochondria and chloroplasts and was comprehensively proposed because of work performed by early scientists, which has been supported by years of molecular and structural evidence.

4-8. The invagination theory suggests the origin of the nuclear membrane and unique membranous organelles, such as the endoplasmic reticulum and Golgi apparatus, occurred in a way similar to plasma membrane invagination in prokaryotic cells.

4-1. History of Cells

Cells are the microscopic structures that are the basis of all living organisms. All living organisms consist of cells, including some organisms with only one cell and others, like plants and animals, consisting of many different and complex types of cells. The term *cell* was first used by English scientist Robert Hooke, in 1665. Hooke used a primitive microscope (Figure 4.2) composed of lenses he created by melting strands of spun glass. He viewed various types of structures under this microscope, including fish scales and insects. However, his greatest observation was that of a thin piece of cork found on the outer bark of an oak tree. Hooke was observing boxes made from the nonliving cell walls of cork, and he called these boxes "cells." These "cells" reminded Hooke of cubicles (*cellulae*), which monks would occupy to study and pray (Figure 4.2).

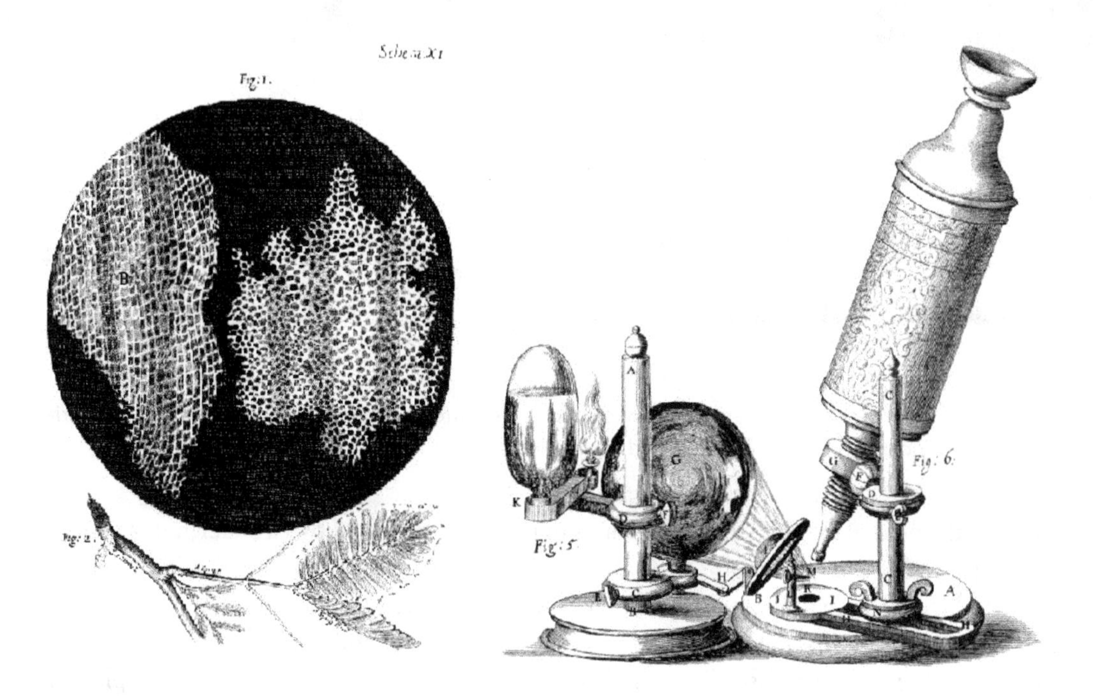

FIGURE 4.2 Upon viewing cork cells under his microscope, Robert Hooke coined the term *cellulae* to describe the "box" shaped structures that resembled cubicles where monks would study and pray.

In 1673, improvements were made to microscope lenses with the invention of the simple microscope by Dutch tailor and naturalist Antonie van Leeuwenhoek (Figure 4.3). Leeuwenhoek's invention appeared more primitive than the microscopes of the time, but his single lens provided a much clearer image and more effective magnification, up to 266 times. He used this microscope to observe different types of living organisms in various types of environments. He was able to view aquatic organisms, which he named "animalcules," (Figure 4.3) along with other types of cells, including blood cells, sperm cells, and the eggs of insects. He went on to build several hundred microscopes over the next few years, and even invited Robert Hooke to use these microscopes in 1674. Leeuwenhoek's invention paved the way for the disciplines of microbiology and cell biology.

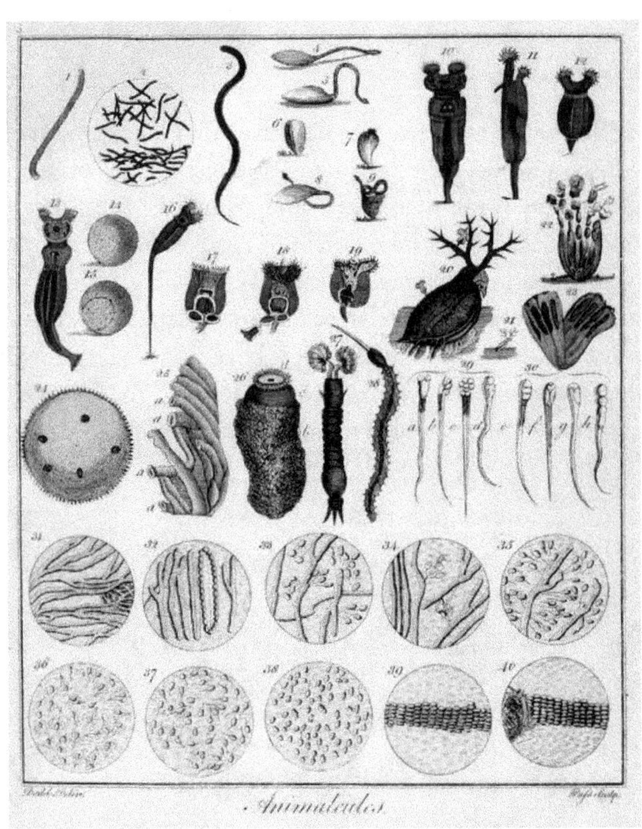

FIGURE 4.3 Leeuwenhoek's simple single lens microscope was effective in allowing him to observe a variety of different of cells, including those that he named "animalicules."

In 1839, the field of cytology began because of studies performed by two German scientists and the development of the cell theory. In 1838, Matthias Schleiden (Figure 4.4 - middle), a German botanist, published an article entitled "Contributions to Our Knowledge of Phytogenesis," wherein he claimed that all plants were composed of cells. In 1839, Theodor Schwann (Figure 4.4, left), a German zoologist, while observing cartilage made the claim that all animals, like plants, were composed of cells. Working alongside Schleiden, Schwann published the *cell theory* in 1839, which stated that all living organisms are composed of cells.

FIGURE 4.4 Three scientists are credited for the development of the cell theory, including Theodor Schwann (left), Matthias Schleiden (middle), & Rudolf Virchow (right).

In 1855, German scientist Rudolf Virchow (Figure 4.4, right) expanded the idea of the cell theory. Virchow was a physiologist and claimed the human body was composed of a sum of individual units, with each unit exhibiting the characteristics of life. This claim was added to the cell theory by stating that all organisms are composed of cells, which are the basic units of both structure and function. Virchow also predicted that every cell originated from another cell similar to itself and that the only source of a living cell was another living cell. Therefore, the remaining piece of the cell theory was added, which stated that cells arise from preexisting cells.

4-2. Prokaryotic Cells

There are two main types of cells that exist, prokaryotic and eukaryotic (which will be discussed in the next section). **Prokaryotic cells** are the smallest and simplest of all cells, and their name is derived from the lack of membrane-bound organelles, including a nucleus. Prokaryotic cells are found in various environments, including in the air, water, and soil, and even living within or on other organisms, including humans.

Prokaryotic cells are divided into two taxonomic domains based on differences in their DNA and RNA sequences—*Bacteria* and *Archaea*—both of which are discussed in more detail in chapter 12. Although prokaryotic cells have many similar features in common with eukaryotic cells, such as a plasma membrane, cytoplasm, DNA, and ribosomes, they also have unique structures that differentiate them and give them the unique ability to interact well within their specialized habitats.

Key Structural Features

Prokaryotes (Figure 4.5) are small organisms whose cellular physiology is contained within a single cell not much larger than 10 μm in diameter, surrounded by a cell wall. The **cell wall**, located directly outside the plasma membrane, functions in giving prokaryotic cells their unique shape

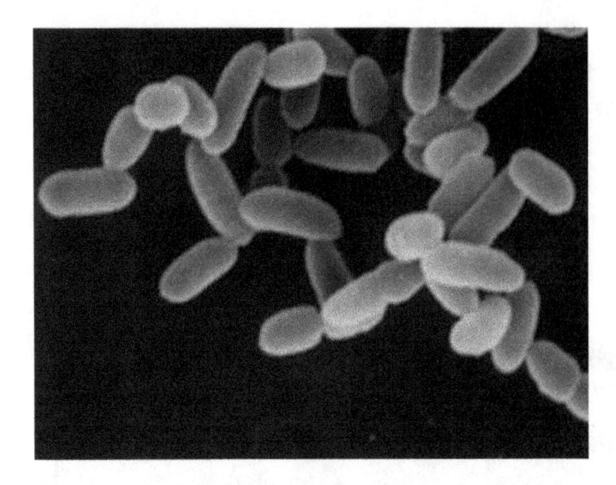

FIGURE 4.5 Prokaryotes are small, simple organisms that lack a nucleus and other membranous organelles. However, these cells have unique, specialized structures that make them successful in their habitats.

(spherical, rod, spiral, rectangular, or square) and acting as a protective barrier to prevent cellular crenation or lysis due to changes to internal cellular volume. Some prokaryotic cells secrete a gelatinous, sticky substance outside the cell wall called a **glycocalyx**, which is more commonly found in bacterial than archaeal cells. The glycocalyx, generally composed of polysaccharides, proteins, or a combination of both, is produced within the cell and secreted externally on its surface. If the glycocalyx is thin and unorganized, it forms a **slime layer**, while a thicker, well-organized glycocalyx is called a **capsule**. Both external layers enable prokaryotic cells to adhere to different surfaces or to each other and offer protection from dehydration. In certain bacterial cells, however, the capsule provides pathogenicity by protecting the cell from a host cell's immune system. Another structure formed by bacterial cells known to cause pathogenicity are **endospores**, which are protective structures that enable cells to remain dormant and viable when unfavorable environmental conditions prevent normal cellular vitality. Finally, some prokaryotic cells have additional external structures extending beyond the surface of the cell wall and glycocalyx that enable movement and attachment, such as flagella, fimbriae, and pili.

Flagella are long, proteinaceous structures that provide movement, while **fimbriae** are short, rodlike fibers extending from the surface of a prokaryotic cell that enable the cells to attach to a variety of substrates or to other prokaryotic cells. A special type of fimbriae, known as **pili** (singular—*pilus*), are external structures often referred to as sex pili, because prokaryotic cells use these structures to attach to other cells to exchange genetic material during a process known as conjugation, which will be discussed shortly.

The internal structures of prokaryotic cells include DNA and ribosomes. Located within the cytoplasm is a nonmembranous region, known as the **nucleoid**, that contains the prokaryotic DNA, which exists as a single circular chromosome. In addition, most prokaryotic cells have plasmids, small, extrachromosomal rings of DNA that typically lack genes essential to their survival. However, in certain prokaryotes, the plasmids may carry genes that confer genetic advantages, such as antibiotic resistance, or assist in the transfer of genetic material from one bacterial cell to another by conjugation. Plasmids are also important structures used in genetic engineering. Finally, prokaryotic ribosomes are responsible for synthesizing proteins within the cell and are usually a primary target in the use of certain antibiotics.

4-3. Eukaryotic Cells

Eukaryotic cells, such as animal and plant cells (Figure 4.6), are a group of cells that differ from prokaryotic cells in many ways. Eukaryotic cells are larger than prokaryotic cells, with diameters usually ranging between 10 to 100 μm in diameter. However, in the human body, the cell with the smallest diameter is a sperm cell (1–3 μm), while the largest is the egg cell (1000 μm or 0.1 mm). In addition, eukaryotic cells are more complex, contributed to by the presence of internal structures found within the cytoplasm responsible for compartmentalizing cellular functions. These structures are membrane-bound organelles, which have specific structures and specialized functions similar to tiny organs. Some organelles are important in providing compartments that house specific biochemical reactions so that they can occur without disrupting the cell. Other organelles perform such functions as deriving energy from nutrients, protecting the cell from harsh chemicals, degrading debris, and helping the cell reproduce. The final distinction between eukaryotic and prokaryotic cells is the presence of a nucleus. The nucleus is the most prominent structure found within the eukaryotic cell and is responsible for housing and protecting the cell's DNA, in addition to communicating with the cell's organelles and coordinating all cellular function responsible for life and reproduction. These group of cells are found in domain *Eukarya* and include the cells of algae, protozoa, fungi, plants, and animals.

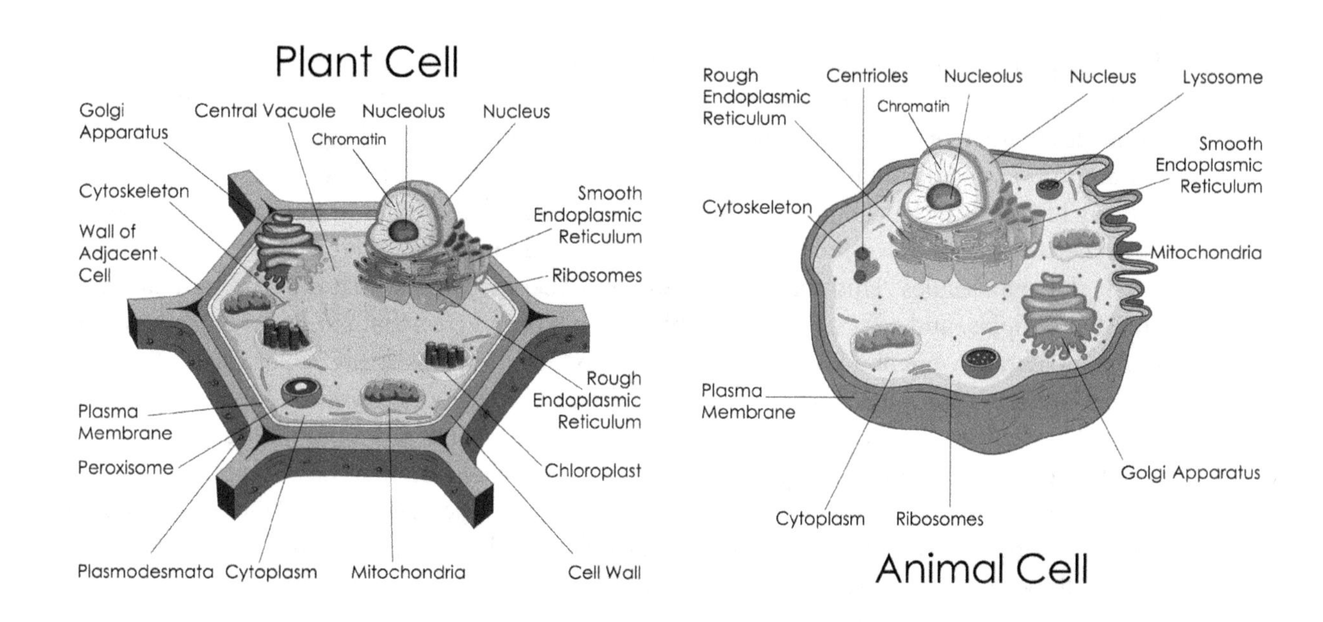

FIGURE 4.6 Animal and plant cells are typical eukaryotic cells that have many of the same types of organelles and cellular structures. However, plant cells contain a cell wall, chloroplasts, and vacuoles, cellular structures not found in animal cells.

4-4. Eukaryotic Cells: Nonmembranous Structures

Glycocalyx

Some eukaryotic cells have glycocalyces, including animal and protozoan cells, and are similar to those structures found in prokaryotic cells, although not as structurally organized. Since most protozoans and animals lack cell walls, the cells may have a glycocalyx anchored to their plasma membranes, which function in anchoring the cells to other cells, strengthening the cell, protecting against desiccation, and cellular communication. Glycocalyces are not present in eukaryotic cells that have cell walls, including fungi and plant cells.

Cell Wall

The cell wall is a rigid structure located outside of the plasma membrane and is found in some protozoan, algae, fungi, and plant cells (Figure 4.6). Cell walls were first observed by Robert Hooke in 1665 as he was viewing the nonliving plant cell walls of cork. These structures have important functions, including providing protection, anchoring cells to one another, and providing shape and structure. The cell walls of eukaryotic cells are not composed of peptidoglycan, as found in prokaryotic cells, but are composed of different types of polysaccharides. The most common types of polysaccharides found in the cell walls of eukaryotic cells include cellulose (plants) and chitin (fungi). Algal cell walls generally consist of a variety of polysaccharides and chemicals, such as cellulose, proteins, agar, and/or calcium carbonate.

Plasma Membrane

The plasma membrane is a complex and dynamic structure encompassing a cell and its contents (Figure 4.6). The plasma membrane has several different functions, including isolating the cell from its outside environment, regulating the movement of substances into and out of the cell, cell-to-cell communication, cellular attachments, and regulating biochemical reactions.

Early ideas about the composition of the plasma membrane began in the early 1900s, when researchers discovered that lipid-soluble molecules could pass through cells more rapidly than water-soluble molecules. This observation led researchers to suggest that the plasma membrane consisted of lipids, and after chemical analysis, it was discovered that the lipids composing the plasma membrane were phospholipids. In 1925, two Dutch scientists, Evert Gorter and François Grendel, extracted phospholipids from a sample of red blood cells and placed them on the surface of water. The scientists discovered that the lipids spread out in a single layer, but the area covered was almost twice the surface area of the red blood cells. They concluded that the phospholipid surface must consist of two layers. Based on the results of their experiment, Gorter and Grendel proposed that the structure was composed of two opposing molecular layers, arranged in a manner in which hydrophilic polar heads were directed outward and hydrophobic nonpolar tails were directed inward.

Additional research on the plasma membrane continued, with the hypothesis that the membrane contained proteins supporting characteristics such as surface tension and permeability. In 1935, two English scientists, Hugh Davson and James Danielli, proposed a "protein-sandwich" model, which helped explain the role proteins played in plasma membranes. The model suggested that the lipid bilayers existed between thin sheets of proteins. In 1957, J. David Robertson, an American scientist and premier electron microscopist of cell membranes, proposed another plasma membrane model introduced as the "unit membrane" model. Robertson's model suggested that all plasma membranes shared a similar structure, one in which a lipid bilayer was covered on both sides with thin sheets of proteins, a structure similar to that proposed by Davson and Danielli in 1935. Unfortunately, Robertson's proposal did not take into consideration that many proteins either penetrated the surface or diffused laterally through the plasma membrane.

The current understanding of how phospholipids and proteins are arranged within the plasma membrane can be explained by the **fluid mosaic model**. This model was first introduced in 1972 by American scientists S. J. Singer and Garth L. Nicolson in their paper entitled "The Fluid Mosaic Model of the Structure of Cell Membranes." In the paper, the scientists suggested that the plasma membrane exhibited a mosaiclike structure, with randomly distributed proteins alternating within sections of the phospholipid bilayer (Figure 4.7). The authors continued by stating that the lipids were of a fluid state, suggesting a dynamic structure containing proteins and lipids laterally flowing within the membrane.

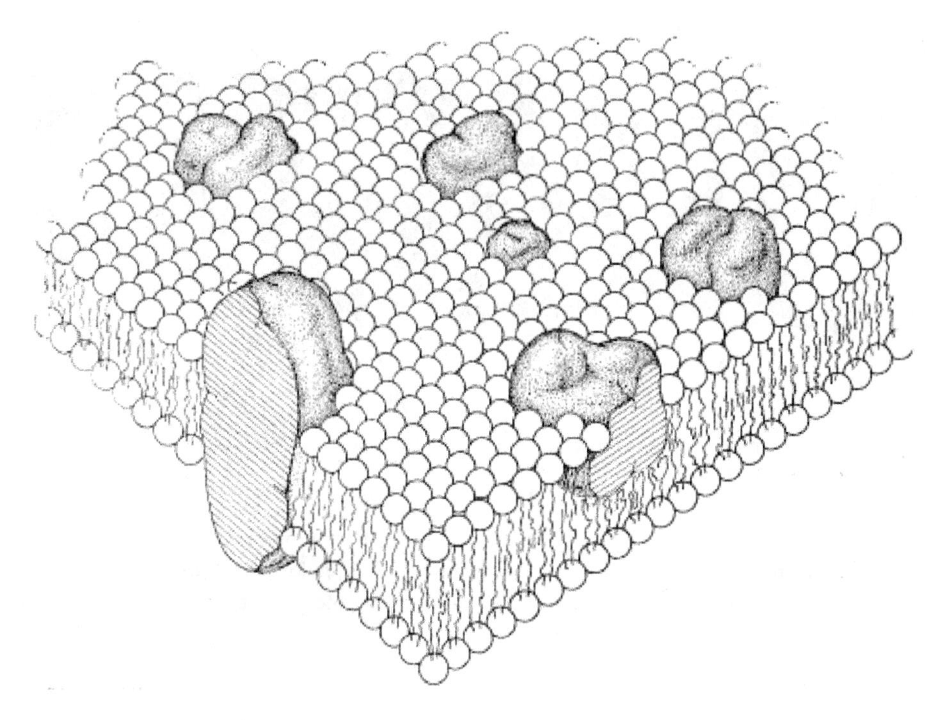

FIGURE 4.7 The eukaryotic plasma membrane, proposed by Singer and Nicolson, shows randomly distributed globular proteins embedded into a phospholipid bilayer representing a mosaic-like structure.

Eukaryotic plasma membranes, as proposed by Singer and Nicolson, are composed of a mosaic of phospholipids and proteins. The phospholipids are arranged in a bilayer consisting of hydrophilic polar heads directed outward and hydrophobic nonpolar tails directed inward. In addition, the phospholipids are not individually bonded to one another, and at times, the phospholipids may contain unsaturated fatty acids, which contain "kinks" that can increase the fluidity of the membrane as the number of unsaturated fatty acids increases. Both characteristics enable the phospholipids to move around within each layer, giving the plasma membrane its fluidlike characteristic, about the consistency of olive oil. Also, these features allow membranes to perform different functions and function better in different types of environments. Generally, the higher the temperature, the more fluid the membrane, while at lower temperatures, the membrane loses fluidity. The phospholipids are primarily responsible for isolating cellular content from the external environment and are considered the "fluid" portion of the membrane.

The plasma membrane of animal cells consists of another type of lipid, cholesterol (Figure 4.8), while plant cells have low levels of cholesterol or other types of related steroid-based lipids.

The number of cholesterol molecules located between the phospholipids may range from a few to an amount that outnumbers the phospholipid molecules. Cholesterol is an important molecule for several reasons. It is important in maintaining plasma membrane strength and flexibility, maintaining membrane fluidity at different temperatures, and making the plasma membrane less permeable to water-soluble molecules, such as ions and simple sugars.

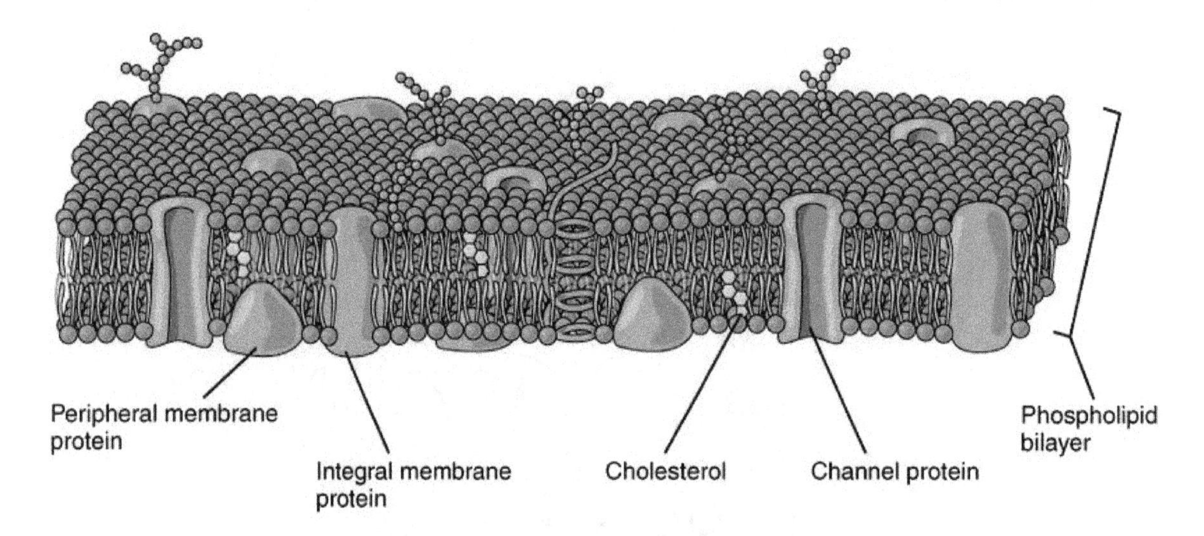

FIGURE 4.8 The plasma membrane consists of cholesterol, an important membrane lipid located between the phospholipids and helps maintain the membrane's strength, flexibility, fluidity, and solubility. The membrane also possesses two types of proteins called integral and peripheral proteins. Integral proteins are embedded, while peripheral proteins are located on the surface.

Proteins, which represent the "mosaic" characteristic proposed by Singer and Nicolson, are located about every 50 to 100 phospholipids, with thousands of proteins either embedded in or attached to the surface of the plasma membrane.

More than 70 percent of the proteins found within the plasma membrane are called **integral proteins** (Figure 4.8). These proteins are globular in shape, with a majority found spanning the entire membrane (**transmembrane proteins**) or only penetrating about half of the membrane. Singer and Nicolson suggested that integral proteins play a critical role in maintaining the structural integrity of the plasma membrane.

The other proteins associated with the plasma membrane are found loosely attached either to the outside or to the inside surface of the plasma membrane. These proteins are called **peripheral proteins** (Figure 4.8), and the number and types of proteins vary within the plasma membrane, producing an irregular, mosaiclike pattern. The proteins are primarily held in place and oriented within the plasma membrane through hydrophobic and hydrophilic interactions. In addition, the proteins can also be attached to various protein fibers either inside or outside the cell or by means of complex carbohydrate molecules, which act as a fingerprint for the cell, enabling it to be recognized by other cells. These attachments help prevent the proteins from moving around within the fluid phospholipid bilayer.

There are four main types of plasma membrane proteins that are responsible for determining most of the membrane's functions.

Receptor Proteins

Receptor proteins can exist as either transmembrane or peripheral proteins and contain a binding site that complements the shape of a specific molecule, like a hormone. When the molecule binds to the binding site, it often results in a change in the protein, initiating a series of chemical reactions that produce a cellular response.

Recognition Proteins

Recognition proteins can either be transmembrane or peripheral proteins and are responsible for identifying specific cells in the body and differentiating them from foreign invaders, such as bacteria or viruses. These types of proteins are important in helping the immune system, for example, target pathogens that invade our body, instead of our body's own cells.

Transport Proteins

Transport proteins are transmembrane proteins that enable substances to pass from one side of the plasma membrane to the other. There are two different types of transport proteins: **channel proteins**, which form channels that allow molecules to pass freely across the plasma membrane, and **carrier proteins**, which contain binding sites where a molecule will attach, changing the shape of the protein and, along with an input of energy, moving the molecule across the plasma membrane to the other side.

Enzyme Proteins

Enzyme proteins are either transmembrane or peripheral and are responsible for increasing the speed of metabolic reactions needed for the synthesis and breakdown of biological molecules. These proteins can initiate these reactions either on the surface of the plasma membrane or inside or outside the plasma membrane.

Movement of Substances Across the Membrane

One of the main functions of the plasma membrane is regulating the movement of substances into and out of the cell. The plasma membrane regulates the movement of necessary nutrients into the cell in order for it to function properly and to produce important molecules, while also removing metabolic wastes and newly formed molecules needed elsewhere in the body. The types of molecules that move into and out of the cell are regulated as a result of the plasma membrane being **selectively permeable** to certain substances. In other words, some substances can move across the membrane, but others cannot. For example, gases, such as carbon dioxide and oxygen can freely cross the plasma membrane, while other materials, like ions, large molecules (glucose, amino acids), and water, move slowly and generally require assistance from proteins within the plasma membrane. The movement of substances across the plasma membrane can occur in a variety of ways.

Passive Transport

Passive transport is the spontaneous movement of substances across the plasma membrane without a direct input of energy from the cell. As substances move across the plasma membrane, they move "down" a concentration gradient, which refers to the substance moving from an area of high concentration to an area of low concentration. Examples of passive transport include diffusion and osmosis.

Diffusion is the movement of molecules across the plasma membrane from higher to lower concentration until dynamic equilibrium has been reached (Figure 4.9). There are two types of diffusion. **Simple diffusion** occurs when molecules freely move without assistance from other types of molecules and can occur even without the presence of a membrane. Only small or lipid-soluble molecules, such as gases and fatty acids, can diffuse through the plasma membrane without issue.

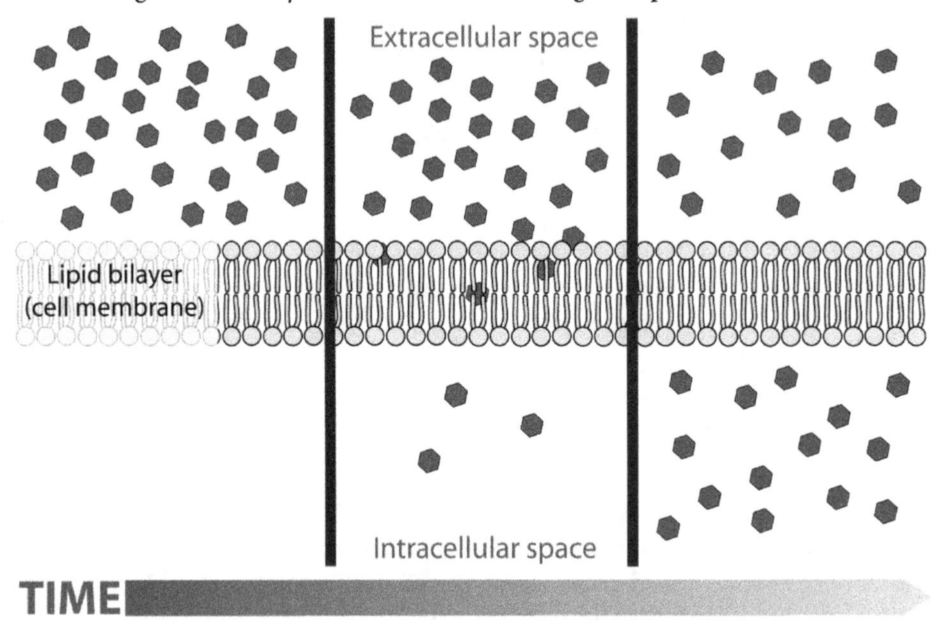

FIGURE 4.9 Diffusion involves the movement of molecules across a selectively permeable membrane from an area of high concentration to an area of lower concentration.

However, larger molecules, such as glucose, proteins, and ions, cannot diffuse through the plasma membrane without assistance; therefore, they rely on a process known as **facilitated diffusion**. Facilitated diffusion is the movement of molecules across the plasma membrane requiring the assistance of transport or carrier proteins (Figure 4.10). There are several factors that influence how quickly diffusion occurs, including temperature, molecular size, and solubility.

Osmosis is another example of passive transport and is the diffusion of water from an area of high concentration to an area of lower concentration (Figure 4.11).

Due to the hydrophobic nature of the phospholipid bilayer, the passive movement of water across the plasma membrane is facilitated by specialized water channel proteins embedded within the membrane known as **aquaporins** (Figure 4.12).

The movement of water across the plasma membrane is also dependent on the relationship between solute (dissolved substances) concentrations inside and outside of the cell, referred to as **tonicity**. A **hypertonic** solution (Figure 4.13, left) is a solution in which the concentration of solutes is higher outside the cell (less water) than the concentration inside the cell (more water).

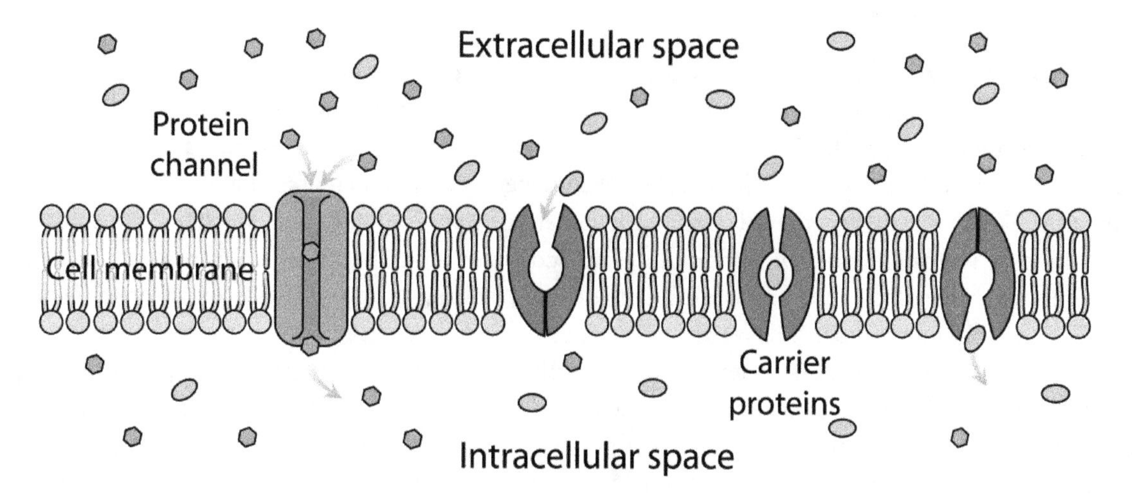

FIGURE 4.10 Facilitated diffusion requires either transport or carrier proteins to move larger molecules, like glucose or proteins, across the plasma membrane.

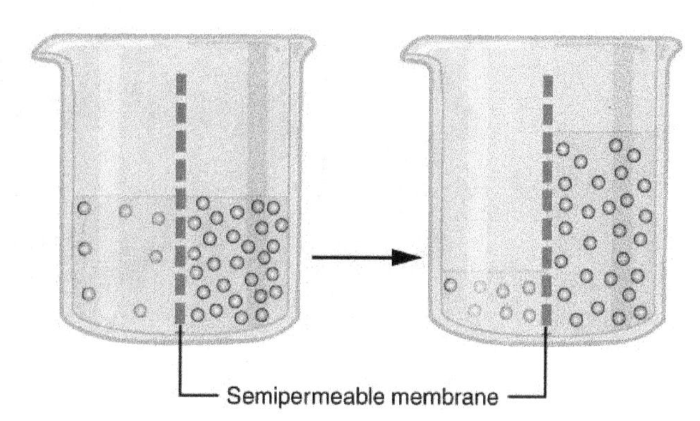

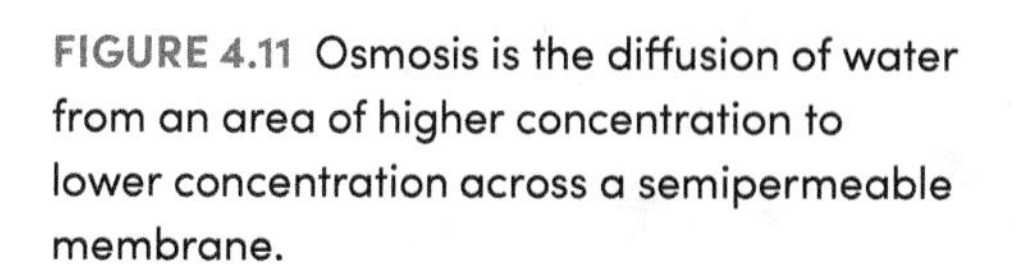
FIGURE 4.11 Osmosis is the diffusion of water from an area of higher concentration to lower concentration across a semipermeable membrane.

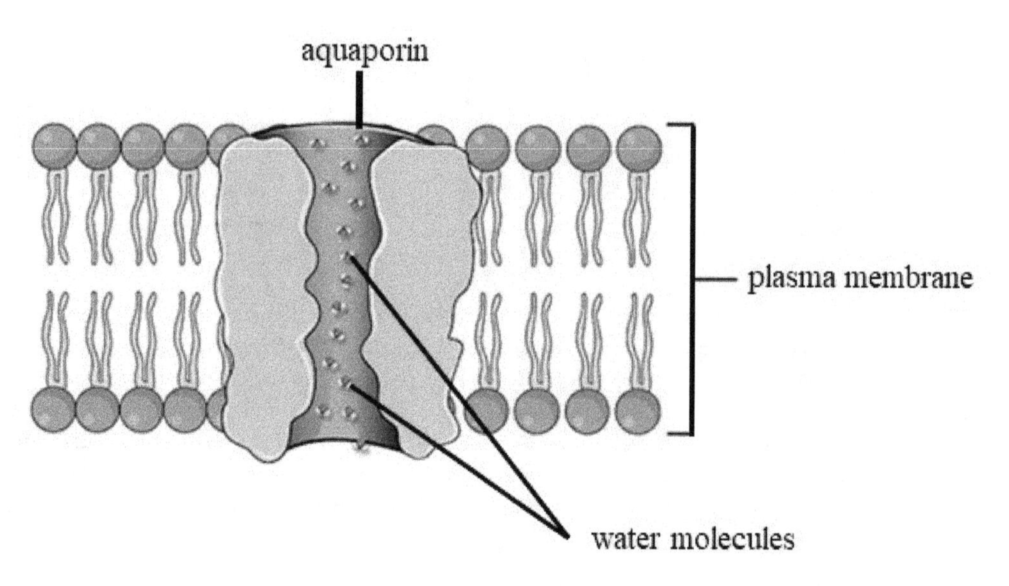

FIGURE 4.12 Aquaporins are water channel proteins required for the movement of water across the plasma membrane.

This results in water leaving the cell, which causes the cell to shrink or shrivel and potentially die. Organisms such as marine animals that are exposed to a hypertonic environment can prevent water loss in their cells by excreting excess salt across their gills or through specialized glands. Plant cells, when exposed to a hypertonic solution, will lose water, causing the plasma membrane to pull away from the cell wall, resulting in death. When the concentration of solutes is equal on both sides of the plasma membrane and there is no net movement of water between the two sides, the solution is said to be **isotonic** (Figure 4.13, middle). If the concentration of solutes is lower outside the cell (more water) than the concentration inside the cell (less water), the solution is **hypotonic** (Figure 4.13, right). Hypotonic solutions result in the movement of water from outside the cell to inside the cell, which can cause a cell to expand and potentially burst due to excess pressure. However, in the case of plant cells, the cell wall protects the cell from bursting, and the increase in pressure is important for plants to maintain their erect position. Other organisms, such as freshwater protozoans, prevent the uptake of excess water by removing the water through specialized structures called contractile vacuoles (described shortly).

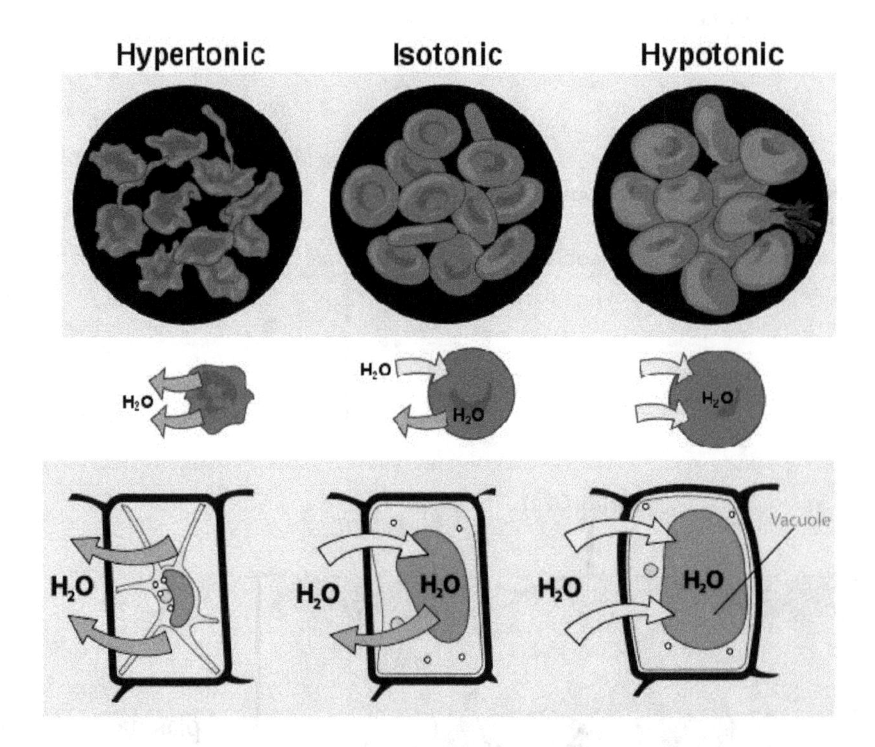

FIGURE 4.13 Osmosis is dependent upon the solute concentration of a particular solution, known as tonicity. Tonicity refers to the solute concentration of a particular solution and greatly affects the movement of water during osmosis, as seen in red blood cells and plant cells. When these cells are exposed to a hypertonic solution (left), water moves out, causing them to shrink and potentially die. There is no net movement of water in an isotonic solution (middle) because the concentration of water and solutes remain equal on either side of the plasma membrane of the cells. A hypotonic solution (right) results in the movement of water into the cells due to high concentrations of solute within, thus causing red blood cells to burst. Alternatively, the cell wall of plant cells protects the cell from bursting.

Active Transport

Active transport is a transport system responsible for moving molecules "against" a concentration gradient or moving molecules from an area of low concentration to high concentration. Active transport requires two essential components: a direct input of energy from ATP and carrier proteins, which work together in unison to move molecules. The carrier proteins are activated by energy from ATP, causing them to function like pumps that can transport molecules against the concentration gradient. The most common example of active transport in cells is the **sodium-potassium pump** (Figure 4.14). The sodium-potassium pump was first discovered in 1957 by Danish scientist Jens Christian Skou, who received the Nobel Prize in Chemistry for his discovery in 1997. The sodium-potassium pump is situated along the outer region of the plasma membrane of all animal cells, specifically muscle and nerve cells, and is responsible for maintaining the concentration gradient of extracellular (outside of cell) sodium and intracellular (inside the cell) potassium by pumping three sodium ions out of the cell and two potassium ions into the cell for every ATP molecule used by the carrier protein. This form of active transport helps maintain the concentration gradient of sodium and potassium across the plasma membrane to ensure cells can initiate crucial physiological processes, such as stimulating muscle contractions, producing nerve impulses, and regulating the osmotic equilibrium of fluids in the kidneys.

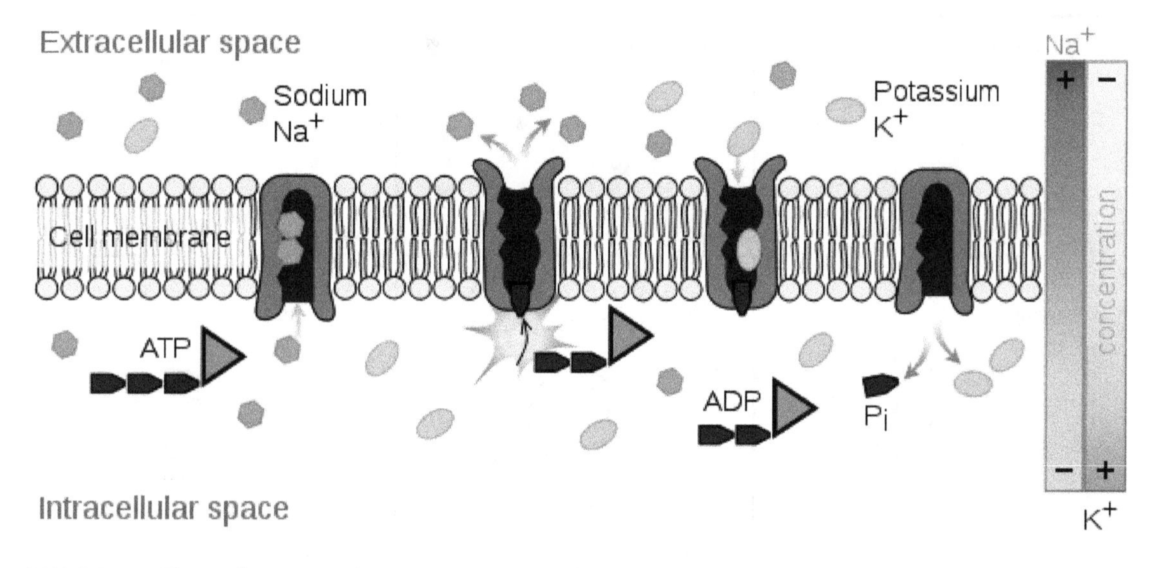

FIGURE 4.14 The sodium-potassium pump maintains the concentration gradient of sodium and potassium ions by using an input of ATP to actively transport three sodium ions out of the cell and two potassium ions into the cell to initiate crucial physiological processes.

Bulk Transport

Bulk transport is a membrane-assisted transportation method used when larger molecules, such as proteins and polysaccharides, are unable to move across the membrane by means of passive or active transport. These large molecules can enter and leave cells through the formation of vesicles that arise from the plasma membrane. There are two different types of bulk transport that exist.

Endocytosis is the process by which a cell uses a portion of its plasma membrane to form a vesicle around a molecule or particle to bring it into the cell. This process can occur in three different ways (Figure 4.15). Cells will use the process of **phagocytosis** to engulf large food particles or another cell. This is a common means of ingestion performed by single-celled protozoans and white blood cells. **Pinocytosis** is another form of endocytosis and involves the formation of small vesicles around liquids or dissolved particles. Pinocytosis is common in blood cells and plant root cells to ingest substances. **Receptor-mediated endocytosis** is a specialized form of pinocytosis in which protein receptors located on the plasma membrane recognize a specific type of molecule, such as a vitamin or hormone. Once the molecule binds to the receptor protein, the plasma membrane will form a vesicle around the molecule, bringing it into the cell. A common example of receptor-mediated endocytosis occurs along liver cells, which contain protein receptors associated with a type of cholesterol carrier called low-density lipoprotein (LDL). The LDL molecule will bind to the receptor proteins, causing the plasma membrane to fold itself around the molecule. Once inside the cell, the LDL molecule is broken down to release cholesterol, which is used in the production of estrogen and testosterone.

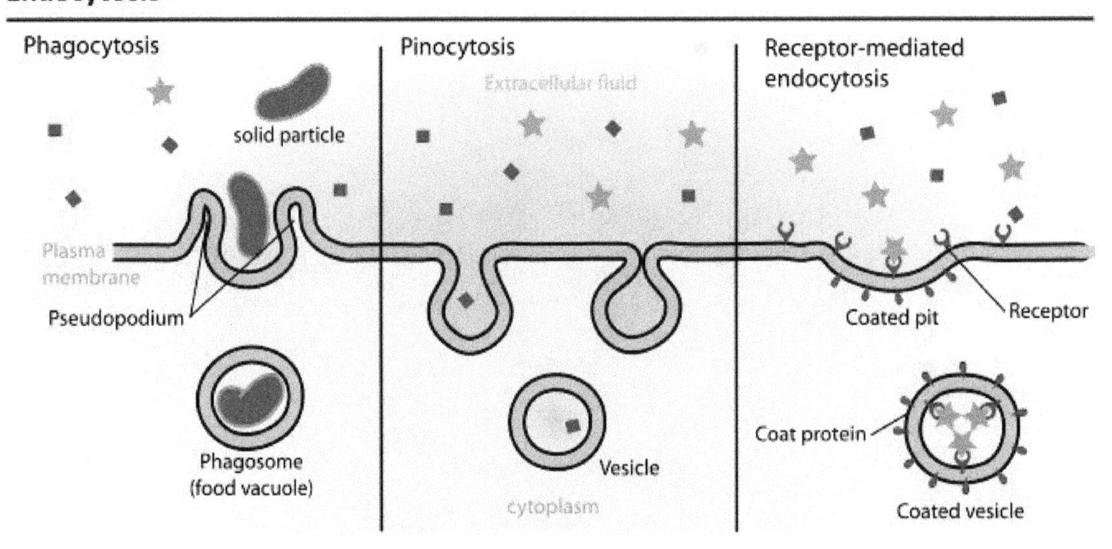

FIGURE 4.15 Bulk transport helps move much larger molecules, such as proteins and carbohydrates into or out of the cell. Endocytosis involves the formation of a plasma membrane-based vesicle around a particle and bringing it into the cell. Endocytosis includes phagocytosis, pinocytosis, and receptor-mediated endocytosis.

Exocytosis occurs when a cell encloses particles, such as metabolic waste or hormones, within vesicles produced by an organelle called the Golgi apparatus to transport the particles out of the cell. The formed vesicle fuses with the plasma membrane, and the contents of the vesicle are expelled outside the cell. There are many common examples of exocytosis including insulin secretion from the pancreas, release of neurotransmitters in the nervous system, and release of proteins and carbohydrates from plant cells to build cell walls.

Cytoplasm

The cytoplasm of eukaryotic cells consists of cytosol, a semi-fluid, gelatinous solution consisting of water, ions, macromolecules, and wastes, and numerous organelles (Figure 4.6). The cytoplasm lies inside the eukaryotic cell between the nucleus and plasma membrane. Most biochemical reactions important for supporting life occur within the cytoplasm, including protein synthesis. This process takes place on ribosomes, which are located within the cytoplasm of cells and on an organelle called the endoplasmic reticulum (discussed shortly).

The ribosomes found in eukaryotic cells are considerably larger than those found in prokaryotic cells. Prokaryotic ribosomes have an approximate size of 70S, while eukaryotic cells are 80S. In addition, eukaryotic ribosomes are composed of two subunits, approximately 60S and 40S in size. These characteristics are important distinctions between prokaryotic and eukaryotic cells.

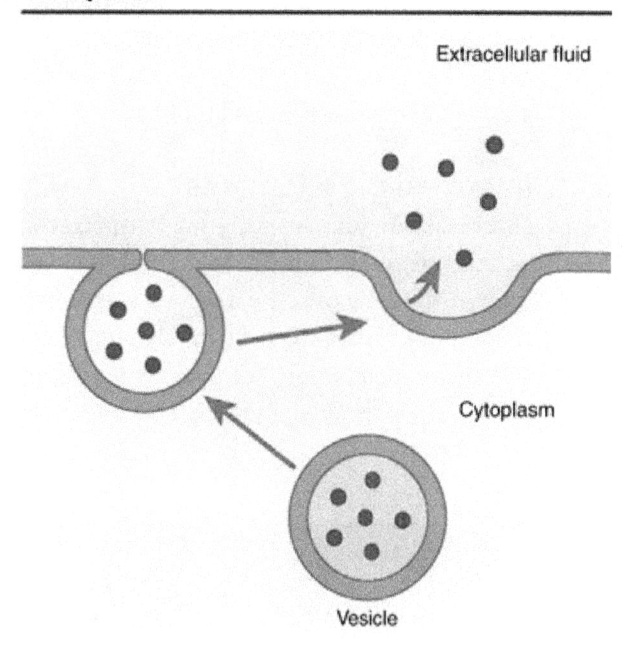

FIGURE 4.16 Exocytosis expels metabolic waste products outside of the cell by means of vesicles, which fuse with the plasma membrane.

Cytoskeleton

The cytoskeleton is an intricate network of interconnected proteinaceous fibers that create an internal scaffolding extending from the nucleus to the plasma membrane (see Figure 4.6 and Figure 4.17). The cytoskeleton performs many different functions: provides cell shape and structure, enables cellular movement, supplies tracks that enable organelles and important molecules to move throughout the cell, and forms the necessary components for proper cellular division.

There are three main types of proteinaceous fibers that make up the cytoskeleton.

Microtubules

Microtubules are thick, hollow tubes composed of a globular protein called tubulin (Figure 4.17, left). These groups of fibers are responsible for helping maintain the shape of the cell and act as tracks to which important molecules and organelles can attach and move within the cell. In addition, microtubules are also important structures involved in the separation of chromosomes during cellular division.

There are two types of microtubule-based structures found in eukaryotic cells, which extend from the plasma membrane—cilia and flagella. Cilia are short, hairlike extensions attached to the surface of some cells, and flagella are more taillike and are usually limited to one or a few per cell. Both structures are composed of nine pairs of microtubules that extend down their entire length, attached just below the plasma membrane. Cilia move in a highly coordinated rhythmic fashion, producing wavelike patterns that can move the cell or propel substances over the surface of the cell. Ciliated cells are important in the respiratory tract, where they sweep away trapped debris within mucus to prevent its entry into the lungs. Flagella produce a whiplike or undulating motion that provides movement to cells, such as single-celled eukaryotes and sperm cells.

Microfilaments

Microfilaments are the smallest of all cytoskeleton proteinaceous fibers and exist as long, thin, solid rods composed of the globular protein actin (Figure 4.17, middle). Microfilaments contain two chains of actin proteins, twisted in a helical structure, and can exist as bundles or mesh-like networks. These group of fibers are responsible for maintaining cell shape, anchoring one cell to another, and permitting cell motility.

Intermediate Filaments

Intermediate filaments are medium-sized structures composed of various fibrous proteins (such as keratins) coiled into a ropelike system (Figure 4.17, right). These types of filaments are found only in the cells of some animals (vertebrates) and are prominent in skin and nerve cells. The numerous overlapping proteins that make up intermediate filaments help provide the cell strength and shape. In addition, the intermediate filaments are responsible for anchoring the nucleus, supporting the plasma membrane, and assisting in the formation of cellular junctions.

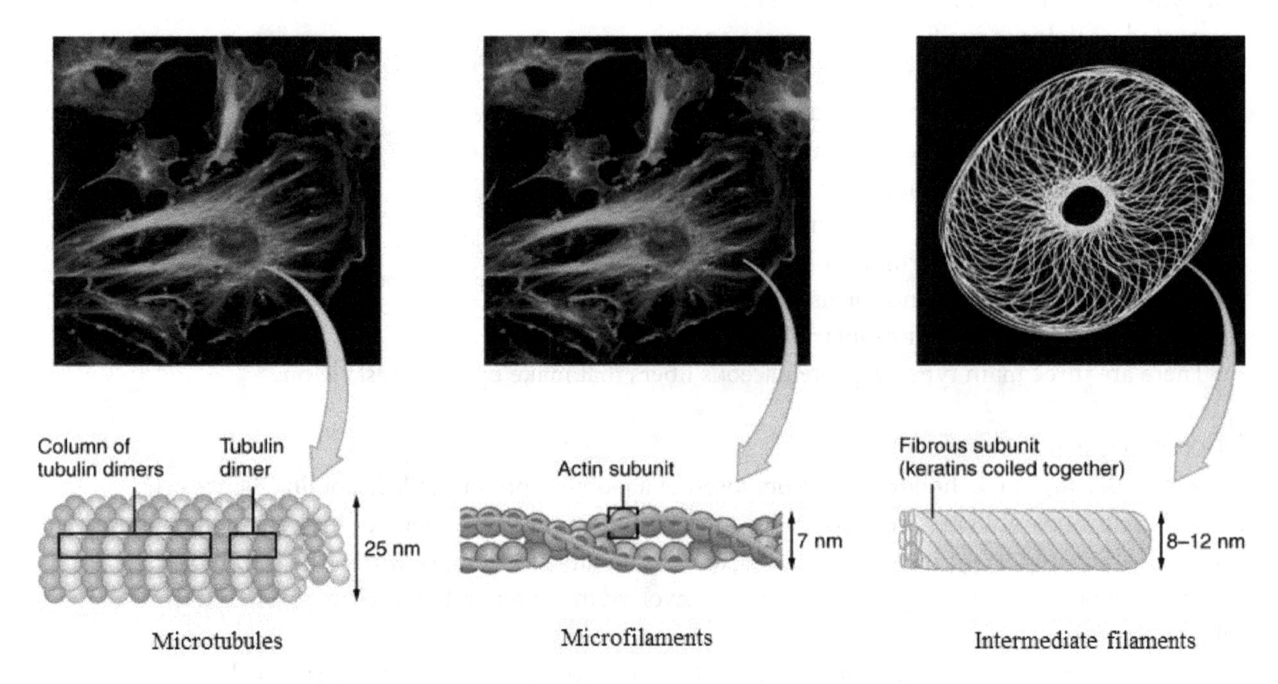

FIGURE 4.17 The cytoskeleton is an internal structure composed of three types of proteinaceous fibers, including microtubules (left), microfilaments (middle), and intermediate filaments (right). The cytoskeleton is an important cellular structure providing many different types of functions.

Cellular Junctions

Cellular junctions exist as points of contact between cells, enabling cells to behave and communicate in a coordinated way. There are three main types of cellular junctions that exist in animal cells. **Gap junctions** are junctions occurring between the cytoplasm of adjacent cells, forming tubular channels (Figure 4.19, left). These channels enable cells to communicate and behave as one unit, as well as exchange cellular contents, such as ions, nutrients, and other small molecules. Gap junctions are important structures found in cardiac and smooth muscle tissues because they allow ions and electrical signals to spread rapidly between the cells. **Desmosomes** are the strongest of all cellular junctions and function like rivets or spot welds (Figure 4.19, middle). These cellular junctions occur between the intermediate filaments of adjacent cells, forming flexible yet strong and sturdy sheets of cells capable of withstanding tension, stretching, and the stresses of sustained movement. Desmosomes are commonly found in the heart, stomach, and skin. **Tight junctions** are cellular junctions that form between the plasma membrane proteins of neighboring cells in a zipper-like fashion (Figure 4.19, right). These junctions hold the cells together so tightly they form sheet-like barriers preventing the movement of fluid between cells. The sheetlike barriers produced by these tight junctions exist within the digestive tract, kidneys, and urinary bladder.

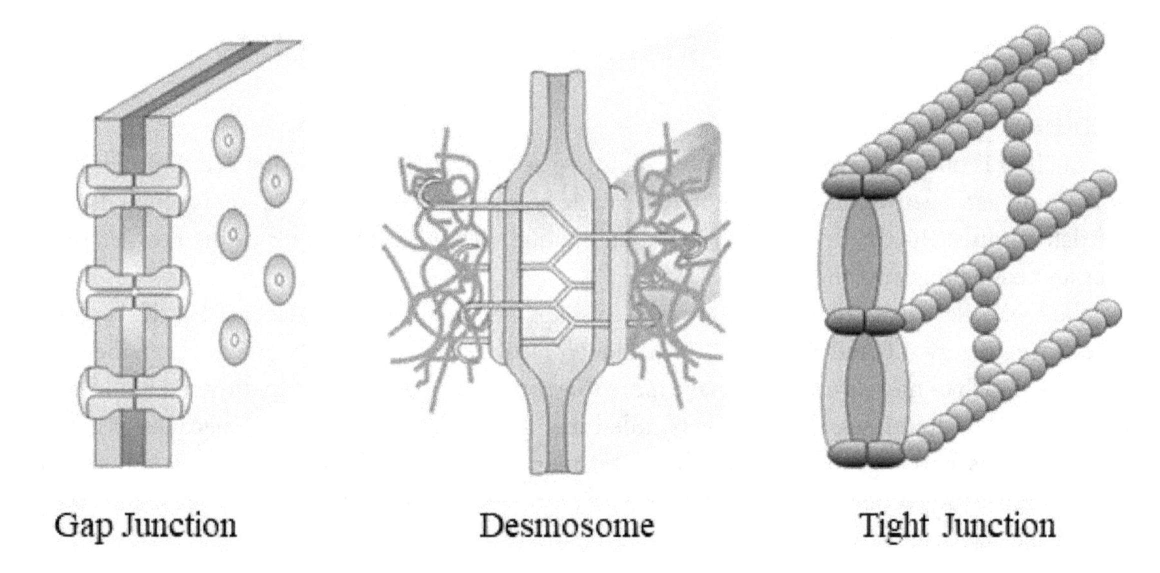

FIGURE 4.18 Cellular junctions are points of contact between cells. There are three main types found in animal cells, including gap junctions (left), desmosomes (middle), and tight junctions (right). All types of cellular junctions enable cells to behave and communicate in a coordinated way.

Alternatively, plant cells are connected to one another by membrane-lined structures known as **plasmodesmata** (plasmodesma, *singular*), which pass through the cell walls of adjacent plant cells creating channels that join their plasma membranes, cytoplasm, and endoplasmic reticula (Figure 4.19). The channels provide a means of cellular communication and a low-resistant pathway for the transportation of water, small molecules, and macromolecules, such as proteins and nucleic acids, between plant cells.

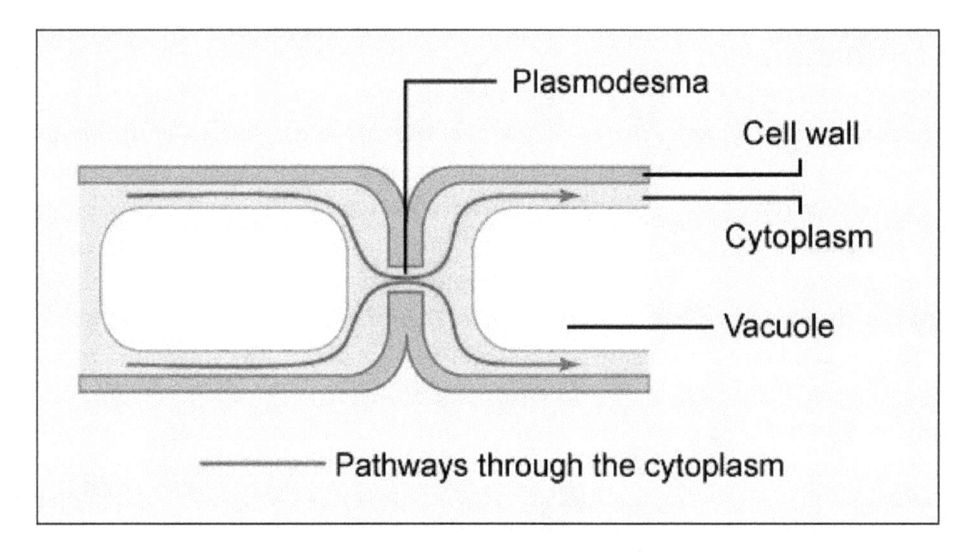

FIGURE 4.19 Plasmodesmata are membrane-lined channels that provide a means of cellular communication and a pathway for the transportation of molecules between adjacent plant cells.

4-5. Eukaryotic Cells: Membranous Organelles

Nucleus

The nucleus is one of the most prominent, defining structures in a eukaryotic cell, first observed by Antonie van Leeuwenhoek in the 1670s. While studying orchids under a microscope in 1831, Scottish botanist Robert Brown discovered an opaque area within the cells of the flower's outer layer and called this area the **nucleus**.

The organelle is usually spherical to ovoid in shape, relatively large, and located in the center of the cell (see Figure 4.6 and Figure 4.20). Although most cells contain only one nucleus, there are some cells, such as muscle cells, that have more than one, while others, like mature red blood cells, lack a nucleus. Oftentimes, the nucleus is called the "control center of the cell," because of its two main functions: housing most of the cell's genetic material (DNA) and directing cellular function by controlling the production of necessary molecules responsible for regulating physiological processes. The nucleus consists of two main parts.

Nuclear Envelope

The **nuclear envelope** is a double membrane structure separating the nucleus from the rest of the cell (Figure 4.20). Each membrane is a phospholipid bilayer, equaling a membrane system consisting of four phospholipid bilayers. The structure is perforated with specialized, protein-lined channels called **nuclear pores** (Figure 4.20), which function in controlling the movement of substances, such as water, ions, and molecules, into and out of the nucleus.

Nucleoplasm

The nucleus consists of a semifluid, gelatinous solution called **nucleoplasm**, which possesses two important components—chromatin and nucleolus (Figure 4.20). **Chromatin** is a grainy-appearing mass of DNA strands that will coil into chromosomes prior to cellular division. Chromatin is wrapped around a group of proteins called *histones*, which perform two functions—tightly packing the DNA within the nucleus and keeping the DNA from getting tangled. The second component located within the nucleoplasm is a dark-staining region of chromatin called the **nucleolus**. Located within the nucleolus is ribosomal RNA (rRNA), proteins, various stages of synthesized ribosomes, and DNA, which are all used for the synthesis of ribosomes, the structures responsible for producing proteins.

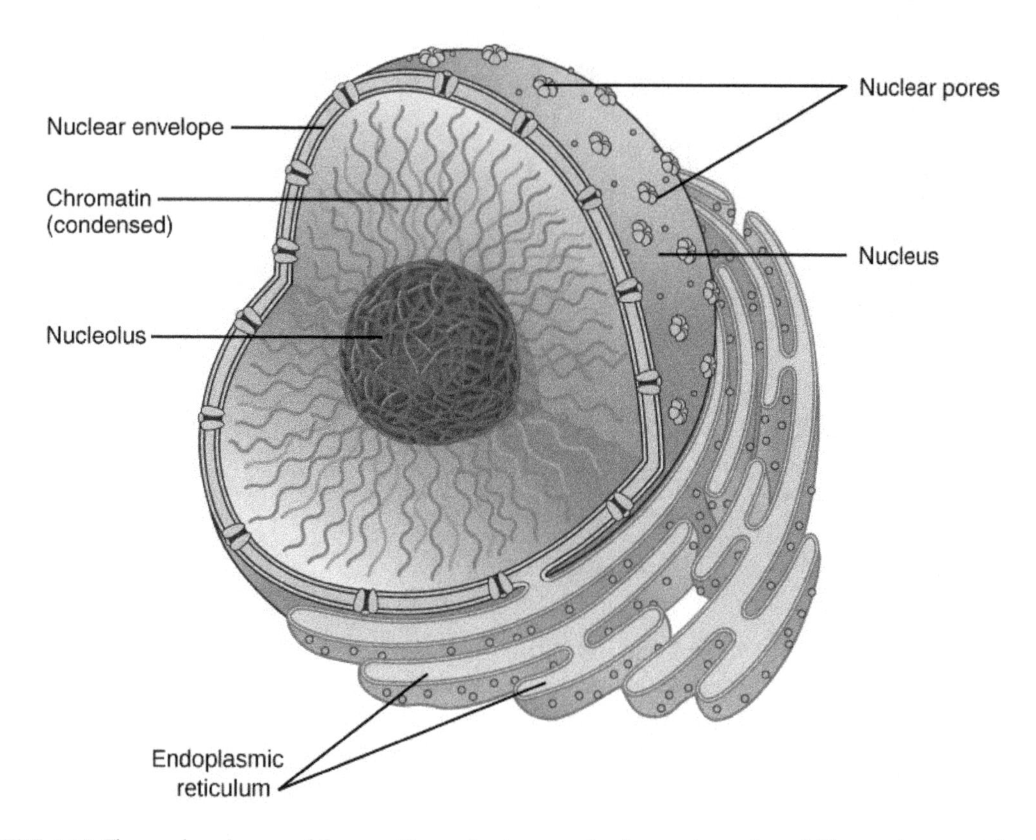

FIGURE 4.20 The nucleus is one of the most important organelles in a eukaryotic cell. The nucleus contains chromatin (DNA) and ribosomal RNA within the nucleolus, both of which are surrounded and protected by a double phospholipid bilayer called a nuclear envelope, perforated with nuclear pores.

Endoplasmic Reticulum

The endoplasmic reticulum is part of the endomembrane system of a cell, first observed by Italian histologist Camillo Golgi in 1898. Advances in microscopy led to its detailed description as a "lacelike reticulum" in a paper entitled "A Study of Tissue Culture Cells by Electron Microscopy" in 1945 by scientists Keith R. Porter, Albert Claude, and Ernest F. Fullam. The name "endoplasmic reticulum" was assigned to the elaborate structure by Porter in his 1953 paper entitled "Observations on a Submicroscopic Basophilic Component of Cytoplasm." What these scientists observed was a network of flattened, highly folded, membranous tubules extending from the nuclear envelope, occupying a large portion of the cell's cytoplasm.

Although each eukaryotic cell contains only one endoplasmic reticulum, there are two forms of the organelle present within the cell (see Figure 4.6 and Figure 4.21).

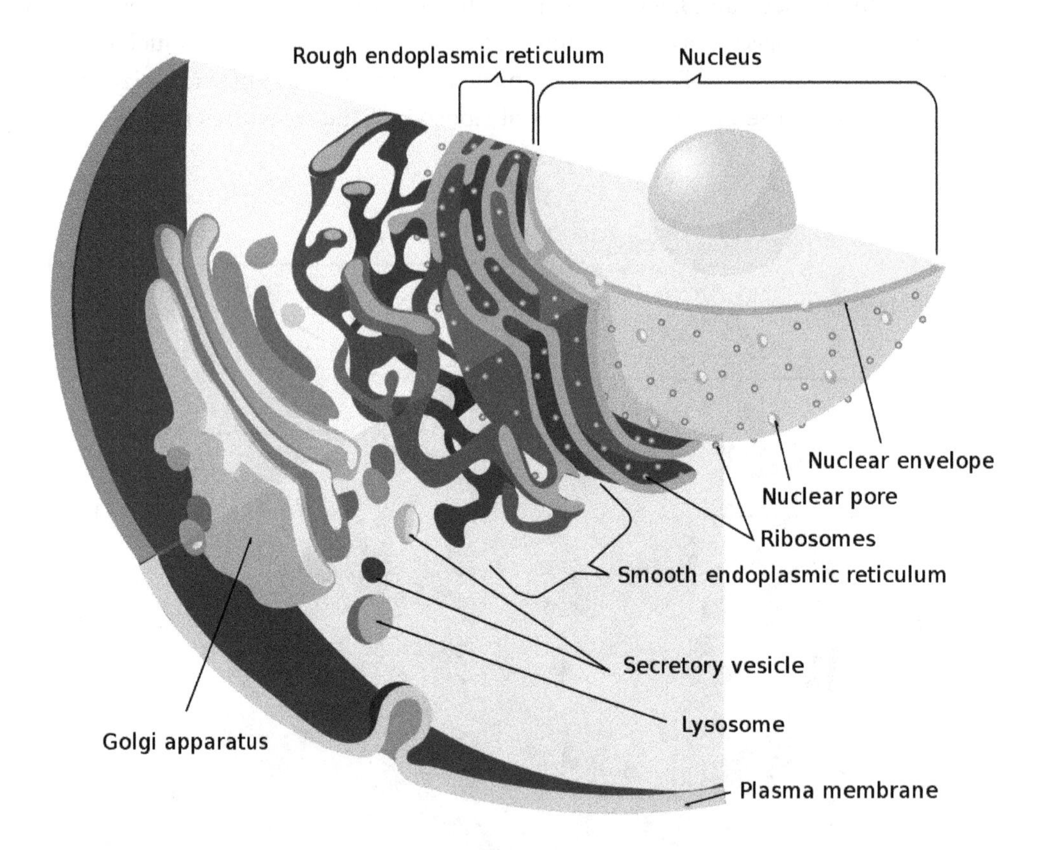

FIGURE 4.21 The endomembrane system includes the endoplasmic reticulum, Golgi apparatus, and lysosomes. The endoplasmic reticulum and Golgi apparatus are responsible for the synthesis, packaging, and transportation of important macromolecules, such as proteins. Lysosomes are responsible for the digestion of macromolecules, organelles, and cellular structures using hydrolytic enzymes.

Rough endoplasmic reticulum is given its name because the outside of the structure is studded with ribosomes, which are the sites of protein synthesis. The ribosomes located on the rough endoplasmic reticulum are responsible for producing various types of proteins, including plasma membrane proteins, digestive enzymes, and protein-based hormones, such as insulin. Once produced, these proteins are packaged and moved along the channels of the endoplasmic reticulum, where they will be used within the cell, on the cell surface, or outside the cell.

Smooth endoplasmic reticulum lacks the ribosomes that characterize the rough endoplasmic reticulum, giving this endoplasmic reticulum a smooth appearance. Smooth endoplasmic reticulum has a variety of functions, all of which are performed by enzymes specific to this structure. The smooth endoplasmic reticulum is responsible for the production of lipids, most notably fatty acids, phospholipids, and steroids. Two of the most common steroids produced by the smooth endoplasmic reticulum are the sex hormones, estrogen, and testosterone. Smooth endoplasmic reticulum is also abundant in liver cells, which consist of detoxifying enzymes responsible for breaking down harmful molecules, such as alcohol and drugs, and metabolic waste, like ammonia. Other smooth endoplasmic reticulum enzymes are responsible for the metabolism of carbohydrates, also located within the liver. The polysaccharide glycogen is a carbohydrate stored in the liver, and enzymes break down the complex sugar into individual glucose molecules used to produce energy. Finally, the smooth endoplasmic reticulum of muscle cells also stores calcium ions, important ions necessary to produce muscle contractions.

Golgi Apparatus

The Golgi apparatus is another part of the endomembrane system, accidently discovered by Italian histologist Camillo Golgi as he was studying Purkinje cells (a group of neurons located in the cerebellum) of the nervous system. Golgi published his discovery in 1898 and called the structure *apparato reticolare interno* ("internal reticular apparatus"). By 1913, the term "Golgi apparatus" had replaced *apparato reticolare interno* in the scientific literature. At the time of Golgi's discovery, many scientists disputed his find as nothing more than a staining artifact, and the debate continued for several decades. However, in 1954, with the development of more powerful microscopes, such as the electron microscope, any controversies about the organelle were solved. Several groups of scientists verified the existence of the Golgi apparatus using electron microscopy; however, A. J. Dalton and M. D. Felix are the two scientists given credit for introducing the scientific community to the organelle. They discovered that the organelle consisted of fine structural components, such as lamellae, vesicles, and vacuoles.

The Golgi apparatus is described today as a specialized organelle consisting of flattened, hollowed, membranous sacs enclosed by a phospholipid bilayer (see Figure 4.6 and Figure 4.21). Often referred to as a "shipping facility," the Golgi apparatus receives, processes, modifies, and packages molecules produced by the rough endoplasmic reticulum and ships the molecules out of the cell through exocytosis in secretory vesicles. The organelle serves such functions as modifying proteins received from the rough endoplasmic reticulum through the addition of phosphates or carbohydrates or breaking larger proteins into smaller ones. In addition, the Golgi apparatus is also responsible for producing carbohydrates, such as cellulose, used in the cell walls of plants. When receiving proteins and lipids from the endoplasmic reticulum, the Golgi apparatus separates the molecules and packages them into vesicles, which are then sent out to their final destination.

Lysosomes

In 1955, the final component of the endomembrane system of a eukaryotic cell, called the lysosome, was discovered by means of centrifugation by a team of scientists led by Belgian scientist Christian de Duve. De Duve's discovery of the lysosome began in 1949, when he and his team began studying how insulin worked on liver cells by determining the location of an enzyme called glucose-6-phosphatase, responsible for regulating blood glucose levels. After repeated experimental attempts, the group of scientists could not effectively purify and isolate the enzyme; however, using centrifugation (the process responsible for separating different cellular components based on size and density), the team was successful in isolating the enzyme and detecting its activity. During their experiments, the scientists used an enzyme called acid phosphatase as a control, but they found that the enzyme's activity was only operating at 10 percent of its normal rate. However, after discovering centrifuged cellular components left in a refrigerator for five days, the scientists measured the enzymatic activity of the components and discovered it was back to normal. The scientists repeated these experiments, and each time, the results remained unchanged. This led de Duve and his team to hypothesize that a membrane-like barrier was restricting the ability of the enzyme to work on the substrate and described it as a "saclike structure surrounded by a membrane ... containing acid phosphatase." In 1955, additional enzymes were discovered in these membrane-like structures, suggesting a new organelle with digestive properties, which de Duve named "lysosomes." In the same year, de Duve and a visiting American scientist, Alex B. Novikoff, used electron microscopy to produce the first pictures of a lysosome and were able to confirm the location of the hydrolytic enzymes, including acid phosphatase, within the organelle.

Lysosomes are membrane-enclosed organelles that contain hydrolytic enzymes, which are produced in the rough endoplasmic reticulum and sent to the Golgi apparatus to be packaged into vesicles (see Figure 4.6 and Figure 4.21). There are about 50 different hydrolytic enzymes found in lysosomes, which are responsible for digesting macromolecules, along with breaking down worn-out organelles and excess cellular membranes and dismantling ingested microorganisms, such as bacteria. The number of lysosomes in a cell varies based on the cell type. For example, white blood cells have several lysosomes to protect an organism from debris and pathogens. A correct balance of enzymes within the lysosomes is important to an organism's health. In the case of Tay-Sachs disease, a human genetic disorder in which an individual lacks a specific lipid-digesting enzyme, lipids begin to accumulate within the lysosomes, resulting in the organelles swelling, bursting, and digesting the cell. An individual begins to exhibit neurological issues within the first four to six months, eventually leading to loss of sight, hearing, and the ability to move, and ultimately death by the age of three or four.

Vacuoles

Vacuoles are large, centrally located, membranous organelles first observed as "star-shaped" structures in protozoans in 1776 by Italian scientist Lazzaro Spallanzani, who suggested that these structures were for respiratory purposes. Another prominent scientist, French biologist Félix Dujardin, has been given credit for naming these "star-shaped" structures *vacuoles* when observing them within protozoans in 1841. In 1842, the term "vacuole" was applied in a description of plant cells by cell theorist Matthias Schleiden as a means of distinguishing this structure from the cytoplasm. Studies in the mid-to-late 1880s by Dutch botanist Hugo Marie de Vries showed that vacuoles were responsible for regulating water content when exposed to pure water and various concentrations of solutes, ultimately acting as osmoregulators.

Vacuoles have a function in the storage and transport of important cellular substances and molecules and are formed from the endoplasmic reticulum and Golgi apparatus. These organelles are found in all eukaryotic cells; however, they are more prominent in plant cells, where they can occupy up to 90 percent of its cellular volume. Vacuoles perform various functions, dependent on the cell in which they are found.

Food Vacuoles

Food vacuoles are primarily found in amoebae and other unicellular protists and act as storage compartments for food particles. The vacuoles form because of phagocytosis, in which the organism engulfs small organisms or particles. The vacuole fuses with a lysosome, and the hydrolytic enzymes within digest the food particles into smaller molecules, which become nutrients for the cell.

Contractile Vacuoles

Contractile vacuoles are "star-shaped" structures found in fresh-water protists, such as *Paramecium.* As water enters the cell, the cell begins to fill and swell. The water is transported from the cell into the contractile vacuole with the help of ATP, and when the vacuole is full, it contracts, expelling water out through a small pore within the plasma membrane. Other variations of vacuoles exist in plants and fungi, which contain hydrolytic enzymes and function in a manner similar to lysosomes in animals. Plants have additional modified vacuoles that store important organic compounds required for growth, toxic compounds for protection against predators, and pigments responsible for flower color to help attract pollinators.

Central Vacuole

The central vacuole is the largest of all vacuoles, occupying a majority of a plant cell's volume (Figure 4.6). The vacuole consists mainly of water and is responsible for helping maintain the water balance within a plant cell. In addition, the central vacuole is also a storage area for important ions, such as potassium and chloride; important molecules like sugars and amino acids; and hazardous metabolic waste products. Finally, the central vacuole is also responsible for maintaining **turgor pressure**, which is the pressure of water pressing against the walls of the central vacuole. Turgor pressure is important in helping maintain cell shape and rigidity, as well as providing support to nonwoody parts of plants.

Mitochondria

Observations of structures resembling mitochondria began shortly after the discovery of the nucleus in the 1840s. In 1890, a German pathologist and histologist, Richard Altmann, was the first to observe granular structures he described as "elementary organisms," capable of living within a cell and carrying out vital metabolic and genetic functions. In his book entitled *The Elementary Organism,* Altmann called the structures he observed "bioblasts," and he also suggested in his book that the organelles may have symbiotic origins. The idea of symbiotic origin of mitochondria was also observed in the 1920s, when American scientist Ivan E. Wallin suggested that the organelles were once free-living bacteria. Wallin published several papers and a book detailing the bacterial origin of mitochondria and their symbiotic relationship with ancient eukaryotic cells, an act he called symbionticism. Wallin's idea was further confirmed with biochemical work performed by several scientists between the 1970s and 1980s (described in section 4-7). The name "mitochondria" was first used in the publication *On Spermatogenesis of Vertebrates and Higher Invertebrates, Part II: The Histogenesis of Sperm* by German scientist Carl Benda in 1898, as he was observing the structures while studying spermatogenesis.

Mitochondria are spherical to elongated organelles present in all eukaryotic cells (see Figure 4.6 & Figure 4.22). While some eukaryotic cells may have only one mitochondrion, other cells, like liver or muscle cells, may have thousands of mitochondria. Mitochondria consist of two phospholipid bilayers constituting a double-membrane system. The first membrane is a smooth outer membrane, while the inner membrane is a highly convoluted structure with many intricate folds, called the **cristae**. Within the cristae is the **mitochondrial matrix**, a semifluid mixture of circular pieces of DNA, 70S ribosomes, and enzymes. The mitochondria are often referred to as the "powerhouse" of the cell because they are responsible for converting energy within sugar molecules into cellular energy called ATP in the presence of oxygen, in a process known as cellular respiration (discussed further in chapter 5). Within the mitochondrial matrix are a variety of enzymes that are responsible for initiating most of the reactions involved in cellular respiration, while other enzymatic reactions occur just outside of the cristae. In human cells, the mitochondria are highly effective energy producers, producing about 90 percent of the energy needed for the cells to function.

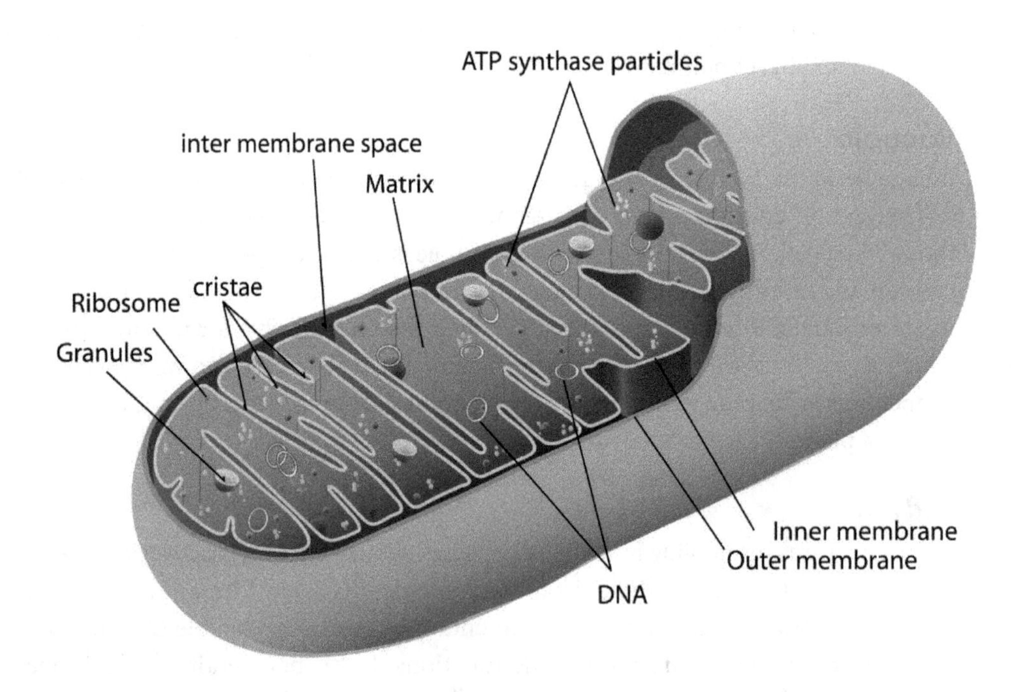

FIGURE 4.22 A mitochondrion consists of a double membrane system, with the inner membrane containing the cristae and the mitochondrial matrix. Within the mitochondrial matrix are circular pieces of DNA, ribosomes, and enzymes, which are responsible for the production of ATP during the process of cellular respiration.

Chloroplasts

The first description of chloroplasts was provided by German botanist Hugo von Mohl in his 1835 paper entitled "About the Propagation of Cells of Plants Through Division." In his paper, von Mohl described chloroplasts as *chlorophyllkörnern*, or "chlorophyll granules." It was not until 1883 that German botanist Andreas Schimper coined the term "chloroplast" in one of his most memorable papers, "On the Development of Chlorophyll Granules and Colored Bodies." It was also in this paper that Schimper suggested that new chloroplasts arose from division and their appearance closely resembled that of free-living cyanobacteria. This paper was the precursor to the symbiogenesis hypothesis laid out by Russian scientist Constantin Merezhkowsky about the origin of chloroplasts

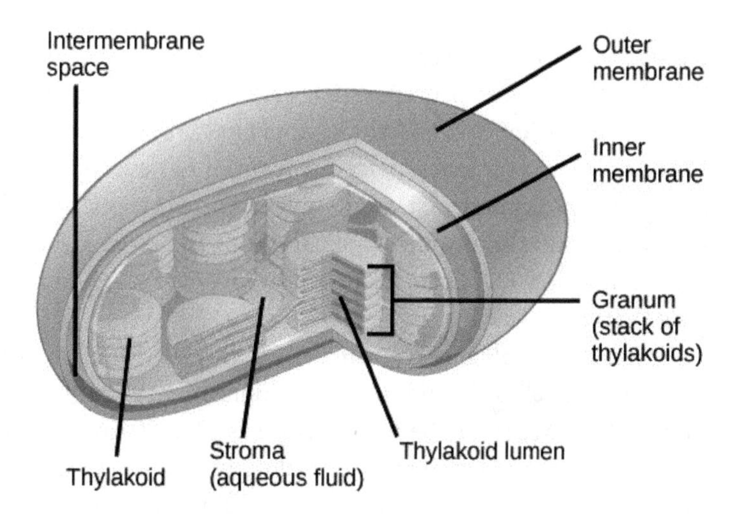

FIGURE 4.23 The chloroplast, like a mitochondrion, consists of a double membrane system. The inner membrane is called the stroma and contains circular pieces of DNA, ribosomes, and various enzymes. Thylakoids are flattened sacs located within the stroma stacked into grana and contain different light trapping pigments responsible for photosynthesis, called chlorophylls.

in plant cells, an idea later supported by means of microscopy and biochemistry in the 1960s and 1970s (described in section 4-7).

Chloroplasts are ovoid, almost flattened light-harvesting organelles, consisting of two phospholipid bilayers, found in photosynthetic eukaryotic cells, such as algae, protozoans, and plants (see Figure 4.6 and Figure 4.23). Some algal cells contain only one chloroplast, while the cells of plant leaves may contain 40 to 50 or as many as 100 chloroplasts. Chloroplasts are the site of photosynthesis (discussed further in chapter 5), a process wherein solar energy is converted to chemical energy used in the production of sugar molecules. The chloroplast double membrane system includes an outer membrane encircling the organelle and an inner membrane containing a gelatinous fluid called the **stroma**, consisting of circular pieces of DNA, 70S ribosomes, and a variety of enzymes. Also located within the stroma are flattened, interconnected sacs stacked upon one another like coins, called **thylakoids**. The stacks of thylakoids (known as grana) contain light-trapping pigments called **chlorophylls**. These pigments are responsible for capturing solar energy and transferring this energy along a variety of molecules to ATP. The energy within ATP is then used outside of the thylakoids to produce sugar from carbon dioxide and water.

4-6. Membrane Contact Sites and Organelle Communication

In 1896, American cytologist Edmund Beecher Wilson drew some of the first cells, depicting isolated and floating organelles within specific compartments of the cytoplasm. However, despite the various cellular discoveries and observations that have occurred since, most of the cellular depictions in biological textbooks, including this one, have remained relatively unchanged. This has caused contention among recent cell biologists suggesting that the images should properly depict the way a cell looks. The contention stems from a recent discovery that organelles are not static structures but rather dynamic entities that move and interact like dancers in a theatrical performance.

In 1990, Jean Vance, a British-Canadian cell biologist, made a unique discovery: while studying mitochondria from rat livers, she found pieces of endoplasmic reticulum stuck to the mitochondria.

Despite the observations made by other scientists prior to this discovery that this was nothing more than contamination or preparation artifact, Vance suggested that there must be a contact point between mitochondria and the endoplasmic reticulum. This contact point, Vance conjectured, would serve as a passageway for the transfer of newly formed lipids from the endoplasmic reticulum to the mitochondria. Despite several presentations given by Vance thereafter, there were still several scientists who were skeptical of this hypothesis. However, within the last couple of years, researchers now recognize how the significance of her discovery suggests how cells can maintain order within a crowded space.

Interactions called **membrane contact sites** (MCS) exist between the plasma membrane and various organelles, including the endoplasmic reticulum, Golgi apparatus, and mitochondria. Today, these MCSs can be seen as close as 10 to 30 nm apart with the help of advances in imaging technologies and microscopy resolution. However, the first interaction between organelles was microscopically seen as early as the 1950s, when French scientists observed junctions occurring between mitochondria and the endoplasmic reticulum in rat cells, like what Vance observed in 1990. In addition, other scientists between 1960 and 1980 also discovered interactions between the endoplasmic reticulum and other organelles, such as the Golgi apparatus and the plasma membrane. Unfortunately, none of these discoveries were supported, and they were deemed biological anomalies.

Within the last decade, scientists have used Vance's 1990 paper as a backdrop for current work performed on the MCSs that exist between organelles. The interactions that occur between the MCSs of different organelles are the result of a group of proteins called *tethers*. In 2009, a team of scientists discovered that a group of four proteins was responsible for creating the tether connecting the endoplasmic reticulum to the mitochondria in yeast cells. Another biologist discovered that interactions between the endoplasmic reticulum and the plasma membrane of yeast cells consisted of six proteins. In other studies, more complex tethering arrangements have been discovered, such as those found between the endoplasmic reticulum, mitochondria, and plasma membrane, which consists of at least two types of tethering systems.

MCSs serve different functions that are important in maintaining normal, healthy cell physiology and homeostasis. One of the first functions suggested by biologists was the transportation of lipids (cholesterol and waxes), ions (calcium), and water-soluble molecules (hydrogen peroxide). Other scientists suggest that MCSs serve an important function in helping regulate organelle positioning and division. This was observed in 2011, when images of yeast cells showed a mitochondrion dividing at an MCS shared with the endoplasmic reticulum.

The interaction between organelles is dynamic in that they connect, separate, and connect again. However, prolonged interaction between organelles can lead to different health issues, including insulin resistance, diabetes, and obesity. In addition, certain diseases have also been linked to the presence of different types of MCS proteins. For example, certain tether proteins are linked to diseases such as Alzheimer's, hereditary spastic paraplegia, amyotrophic lateral sclerosis, Charcot-Marie-Tooth disease, tubular aggregate myopathy, and retinal dystrophy. Although evidence suggests that contact sites can affect health, there is still uncertainty as to whether the diseases are the direct or indirect result of defects in the function of the contact sites.

4-7. Endosymbiotic Theory

The **endosymbiotic theory** states that the presence of mitochondria and chloroplasts within cells was the result of ingestion by early ancestors of eukaryotic cells (Figure 4.26). Instead of these cells being digested by the early eukaryotic cell, they formed a symbiotic relationship, labeled endosymbiosis, in which one organism resides within the cell of another, becoming part of that organism because of years of evolution. This theory was proposed by Lynn Margulis in her famed paper "On the Origin of Mitosing Cells" in 1967. A more formal proposal of her theory was outlined in her 1981 book, entitled *Symbiosis in Cell Evolution*. It should be noted, however, that although Margulis has been given credit for proposing the endosymbiosis idea, she was not, in fact, the first, but rather a contemporary, who pulled together the ideas of other scientists who predated her.

In 1905, Russian scientist Constantin Merezhkowsky (Figure 4.24), was studying chloroplast under a microscope and proposed these organelles resembled bacteria. This led him to the idea that chloroplasts were once ingested bacteria, found not only in algae cells but within plants as well, which performed a form a symbiosis he named symbiogenesis. Upon publishing his work in 1905, he suggested that cyanobacteria (Figure 4.24) were the bacteria ingested, and he continued to publish papers over the next 15 years providing evidence to support this theory. Unfortunately, Merezhkowsky never received scientific recognition nor acceptance of this hypothesis.

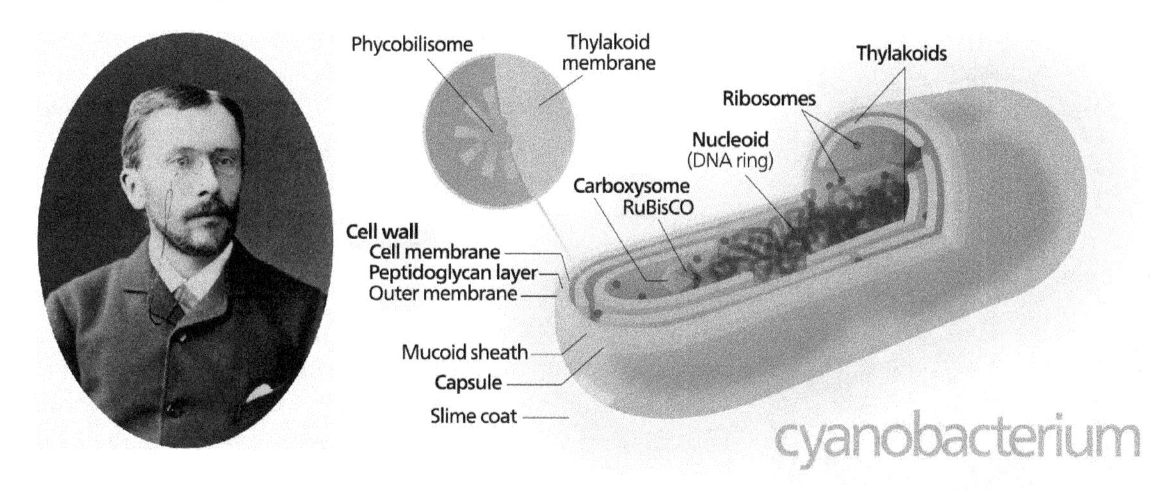

FIGURE 4.24 Constantin Merezhkowsky first proposed the bacterial origin of chloroplast in 1905. He suggested that the bacterial progenitor to chloroplast was cyanobacteria but never received recognition for his hypothesis. However, during the 1960s and 1970s, multiple scientists used microscopy and biochemistry to confirm Merezhkowsky's hypothesis.

In the 1960s, two scientists from the University of Wisconsin began studying the chloroplast by means of microscopy and biochemistry in inquiry of their origin. Using electron microscopy and biochemical staining, Hans Ris and Walter Plaut discovered the presence of DNA within green algae. This discovery implied the idea of endosymbiosis. Upon further study of the chloroplasts under electron microscopy, both scientists discovered similarities between the green alga chloroplasts and bacteria, including DNA fibrils and a double-layered membrane. These features matched one specific group—cyanobacteria—which suggested that these groups of bacteria were ingested by an ancient eukaryotic cell, and instead of being rejected, the bacteria survived, replicated, and began serving the functional role of photosynthesis. What Ris and Plaut discovered closely correlated and agreed with the hypothesis proposed by Merezhkowsky in 1905.

In 1975 and 1976, Linda Bonen and Ford Doolittle published two papers confirming the endosymbiotic theory of chloroplasts as proposed by Margulis and that cyanobacteria were the progenitor of modern-day chloroplasts. Using molecular data, the researchers compared ribosomal RNA from red alga cytoplasm and red alga chloroplasts, along with several other kind of bacteria, discovering that the ribosomal RNA of the chloroplast differed significantly from that of the ribosomal RNA within the cytoplasm of the red algae and was more closely related with the ribosomal RNA of the bacteria. They confirmed the idea of endosymbiosis for that of chloroplasts as once-bacterial cells ingested by an ancient eukaryotic cell. Additionally, Bonen and Doolittle published another paper providing substantial evidence that the bacterial progenitor was cyanobacteria.

The idea of mitochondrial endosymbiosis began around 1920, when American scientist Ivan E. Wallin argued that mitochondria, like chloroplasts, were also once free-living bacteria. From 1922 to 1927, Wallin published nine papers, along with a book (*Symbionticism and the Origin of Species*) detailing his argument that mitochondria were of bacterial origin and existed within a symbiotic relationship with that of the ancient eukaryotic cell, in a phenomenon he coined symbionticism.

FIGURE 4.25 Ivan Wallin first proposed the bacterial origin of mitochondria during the 1920s, publishing a book about symbionticism in 1927. During the 1970s and 1980s, various scientists began studying ribosomal RNA to determine the bacterial progenitor to mitochondria. It was not until 1985 when Carl Woese (pictured) determined that mitochondria closely resembled an alpha-proteobacterium by the name of *A. tumefaciens*.

In 1977, Linda Bonen, along with other scientists, including Ford Doolittle, Michael W. Gray, and Scott Cunningham, took on the task of confirming the idea that mitochondria were a byproduct of the theory of endosymbiosis. As before, this group of researchers employed the method of studying ribosomal RNA, but this time that of the wheat plant. What they discovered was that the ribosomal RNA of the mitochondria did not match that of the wheat ribosomal RNA. Therefore, researchers confirmed that mitochondria resembled bacteria and must have followed the same path as cyanobacteria, becoming ingested, surviving, and propagating to provide an important function in the overall success of the ancient eukaryotic cell. There was one problem with the research, in that they were unable to designate a progenitor to mitochondria. However, in 1985, Carl Woese (Figure 4.25) and a group of researchers discovered that the progenitor was that of a group of bacteria known alpha-proteobacterium. The group of scientists compared the ribosomal RNA genes of seven different organisms, including six bacterial organisms, and the ribosomal RNA from wheat, originally observed by the Bonen team. Several of the bacteria of choice included *E. coli*, *Anacystis nidulans* (cyanobacteria), and *Agrobacterium tumefaciens* (alpha-proteobacterium). When comparing the ribosomal RNA of the prokaryotic organisms to that of the wheat mitochondrial ribosomal RNA, the Woese group of researchers discovered that the bacterial ancestor of mitochondria more closely resembled that of *A. tumefaciens*, an alpha-proteobacterium.

Upon publishing "On the Origin of Mitosing Cells," Lynn Margulis cited the works of Merezhkowsky, Wallin, and Ris and Plaut, and she knew the ideas that she was proposing were those of other individuals. However, Margulis's sole contribution to the endosymbiotic theory was the proposal that flagella, cilia, and centrioles were once spirochetes, a group of bacteria shaped like corkscrews that move by way of twisting motions. She hypothesized that these types of bacteria were engulfed by the ancient eukaryotic cell, experiencing a fate like that of chloroplasts and mitochondria, or attached themselves to the outside of the cell. The idea of the spirochetes attaching themselves to the cell suggested greater mobility and complexity, and the addition of centrioles, Margulis thought, led to the development of the cellular reproductive processes of mitosis and meiosis. This idea seemed plausible given structural evidence that Margulis observed when using electron microscopy. She observed an arrangement of nine tiny tubules around a central ring in all three structures—flagella, cilia, and centrioles—which she suggested was inherited from an ancient spirochete. She also observed two additional tubules within the ring of nine in flagella and cilia, and she designated these as having a (9 + 2) arrangement, but she discovered that centrioles lacked these two additional tubules and therefore labeled these as (9 + 0). These arrangements, she hypothesized, were the byproduct of a spirochete symbiont or evolved later in the ancient eukaryote. Because of the difference in bacterial flagella and eukaryotic flagella, she referred to eukaryotic flagella and cilia as undulipodia ("little waving feet"). At the time of her assertion that flagella, cilia, and centrioles were once spirochetes, microscopy was the basis of her determination. Molecular evidence confirmed the claims that chloroplasts and mitochondria were of bacterial origin; however, no molecular evidence was found linking Margulis's claim that flagella, cilia, and centrioles were of bacterial origin. Microscopy evidence was not enough to support her claims, especially with the advances in molecular testing at the time.

The endosymbiosis theory is generally accepted by most biologists today, especially regarding the bacterial origin of chloroplasts and mitochondria. Besides molecular evidence linking both organelles to bacteria, other structural features and important discoveries support this theory. First, both organelles are surrounded by a double-layered membrane, with the outer membrane being like that of a plasma membrane. Both organelles have circular, naked DNA, similar RNA molecules, and prokaryotic-type and -size ribosomes (70S). Other features include similar membrane lipids and proteins to that of prokaryotic cells, and both organelles appear to independently divide.

4-8. Invagination Theory

The origin of other eukaryotic organelles is currently described by the **invagination** theory. This theory suggests that the plasma membrane folded in on itself, creating inner compartments that underwent further modifications and became specialized for cellular function (Figure. 4.26). The invagination theory has been used to explain the formation of the nuclear membrane, which surrounds the cell's DNA, and the membranes of the endoplasmic reticulum, along with the formation of other membrane-bound organelles within the endomembrane system, such as the Golgi apparatus, as the

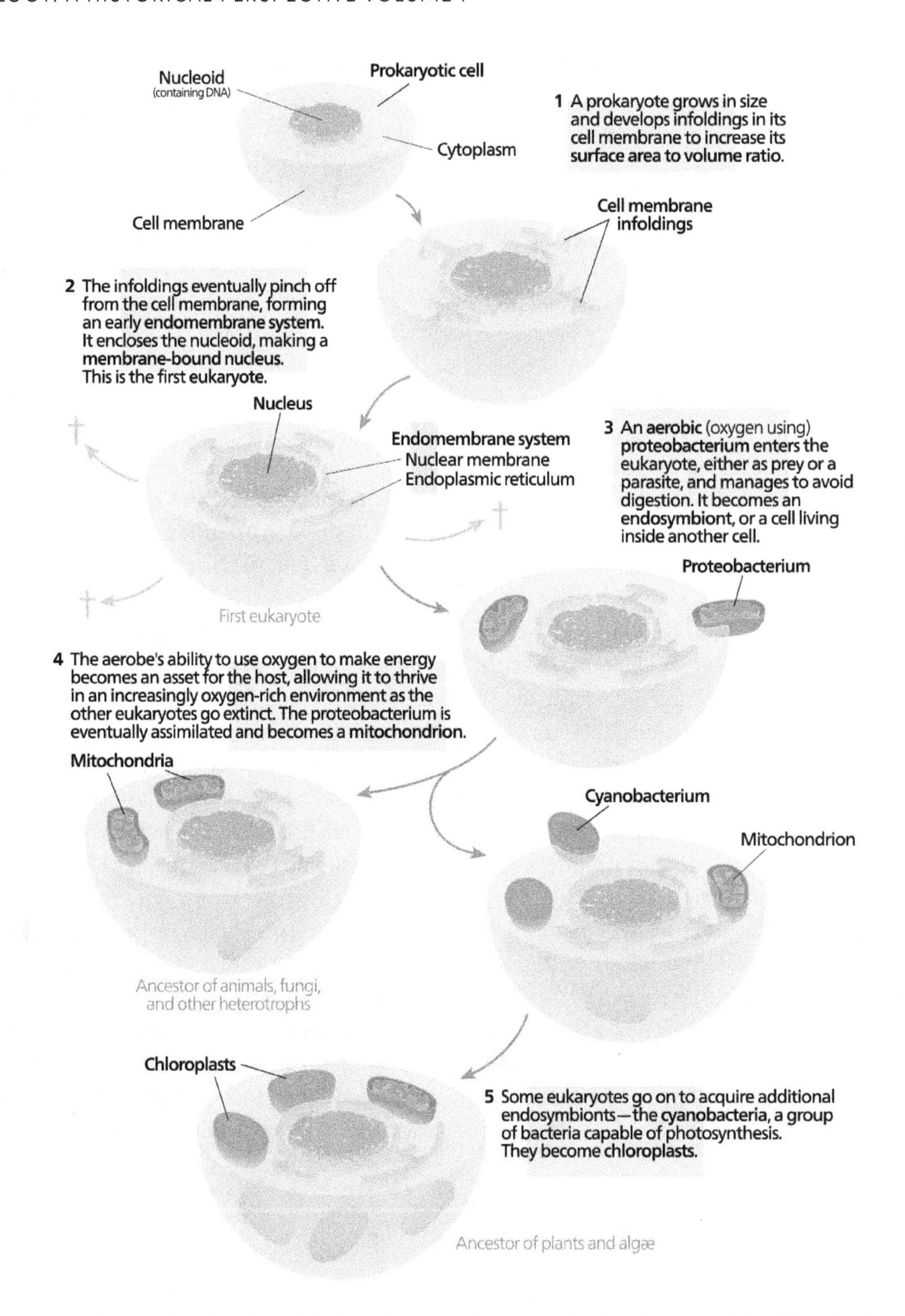

FIGURE 4.26 The theories of endosymbiosis and invagination currently describe the origin of the mitochondria, chloroplast, and endomembrane system of eukaryotic cells. The endosymbiotic theory suggests that mitochondria and chloroplast are of bacterial origin and were ingested by ancestral eukaryotic cells and after many years of evolution, formed a symbiotic relationship with one another. The invagination theory suggests the organelles of the endomembrane system originated from the infolding of the plasma membrane, creating inner compartments with specialized functions.

cells became larger. These newly formed compartments created sufficient spaces for cellular processes to evolve and enabled membrane-bound organelles to function as independent entities within the cell. Although invagination is merely a theory, invagination as seen in prokaryotic cells provides observational evidence supporting this idea. Most bacteria have plasma membrane invaginations, creating compartments that enable the bacteria to process molecules and large macromolecules.

Chapter Summary

Below is a summary of the main ideas from each section of the chapter.

4-1. History of Cells

- All living organisms are composed of microscopic structures called cells. The term *cellulae* was first used by Robert Hooke to describe his observation of cork cells in 1665, from which the word *cell* originated. Microscopic research of cells was introduced and advanced by Antonie von Leeuwenhoek, with his improvements to the simple microscope. Cytology is the study of cells, which began with the development of the cell theory in 1839 by Matthias Schleiden and Theodor Schwann suggested that all living organisms are composed of cells. The cell theory was expanded in 1855 by Rudolf Virchow, who added that all cells arise from preexisting cells.

4-2. Prokaryotic Cells

- Prokaryotic cells are the simplest of all cells, lacking membrane-bound organelles and a nucleus. Prokaryotic cells are found in a variety of environments and habitats, including air, water, and on the human body. Two separate domains of prokaryotic cells exist, *Bacteria* and *Archaea*, which differ in their DNA and RNA base sequences.
- Prokaryotic cells have a variety of key structural features that make them unique. Prokaryotes are small organisms, ranging in size up to around 10 μm in diameter. The cell wall is a rigid structure located outside the plasma membrane, which provides shape and protection. The glycocalyx is a gelatinous, sticky substance composed of either proteins or polysaccharides, or it may exist as a combination of these two molecules, which is produced and secreted by the cell. A thin, unorganized glycocalyx is called a slime layer, and a more organized and thicker glycocalyx forms a capsule. Glycocalyces provide adhesion for prokaryotes to attach to different surfaces, protection, and a means of pathogenicity. An endospore is a protective structure formed when environmental conditions are unfavorable and provides another means of pathogenicity. Flagella are proteinaceous extensions that provide movement; fimbriae are short, rodlike extensions that provide a means of attachment; and pili are a form of fimbriae that allow cells to exchange genetic material. The nucleoid is an internal region of the prokaryotic cell that houses the single, circular chromosome containing the organism's DNA. Plasmids are circular pieces of DNA that are not part of the chromosome but may carry genes that confer genetic advantages or assist in the exchange of genetic material from one cell to another. Prokaryotic ribosomes synthesize proteins.

4-3. Eukaryotic Cells

- In comparison to prokaryotic cells, eukaryotic cells are larger and more complex, containing membrane-bound organelles and having a nucleus. Eukaryotic cells are part of the *Eukarya* domain and include algae, protozoans, fungi, plants, and animals.

4-4. Eukaryotic Cells: Nonmembranous Structures

- Eukaryotic cells contain a variety of nonmembranous structures. A glycocalyx exists in eukaryotic cells that lack a cell wall, including protozoan and animal cells. The eukaryotic glycocalyx is not as structurally organized as a prokaryotic glycocalyx. The glycocalyx is anchored to the plasma membrane and functions in cellular anchorage, strength, protection, and communication. Fungal and plant cells lack a glycocalyx because they have cell walls.
- The eukaryotic cell walls are rigid structures surrounding the cell. Cell walls exist in some protozoans, algae, fungi, and plants. The cell wall provides different functions, including protection, cellular anchorage, shape, and structure. The cell wall is primarily composed of different types of polysaccharides or a mixture of polysaccharides and chemicals, including cellulose (plants and algae), chitin (fungi), and calcium carbonate (algae).
- The plasma membrane encompasses the cell and isolates the cellular contents of eukaryotic cells from the outside environment, regulates the movement of substances into and out of the cell, enables cellular communication, provides cellular attachments, and regulates biochemical reactions. The plasma membrane consists of a double layer of phospholipids arranged in accordance with their preference for water (hydrophilic) or lack thereof (hydrophobic) and exist in a fluid state, as described in the fluid mosaic model by Singer and Nicolson (1972). The plasma membrane contains proteins that provide surface tension and permeability and are randomly distributed throughout the phospholipid bilayer in a mosaic fashion, as described by the fluid mosaic model by Singer and Nicolson (1972). There are two major groups of plasma membrane proteins, integral and peripheral proteins. Integral proteins are embedded within the plasma membrane, while peripheral proteins are located on the surface of the membrane. There are four different types of proteins that are responsible for the overall function of the plasma membrane. Receptor proteins are proteins that consist of a binding site that complements a specific type of molecule. Recognition proteins are proteins that help identify specific cells in the body. Transport proteins help transport substances across the membrane and can exist as channel proteins or carrier proteins. Enzymatic proteins initiate metabolic reactions responsible for the synthesis and breakdown of biological molecules. The plasma membrane contains cholesterol, which is important in helping maintain the strength, flexibility, and fluidity of the membrane at different temperatures.
- One of the functions of the plasma membrane is the movement of substances into and out of the cell. The plasma membrane is said to be selectively permeable, in that some substances, such as different types of gases, can cross the membrane, while others, like glucose, cannot. Movement can occur by means of passive transport, active transport, or bulk transport. Passive transport is the movement of substances down a concentration gradient without the

need for an input of energy. Diffusion is the movement of substances from an area of high concentration to low concentration. Osmosis is the movement of water from an area of high concentration to low concentration. The movement of water across a plasma membrane is facilitated by aquaporins and is dependent on tonicity, the concentration of solutes or dissolved substances inside and outside of the cell. Isotonic refers to an equal amount of solute on both sides of the membrane and no movement of water. Hypotonic refers to a concentration of solutes lower on the outside of the cell than on the inside and the movement of water into the cell. Hypertonic refers to a concentration of solutes higher on the outside of the cell than on the inside and the movement of water out of the cell. Active transport is the movement of substances against a concentration gradient, requiring a direct input of energy. A common example of active transport is the sodium-potassium pump that helps maintain the concentration gradient of sodium and potassium. Bulk transport requires the formation of a vesicle, arising from the plasma membrane, to help transport large molecules into and out of the cell. Endocytosis transports large molecules into the cell. Three different types of endocytosis include phagocytosis, allowing large particles to be ingested; pinocytosis, in which liquids and dissolved particles are ingested; and a specialized pinocytosis, known as receptor-mediated endocytosis, that allows for the ingestion of specific types of molecules, such as hormones. Exocytosis transports large molecules out of the cell.

- The cytoplasm consists of a semifluid, gelatinous solution (cytosol) and organelles found within the membranes of eukaryotic cells. The cytoplasm houses different biochemical reactions, including protein synthesis, a process that takes place on ribosomes.
- A cell's cytoskeleton is composed of an intricate network of interconnected protein-based fibers. Microtubules are thick, hollow tubes composed of tubulin and are responsible for helping maintain the shape of the cell, the movement of organelles, and cell division. Cilia are microtubule-based short, hairlike structures on the surface of cells that produce wavelike patterns of movement, propelling substances into and away from the cell. Flagella are microtubule-based taillike structures extending from the plasma membrane that undulate, providing cellular movement. Intermediate filaments are structures composed of overlapping, fibrous proteins coiled together like a rope, which provide cellular shape and structure and help anchor and support organelles. Microfilaments are long, thin, solid fibers composed of the protein actin and are responsible for maintaining cell shape, anchoring cells to other cells, and permitting cellular motility.
- Cellular junctions are points of contact between cells that allow cells to behave in a coordinated manner. Gap junctions are cellular junctions in which tubular channels form between the cytoplasm of adjacent cells and function in communication and cellular exchange. Desmosomes are cellular junctions that occur between the intermediate filaments of cells and provide increased strength for sheets of cells under tension and stretching. Tight junctions are cellular junctions that form between plasma membrane proteins of adjacent cells and produce sheetlike barriers preventing the movement of fluid between cells. Plasmodesmata are membrane-lined channels that pass through the cell walls of adjacent plant cells. The structures provide cellular communication and help transport water and molecules between plant cells.

4-5. Eukaryotic Cells: Membranous Organelles

- Eukaryotic cells contain a variety of membranous structures.
- The nucleus is a prominent, spherical-to-ovoid-shaped, centrally located organelle in the cell. The nucleus has a function in storing genetic material and directing cellular function. There are two main parts of the nucleus. The nuclear envelope is a double phospholipid membrane that separates the nucleus from the rest of the cell. The nuclear envelope contains specialized structures called nuclear pores, which control the movement of substances into and out of the nucleus. The nucleoplasm is the semifluid, gelatinous fluid found within the nucleus. A mass of DNA that coils into chromosomes, known as chromatin, is found within the nucleoplasm. Chromatin is wrapped around histones, proteins that help tightly pack DNA into the nucleus and prevent the DNA from tangling. Dark-staining regions of chromatin called nucleoli are also found within the nucleoplasm and consist of important materials for the synthesis of ribosomes.
- The endomembrane system consists of the endoplasmic reticulum, Golgi apparatus, and lysosomes. The endoplasmic reticulum is a highly folded, membranous organelle that occupies most of the cell's cytoplasm. There are two forms of the endoplasmic reticulum. The rough endoplasmic reticulum contains highly functional ribosomes that are responsible for the synthesis of a variety of different proteins used throughout the cell. The smooth endoplasmic reticulum lacks ribosomes but consists of important enzymes responsible to produce lipids, metabolism of carbohydrates, and storage of ions. The Golgi apparatus consists of flattened, hollowed, membranous sacs. The Golgi is responsible for receiving, processing, modifying, and packaging molecules produced by the rough endoplasmic reticulum. Other functions of the Golgi apparatus include protein modification and metabolism and carbohydrate production. Lysosomes are membranous organelles that contain hydrolytic enzymes that digest macromolecules, worn-out organelles, and microorganisms.
- Vacuoles are large, centrally located, membranous organelles found in all eukaryotic cells, but they are most prominent in unicellular protists and plant cells. Vacuoles have a variety of functions. Food vacuoles are formed by phagocytosis and are found in unicellular protists, which function as storage compartments for food particles until the particles are digested by lysosomes. Contractile vacuoles are "star-shaped" structures primarily found in freshwater protists to help maintain water content. Contractile vacuoles are also found in plants and fungi containing hydrolytic enzymes. Modified contractile vacuoles found in plants store organic compounds, toxic compounds, and flower color pigments. Central vacuoles are large vacuoles occupying most a plant cell's volume. The central vacuole has a function in maintaining water balance, storing ions, and maintaining turgor pressure (the pressure of water against the walls of the central vacuole).
- Mitochondria are spherical to elongated organelles present in all eukaryotic cells. Mitochondria consists of a double phospholipid bilayer. The inner layer is characterized by its highly convoluted folds, known as the cristae, which is surrounded by a smooth outer layer. The cristae consist of a semifluid mixture called the mitochondrial matrix, which contains circular pieces of DNA, 70S ribosomes, and enzymes. Mitochondria are specialized organelles responsible for the production of energy in the form of ATP during a process known as cellular respiration.

- Chloroplasts are ovoid to flattened organelles present in photosynthetic eukaryotic cells. Chloroplasts consist of a double phospholipid bilayer. The inner layer consists of a gelatinous fluid known as stroma containing circular pieces of DNA, 70S ribosomes, and enzymes, all enclosed by an outer membrane. The stroma contains interconnected sacs known as thylakoids (stacks of thylakoids are called grana), where light-trapping pigments called chlorophylls are located. Chlorophyll pigments capture solar energy and transfer this energy to make ATP, which is used to synthesize sugar, in a process known as photosynthesis.

4-6. Membrane Contact Sites and Organelle Communication

- Organelles are dynamic structures that consistently move within the cell. Membrane contact sites are interactions that exist between the plasma membrane and different types of organelles, such as the endoplasmic reticulum, Golgi apparatus, and mitochondria. The interactions between membrane contact sites of different organelles occur due to varying numbers of proteins called tethers. Membrane contact sites maintain normal, healthy cell physiology; function in the transportation of different molecules, such as lipids and ions; and regulate organelle positioning and division. The interactions occurring between organelles are constantly connecting, separating, and connecting again. Prolonged interactions between organelles and different types of membrane contact proteins have been linked to different health issues and certain disease.

4-7. Endosymbiotic Theory

- The endosymbiotic theory describes the origin of mitochondria and chloroplasts in eukaryotic cells. The endosymbiotic theory states that mitochondria and chloroplasts were free-living bacteria ingested by ancient eukaryotic cells, which formed a symbiotic relationship with each other (endosymbiosis) after years of evolution. Lynn Margulis proposed the endosymbiotic theory in 1967 after pulling together ideas from early scientists.
- Several early scientists proposed hypotheses on the idea of endosymbiosis, which were later confirmed through various microscopy, biochemical, and molecular studies. Constantin Merezhkowsky (1905) was the first to propose that chloroplast resembled ingested cyanobacteria. Microscopy and biochemistry work performed by Hans Ris and Walter Plaut (1960s) discovered similarities between chloroplasts and cyanobacteria, supporting the hypothesis proposed by Merezhkowsky. Linda Bonen and Ford Doolittle (1975, 1976) used molecular data and ribosomal RNA studies to confirm the endosymbiotic theory of chloroplasts proposed by Margulis and hypothesized that cyanobacterium was the bacterial progenitor. Ivan E. Wallin (1920s) was the first to propose that mitochondria were once free-living bacteria. Linda Bonen (1977) and other scientists used molecular data and ribosomal RNA studies to confirm that mitochondria were of bacterial origin, but they were unable to designate a bacterial progenitor. Additional research led by Carl Woese (1985) compared ribosomal RNA genes from several different organisms to discover that the bacterial progenitor of mitochondria was *Agrobacterium tumefaciens*, an alpha-proteobacterium.

- Although the original endosymbiotic theory, as proposed by Margulis, cited works of other scientists, she solely proposed that flagella, cilia, and centrioles were once spirochetes, or corkscrew-shaped bacteria. Margulis proposed that spirochetes were engulfed by ancient eukaryotic cells or attached themselves to the outside of the cell, which allowed for greater mobility and complexity and led to the development of mitosis and meiosis. Using electron microscopy, Margulis observed that flagella, cilia, and centrioles contain tiny tubules arranged around a central ring, which became the structural evidence she used to support her idea of spirochete origin. Molecular testing has not confirmed that flagella, cilia, and centrioles are of bacterial origin.
- The endosymbiotic theory and bacterial origins of chloroplasts and mitochondria are accepted by most biologists. In addition to molecular evidence, structural evidence supports the idea of bacterial origin, including double-layered membranes; circular, naked DNA; similar RNA molecules; 70S ribosomes; similar prokaryotic membrane lipids and proteins; and independent division.

4-8. Invagination Theory

- The invagination theory describes the origin of eukaryotic structures and organelles such as the nuclear membrane, endoplasmic reticulum, and Golgi apparatus. The invagination theory states that membranous organelles arose from the folding of the plasma membrane in on itself, creating inner compartments that underwent further modifications and became specialized for cellular function. Invagination observed in bacterial cells, which creates compartments for molecular processing, supports this idea.

End-of-Chapter Activities and Questions

Directions: Please refer back to what you learned in this chapter to complete the following activities.

Define Each Term in Your Own Words

1. *Cellulae*
2. 70S Ribosomes
3. *Eukarya*
4. Membrane Contact Sites
5. *Agrobacterium tumefaciens*

Chapter Review

1. What are the major principles outlined in the cell theory?
2. Contrast the following bacterial and archaeal structures: glycocalyx, flagella, cell wall, plasma membranes, and cytoplasm.
3. Outline the major differences between prokaryotic and eukaryotic cells.
4. Identify and describe the different membranous organelles found within a eukaryotic cell. Make sure to include in your description the scientists who first observed or named these structures and a general function of each.
5. Compare and contrast the endosymbiotic theory and the invagination theory.

Multiple Choice

1. The cell theory was proposed by which group of scientists?
   - **a.** Carl Woese, George Fox, Linda Bonen
   - **b.** Matthias Schleiden, Theodor Schwann, Rudolf Virchow
   - **c.** A. J. Dalton, Christian de Duve, Matthias Schleiden
   - **d.** Richard Altmann, Ivan E. Wallin, Constantin Merezhkowsky

2. In what ways do archaeal cells differ from bacterial cells?
   - **a.** rRNA base sequences
   - **b.** structural features
   - **c.** habitat
   - **d.** all the choices are correct

3. Which of the following structures is only unique to eukaryotic cells?
   - **a.** cell wall
   - **b.** nucleus
   - **c.** flagella
   - **d.** ribosomes

4. The endomembrane system of eukaryotic cells contains all the following except _______.
   - **a.** mitochondria
   - **b.** lysosomes
   - **c.** endoplasmic reticulum
   - **d.** Golgi apparatus

5. What structural evidence supports the bacterial origin of mitochondria and chloroplasts as outlined by the endosymbiotic theory?
   - **a.** circular, naked DNA
   - **b.** 70S ribosomes
   - **c.** independent division
   - **d.** all the choices are correct

Image Credits

CHAPTER

5

Metabolism and Energy

PROFILES IN SCIENCE

C. B. van Niel was born in Haarlem, Netherlands, on November 4, 1897, and died in Carmel, California, in 1985. After his father's death, van Niel's early education was dictated by family tradition, and he began preparing himself to succeed his father in business. However, after a summer vacation in Holland at age 15, van Niel was introduced to agricultural research, sparking his interest in science. Upon his family's return home, van Niel transferred to a college preparatory high school, where he became interested in analytical chemistry, even setting up his own chemistry lab at home to test and analyze different fertilizer samples. In 1916, he enrolled at Delft University, and after serving one year in the Dutch Army during World War I, he graduated from Delft in 1922 with a chemical engineering degree. After graduation, van Niel became an assistant to Albert Kluyver (founder of the study of comparative biochemistry), maintaining microbial collections, preparing lectures, and publishing papers with Kluyver on yeast and spore-forming bacteria. During his time as Kluyver's assistant, van Niel's individual studies on the biochemistry of propionic acid bacteria led to a doctorate in 1928. Months after receiving his doctorate, van Niel took a job in the United States as an associate professor of general microbiology and comparative biochemistry at Hopkins Marine Station of Stanford, a job he would hold until his retirement in 1962. While at Hopkins Marine Station, he continued his work on photosynthetic bacteria, from which he demonstrated that oxygen was derived from the splitting of water molecules during photosynthesis in 1931. This was a defining moment in the understanding of photosynthetic chemistry, as it contradicted previous thoughts that oxygen production during photosynthesis was from the splitting of carbon dioxide molecules. His contributions to the study of photosynthesis, comparative biochemistry of microorganisms, and excellence in teaching were recognized in 1963 as he became the first recipient of the American National Medal of Science. In 1964, van Niel accepted a position as a visiting professor at the University of California–Santa Cruz, leaving the position after four years. In 1972, van Niel resigned fully from teaching and research, throwing out his entire scientific library and collection of reprints.

Introduction to the Chapter

The sun plays an important and central role in sustaining life on Earth by providing the energy required to perform carefully controlled and well-orchestrated metabolic reactions necessary to maintain the complex systems of living organisms. Energy from the sun is first captured and transformed by the cells of photosynthetic organisms to produce organic molecules. When photosynthetic organisms are consumed by other organisms, it results in a continuous flow of energy between all organisms. As energy flows between organisms, it is continually converted, extracted, and utilized by living cells to perform basic biological functions. In this chapter, we will introduce metabolism and the different pathways organisms use to synthesize and break down large, complex molecules. After presenting basic information on energy, we will detail the process of photosynthesis by focusing on visible light and the important structures and chemicals solely responsible for housing and driving the reactions of this metabolic pathway. We will then discuss how these reactions convert solar energy to chemical energy, which is used to generate organic molecules. We will also discuss alternative photosynthetic pathways that have resulted from years of evolutionary adaptations. The chapter will also introduce information about cellular respiration, a metabolic pathway that uses oxygen produced from photosynthesis to drive the chemical reactions required to extract energy from nutrients in order to generate ATP, a cell's main energy source. Finally, an anaerobic metabolic process known as fermentation will be detailed, and two different types will be defined.

Chapter Objectives

In this chapter, students will learn the following:

- **5-1.** Metabolism is the sum of all chemical reactions that occur in a cell, and the reactions exist as two different types of metabolic pathways: anabolism and catabolism.
- **5-2.** The primary sources of energy on Earth are the sun and inorganic chemicals, which can exist as either potential or kinetic energy, flowing and being transformed through organisms as explained by the laws of thermodynamics, while ATP provides the necessary energy for a cell to perform work.
- **5-3.** Photosynthesis is a metabolic process that involves the conversion of solar energy to chemical energy and uses this energy, along with different types of specialized structures, molecules, and reactions, to synthesize organic molecules.
- **5-4.** Cellular respiration is a metabolic, aerobic process that utilizes different types of specialized structures, molecules, and reactions to transfer energy from the bonds of glucose molecules to ATP molecules.
- **5-5.** Fermentation is a metabolic, anaerobic process that utilizes a different final electron acceptor during the breakdown of glucose to produce either lactic acid or alcohol.

5-1. Metabolism

Metabolism is the sum of all chemical reactions responsible for changing or transforming energy within a cell. These reactions occur in a sequential manner known as **metabolic pathways** and may exist as straight or branched chains or as cycles. Each step of the metabolic pathway utilizes an enzyme that catalyzes reactions resulting in the synthesis and/or breakdown of different types of molecules. There are two different types of metabolic pathways that exist in the cell: catabolism and anabolism.

Catabolism is the metabolic pathway responsible for breaking down large, complex molecules within the cell (Figure 5.1). The breakdown of complex molecules requires the presence of enzymes and produces two important components: small, simple molecules and energy. Energy is released when the bonds holding the large molecule together are broken, and this energy is used within the cell to do work. An example of a catabolic pathway would include breaking down proteins into amino acids.

Anabolism is the metabolic pathway responsible for synthesizing new, complex molecules within the cell (Figure 5.1). The synthesis of complex molecules requires three important components: small, simple molecules; energy; and enzymes. Enzymes, along with the direct input of energy, form bonds between small, simple molecules in a variety of ways to produce different types of larger, more complex molecules. An example of an anabolic pathway would include the synthesis of proteins from amino acids.

Although both of these metabolic pathways are functionally opposite, they do function simultaneously as well. The energy and molecules produced and released in a catabolic pathway are utilized within an anabolic pathway, and vice versa.

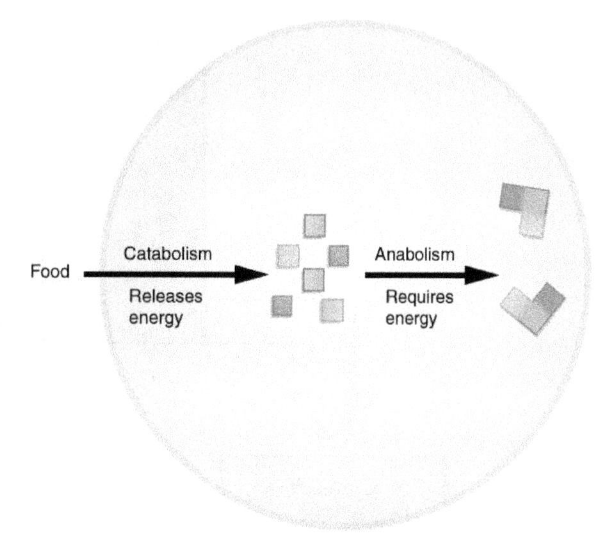

FIGURE 5.1 There are two different metabolic pathways that occur within a cell. Catabolism breaks down large, complex molecules, releasing energy. Anabolism utilizes energy to synthesize new, complex molecules.

5-2. Energy

Energy is defined as the ability to do work, resulting in the movement of matter against an opposing force, such as gravity and/or friction. Energy is measured in calories. A calorie is the measure of the amount of energy needed to raise one gram of water one degree Celsius. However, this term should not be confused with **kilocalorie**, which is used to measure the amount of energy in food and the amount of heat released by an organism. The number of calories on a food label is actually listed in kilocalories. For example, the food label on a soda that reads 250 calories is actually 250 kilocalories, or 250,000 calories (1 kilocalorie equals 1,000 calories).

There are two primary sources of energy found on Earth: energy provided by the sun, known as light or solar energy, and energy provided by inorganic chemicals, known as geothermal energy. Energy flows through various different types of organisms (Figure 5.2) and is important for development, growth, repair, and reproduction. Energy flowing from the sun is used by a group of organisms called **photoautotrophs**, or producers, and includes plants, algae, and a group of bacteria known as cyanobacteria. Energy is captured and utilized to convert carbon dioxide and water into energy-rich organic molecules, during photosynthesis, and are used for important cellular functions. **Chemoautotrophs**, on the other hand, such as bacteria and archaea, obtain their energy from inorganic chemicals, like iron, sulfur, and/or ammonia, to synthesize important organic molecules. **Heterotrophs**, or consumers, are organisms that consume other organisms and are highly dependent on photoautotrophs and chemoautotrophs to supply their energy needs. As a result, energy is continuously flowing from one organism to another.

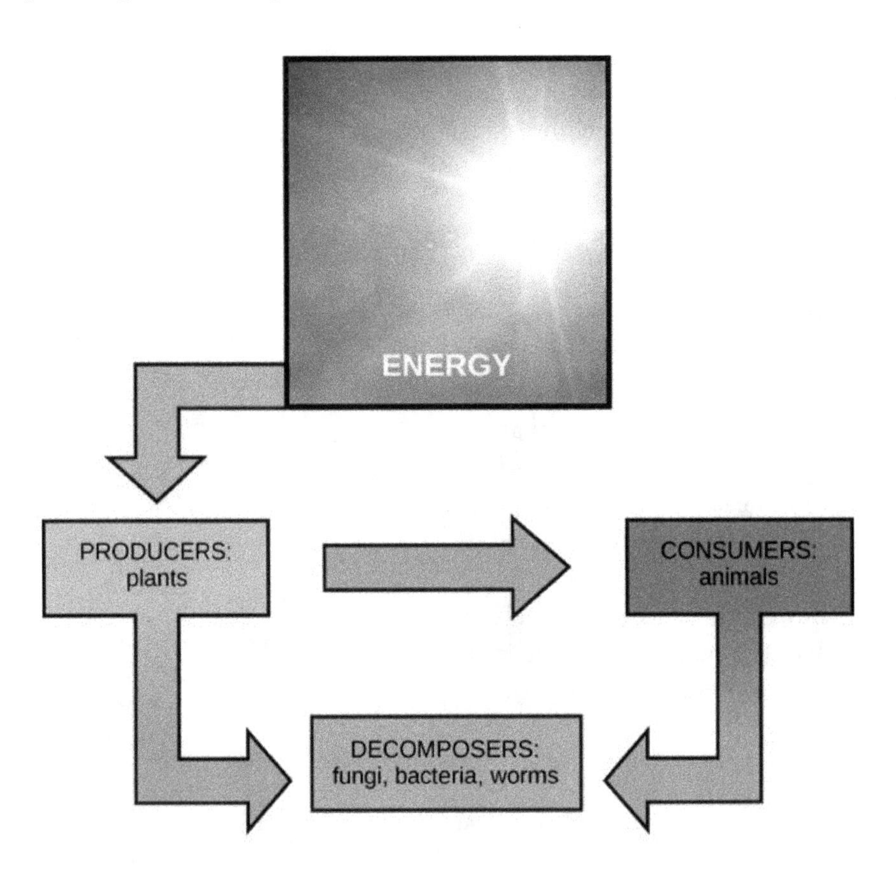

FIGURE 5.2 A majority of living organisms obtain their energy from the sun. Energy from the sun is captured by photoautotrophs and the energy is used to make organic molecules. Heterotrophs are organisms that obtain their energy through the consumption of other organisms. Energy flowing between living organisms is continuous.

Energy exists in two basic forms: potential energy and kinetic energy. **Potential energy** is stored energy, or energy available to do work based on a stationary location or position. Chemical energy is a type of potential energy and is stored in the chemical bonds of molecules. Common examples of potential energy include water behind a dam (Figure 5.3, left), a tablespoon of sugar, or a skier on top of a mountain. **Kinetic energy** is energy being used to move an object. Various types of kinetic energy exist based on the movement of molecules, such as thermal energy (movement of molecules within matter), solar energy (movement of photons), and sound energy (movement of vibrational waves). Common examples of kinetic energy include water flowing through a dam (Figure 5.3, right), a tablespoon of sugar burning, or a skier skiing down a mountain.

FIGURE 5.3 There are two basic forms of energy: potential energy and kinetic energy. A common example of potential energy is water behind a dam (left). Water flowing through a dam is an example of kinetic energy (right).

In order to understand how energy flows through different organisms and is transformed from one state to another (potential to kinetic), it is important to know the laws that govern these events, known as the **laws of thermodynamics**. The **first law of thermodynamics** states that energy cannot be created or destroyed, but it can be converted from one form to another. An example of the first law of thermodynamics involves the conversion of chemical energy, found in the chemical bonds of molecules, into kinetic energy (Figure. 5.4, left). Another example is when plants capture solar (light) energy from the sun, this energy is converted to chemical energy by the process of photosynthesis (Figure 5.4, left). The plant has not created nor destroyed the solar energy, but rather converted the energy into a usable form for its benefit.

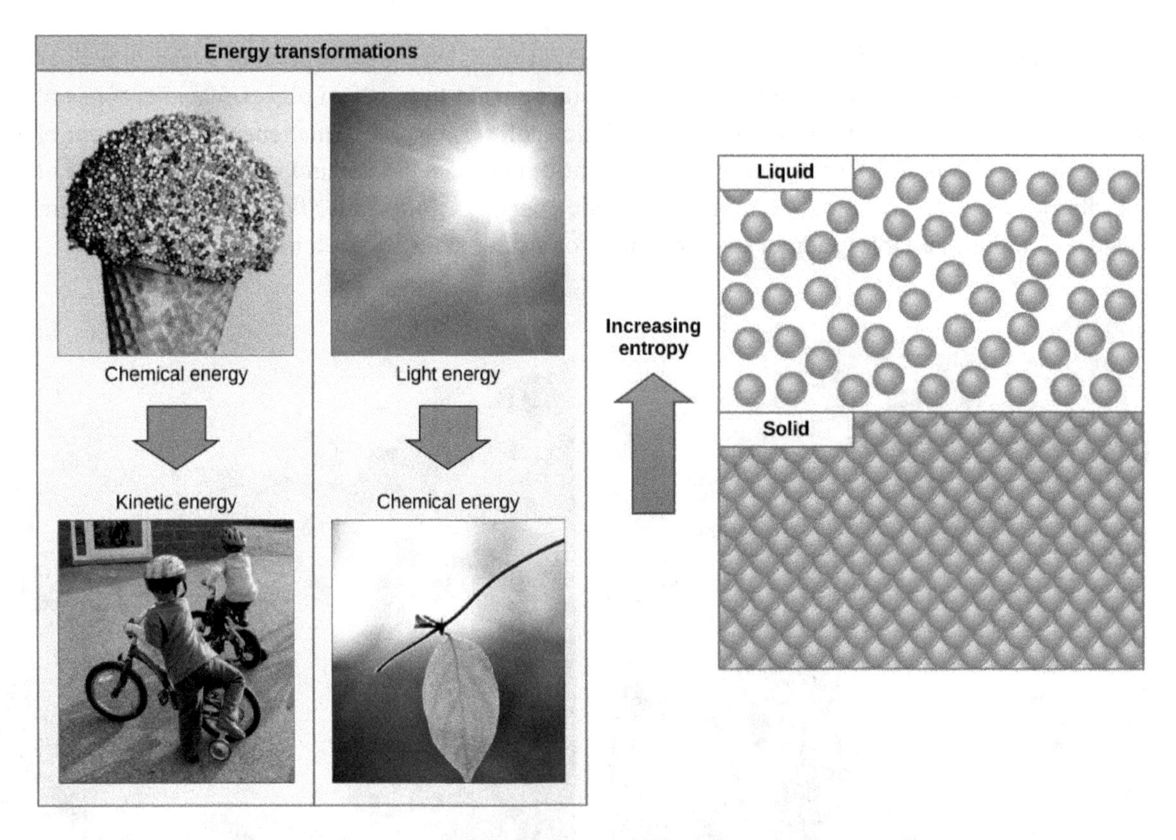

FIGURE 5.4 Energy can be converted from one form to another, as defined by the first law of thermodynamics. An example of energy conversion occurs when chemical energy is converted to kinetic energy or when plants convert solar energy to chemical energy (left). When energy is converted, heat is lost, increasing entropy, or disorder, in the universe, which is explained by the second law of thermodynamics (right).

The **second law of thermodynamics** states, the conversion of energy from one form to another is considerably inefficient and results in the loss of usable energy in the form of heat. Although most energy can be converted from one form to another, heat can never be converted back to any other form of energy. As the amount of heat increases, it increases disorder in the universe, which scientists call entropy (Figure 5.4, right). An increase in entropy leads to decreases in the amount of energy available to do work. For example, an automobile only uses 25 to 30 percent of the potential energy stored in gasoline to produce kinetic energy, while the remaining 70 to 75 percent of energy is lost as heat. Inefficiency of energy conversion and heat loss are also observed in cells, where only 40 percent of energy converted is utilized and 60 percent is lost as heat.

The main energy source in a cell is **adenosine triphosphate** or **ATP** (Figure 5.5). ATP is responsible for providing the energy currency for many different types of reactions that occur within the cell. An ATP molecule is composed of three main molecules: a nitrogenous base (adenine), five-carbon sugar (ribose), and three phosphate groups. The main source of energy that ATP provides is located within the covalent bonds between each of the three phosphate groups (Figure 5.5). The energy in these bonds comes from energy released during the catabolism of nutrients in animals or from the solar energy captured by plants. When the covalent bond holding the last phosphate group to the molecule is broken during the process of hydrolysis (a reaction involving the addition of a water molecule), the reaction releases roughly 7.3 kilocalories of energy, sufficient for many biological reactions, such as protein synthesis or muscle contractions. When a phosphate group is removed from ATP, it becomes adenosine diphosphate, a molecule containing only two phosphate groups, as shown below:

$$ATP \rightarrow ADP + P_i + energy$$

Working cells are rapidly using ATP on a continuous basis; however, the molecule is quickly resynthesized when a phosphate group is added back to the ADP molecule, with the help of an input of energy released during the catabolism of glucose in the mitochondria (Figure 5.5).

$$ADP + P_i + energy \rightarrow ATP$$

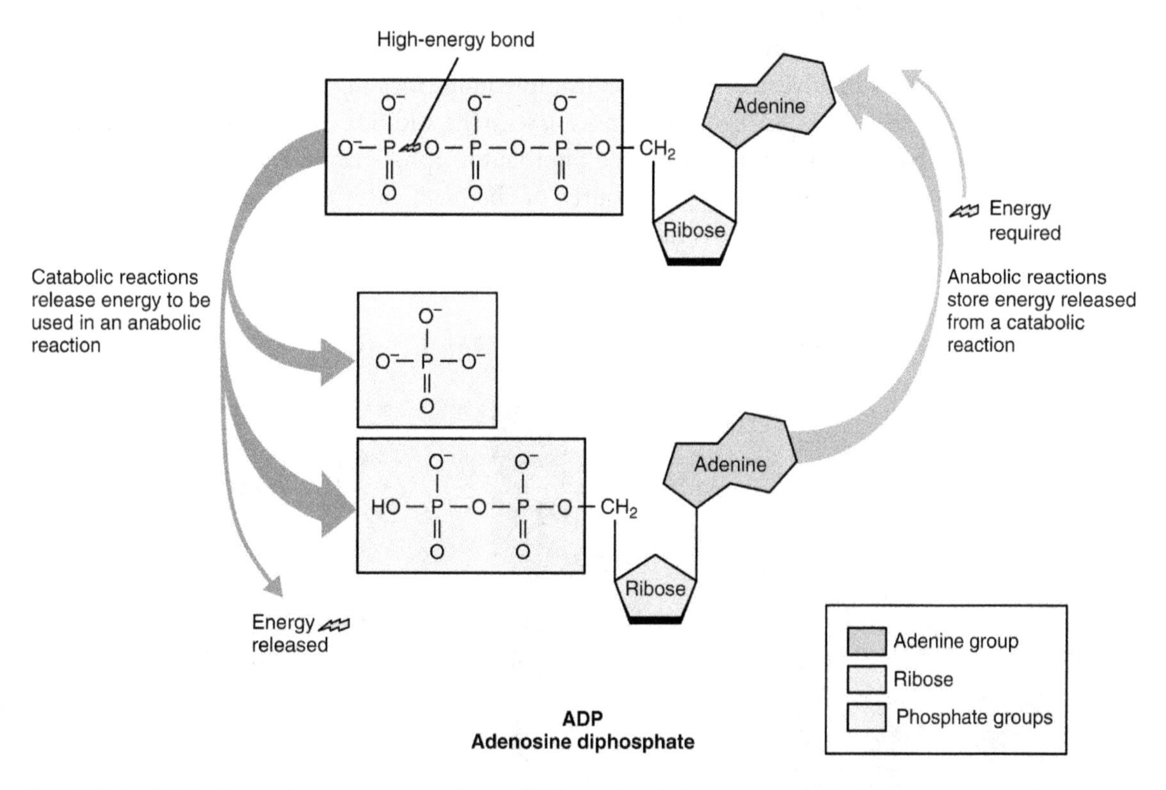

FIGURE 5.5 ATP is the main energy source in a cell. The molecule is composed of adenine, ribose, and three covalently bonded phosphate groups, which contains the majority of the energy within the molecule. When the high-energy bond holding the second and third phosphate group together is broken, energy is released and ATP becomes ADP. The ATP molecule is resynthesized with an input of energy and another phosphate group.

ATP is a reliable and effective source of energy for three reasons. First, ATP is constantly being resynthesized; therefore, it is readily available in the cell when needed. In addition, even energy located within the chemical bonds of lipids and complex carbohydrates can be converted to ATP prior to being used. Next, the bond holding the last phosphate group to the molecule is highly unstable, which makes it easy to break to release energy. Finally, the amount of energy released when the third phosphate bond is broken is almost twice the amount of energy needed to supply biological reactions. The energy not used is released as heat.

Finally, ATP provides the necessary energy for a cell to perform three main functions: **chemical work**—ATP supplies the cell with the energy it needs in order to synthesize important biological molecules, such as proteins and carbohydrates; **transport work**—ATP supplies the energy necessary for pumping substances across the plasma membrane during active transport; and **mechanical work**—ATP provides the cell with energy to perform mechanical functions, such as providing energy for cilia to beat and for muscles to contract.

5-3. Photosynthesis

Photoautotrophs are organisms, such as plants, algae, and cyanobacteria (Figure 5.6), capable of converting solar energy into chemical energy, which is used to synthesize organic molecules. This metabolic process is known as photosynthesis and is responsible for helping sustain life on Earth. **Photosynthesis** involves converting the simple molecules of carbon dioxide (CO_2) and water (H_2O) into glucose ($C_6H_{12}O_6$) and oxygen (O_2), with the addition of solar energy. Organisms such as heterotrophs, organisms that consume photoautotrophs, are reliant on photosynthetic organisms to provide them with glucose, a source of chemical energy, and oxygen, a molecule needed for cellular respiration. The overall process of photosynthesis can be expressed by using the following equation:

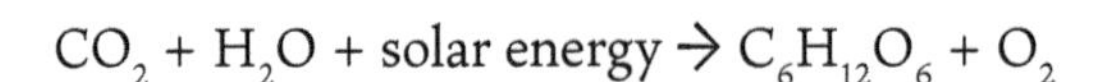

$$CO_2 + H_2O + \text{solar energy} \rightarrow C_6H_{12}O_6 + O_2$$

FIGURE 5.6 Photosynthesis is a metabolic process in which organisms known as photoautotrophs, which include plants (left), algae (middle), and cyanobacteria (right), capture solar energy and convert it to chemical energy to synthesize organic molecules.

Photosynthesis occurs in chloroplasts, located within the green parts of plants, primarily the leaves. The chloroplasts are located in a group of cells called mesophyll cells, with one mesophyll containing up to 200 chloroplasts. Located on either side of the mesophyll cells is an upper epidermis consisting of a waxy substance called a cuticle and a lower epidermis containing small pore openings called **stomata**. Stomata are responsible for opening and closing throughout the day to allow carbon dioxide (CO_2) to enter and oxygen (O_2) to exit the leaves.

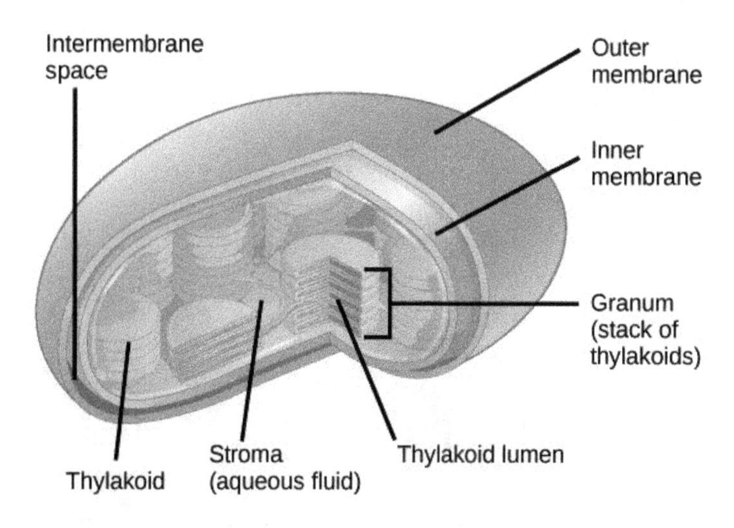

FIGURE 5.7 Chloroplasts are specialized organelles where the process of photosynthesis takes place. Within the inner membrane of the chloroplast is the stroma, a gelatinous-like fluid containing important photosynthetic enzymes and flattened sacs called thylakoids, which consist of different types of chlorophylls, photosynthetic pigments responsible capturing solar energy.

Chloroplasts (Figure 5.7) are a group of specialized organelles consisting of important structures and chemicals that help maximize the efficiency of photosynthesis. Chloroplasts contain a double-membrane system that surrounds a gelatinous fluid consisting of various types of photosynthetic enzymes, called the **stroma**. Within the stroma are membranous structures called **thylakoids**, stacked ten to 20 high, like coins, into grana. Inside of the thylakoids are important photosynthetic pigments called **chlorophylls**, which are responsible for capturing solar energy. Energy captured by chlorophylls is used to drive the photosynthetic reactions to produce sugar and oxygen.

Visible Light

Energy produced by the sun is called electromagnetic radiation. Electromagnetic radiation travels in waves toward Earth; however, most of the radiation is trapped by the ozone layer or by water vapor and carbon dioxide before it reaches Earth's surface. Of the radiation produced by the sun, only 1 percent reaches Earth's surface, but it provides a sufficient amount of energy to power photosynthesis. This type of radiation is called **visible light**.

Our understanding of visible light was introduced in experiments published by Sir Isaac Newton in 1672. Newton ran several experiments involving prisms and white light and discovered that in directing white light through the prism, the light refracted into various different colors, ranging from violet to red. In order to confirm that the prism was not responsible for coloring the light, he ran another experiment demonstrating that the refracted light reproduced white light as it passed through another prism. In an 1865 publication entitled "A Dynamical Theory of the Electromagnetic Field," Scottish mathematician James C. Maxwell outlined by means of mathematical equations that visible light makes up only a small portion of a much larger spectrum of radiation called the **electromagnetic spectrum**. The electromagnetic spectrum organizes the different types of electromagnetic radiation, from short-wavelength gamma rays (measured in nanometers [nm]) to long-wavelength radio waves (measured in kilometers), with visible light's narrow band falling in between and spanning from about 400 nm to 700 nm (Figure 5.8).

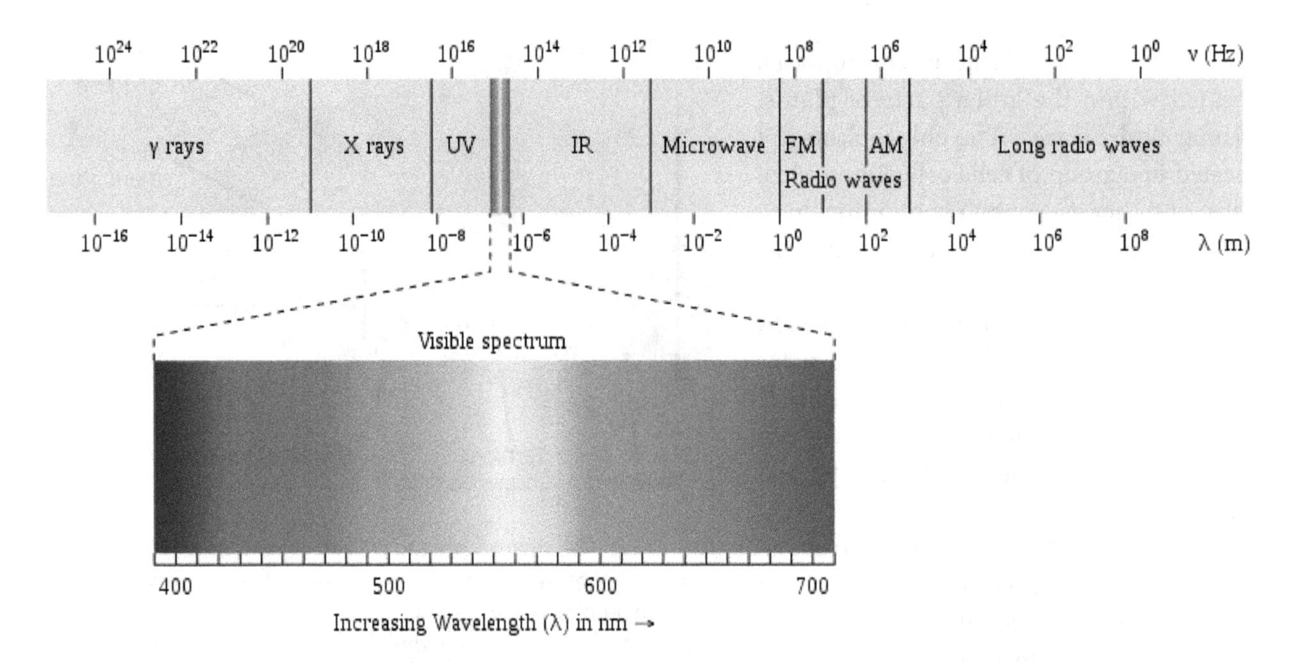

FIGURE 5.8 When visible light is refracted through a prism, it produces a variety of colors ranging from violet to red. Although visible light only makes up a small portion of the electromagnetic spectrum (400 nm to 700 nm), it produces enough energy to drive photosynthesis.

In 1905, four papers written by German physicist Albert Einstein, known as the *Annus Mirabilis* papers, were published in the *Annalen der Physik* (*Annals of Physics*) scientific journal. In one of these papers, "On a Heuristic Viewpoint Concerning the Production and Transformation of Light," Einstein proposed the **quantum theory of light**, which suggested that light is composed of small, individual packets of energy. It was not until 1926 that American chemist Gilbert N. Lewis called these energy packets **photons**.

Photons have a variety of important characteristics:

1. The number of photons absorbed per unit of time determines the brightness of light. For example, the higher the number of photons absorbed per unit of time, the brighter the light.
2. The amount of energy within a photon is fixed and is determined by how fast or slow the photon moves when it vibrates. The amount of energy within a photon is directly correlated to its speed, which means the faster the photon moves, the more energy the photon contains.
3. When a photon vibrates, it travels in waves, which is called a **wavelength**. The distance of a wavelength is measured using nanometers (nm) and typically falls between 380 and 750 nm for visible light.
4. The wavelength of a photon and its energy are inversely related. Therefore, the longer the wavelength, the less energy the photon contains.

When visible light reaches the Earth's surface, one of three fates occurs when it strikes an object. If the light bounces off an object, the light is **reflected** (Figure 5.9, left), which can determine the color of the object. For example, leaves are green during the summer because they reflect the green wavelengths of visible light. If an object reflects all wavelengths of visible light, the object will appear white, while an object that reflects none of the wavelengths will appear black. Visible light can also be **transmitted** (Figure 5.9, middle), or pass through an object. As with reflected light, transmitted light helps determine the color of the object. Finally, light can be captured or **absorbed** (Figure 5.9, right) by an object, which is important in driving the processes of photosynthesis.

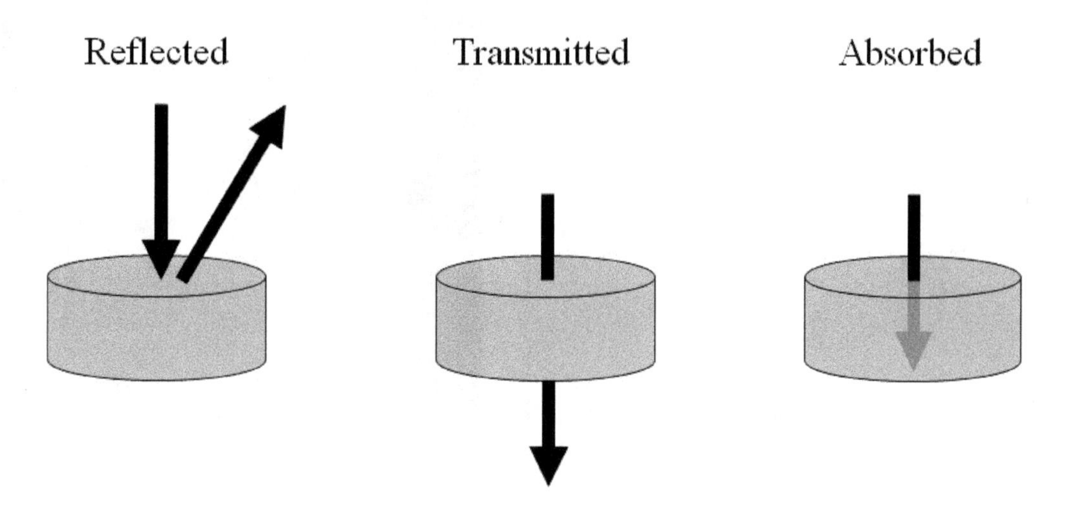

FIGURE 5.9 When visible light reaches Earth, it is either reflected (bounces off an object), transmitted (passes through an object), or absorbed (captured by the object).

Photosynthetic Pigments

Various wavelengths of light (primarily violet, blue, orange, and red) are absorbed by photosynthetic pigments found within the thylakoids of chloroplasts called **chlorophylls**. Plants produce three different types of photosynthetic pigments. The most abundant and important photosynthetic pigment is called chlorophyll *a* and functions as the main light-capturing pigment in the chloroplasts. *Chlorophyll a* (Figure 5.10a) primarily absorbs violet and blue wavelengths of light, around 400 to 500 nm, but can also absorb red and orange wavelengths, between 600 and 700 nm. However, chlorophyll *a* cannot absorb green wavelengths; instead it reflects or transmits this color, giving leaves their characteristic green color. Another important photosynthetic pigment, much smaller and with a slightly different structure than that of chlorophyll *a*, is chlorophyll *b*. Because of the slight difference in structure, chlorophyll *b* absorbs various different blue and red-orange wavelengths. *Chlorophyll b* (Figure 5.10b) reflects or transmits green to yellow wavelengths, again contributing to the color of a plant's leaves. The final group of photosynthetic pigments are the carotenoids. *Carotenoids* are composed of carotenes (Figure 5.10c) and xanthophylls and are responsible for the variety of colors seen in carrots, tomatoes, and some animals, such as fish and flamingos. These pigments are responsible for absorbing violet and blue-green light during photosynthesis, while reflecting yellow, orange, and red wavelengths of light. Within the chloroplast, the color of carotenoids is masked by the dominant green-colored chlorophyll *a* and chlorophyll *b*; however, as the weather begins to change in late summer, the chlorophyll in the leaves begins to break down before the carotenoids do, giving rise to the fall colors of autumn.

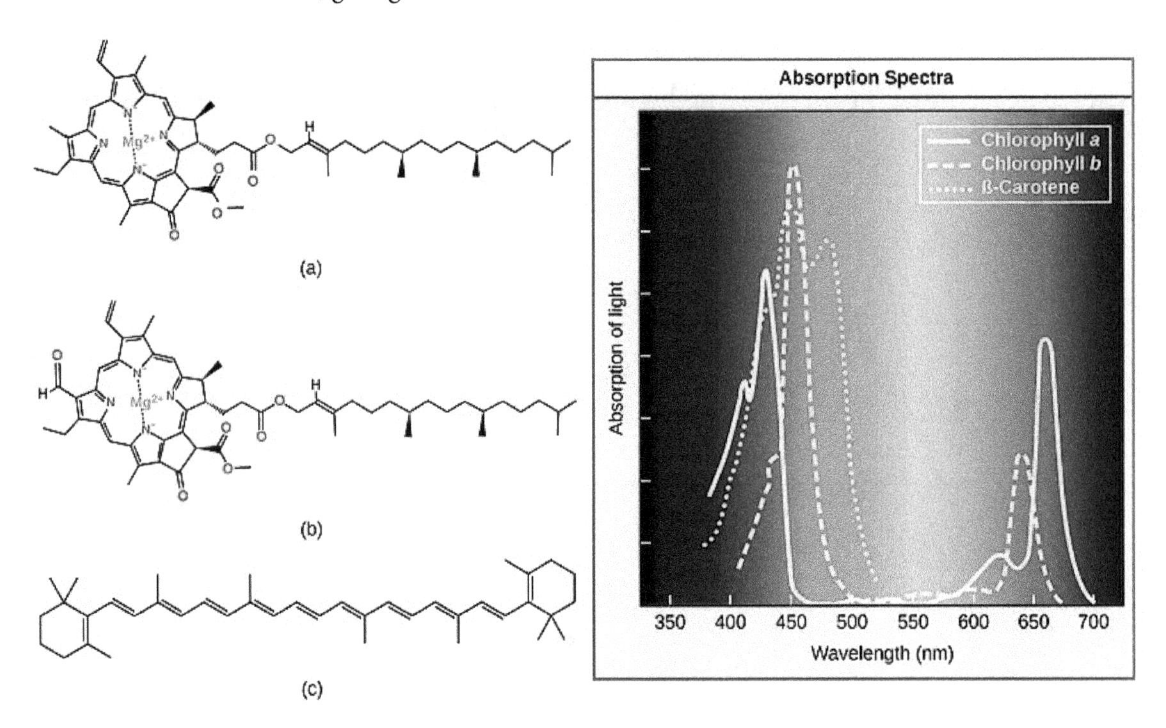

FIGURE 5.10 Photosynthetic pigments are specialized molecules found within the thylakoids of chloroplasts and function in absorbing various different colors of light to help drive photosynthesis. The different types of photosynthetic pigments include (a) chlorophyll a, (b) chlorophyll b, and (c) carotenoids (β-carotene).

Light-Dependent Reactions (Photo Reactions)

The **light-dependent reactions**, also known as the photo reactions, are a group of reactions in which solar energy from the sun is absorbed and converted into chemical energy within the thylakoids of the chloroplasts. The chemical energy is stored in two high-energy carrier molecules, ATP and NADPH, which provide the necessary energy to synthesize important organic molecules, such as glucose, in the light-independent reactions of photosynthesis. In addition to the production of high-energy carrier molecules, oxygen gas is also produced and released and will either enter the atmosphere through a plant's stomata or be used by the plant during the process of cellular respiration (discussed later).

Within the thylakoids are a group of specialized units called photosystems, which consist of two main regions: *light-harvesting pigment complexes* and a *reaction center* (Figure. 5.11). The **light-harvesting pigment complexes**, which contain up to 400 photosynthetic pigments, including chlorophyll *a*, chlorophyll *b*, and carotenoids, bound to proteins, and the **reaction center**, surrounded by the light-harvesting pigment complexes, consisting of a special pair of chlorophyll *a* molecules and various proteins. There are two different types of photosystems found in the thylakoids, photosystem I (PSI) and photosystem II (PSII). It is important to note that the names of the photosystems do not reflect the order in which they proceed, but rather the order in which they were discovered (Figure 5.11).

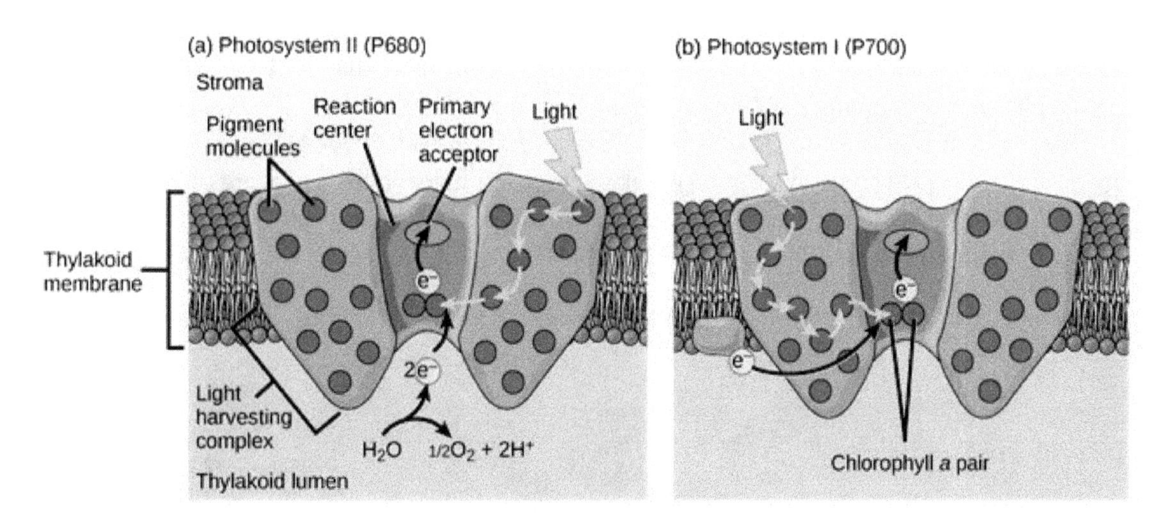

FIGURE 5.11 The photosystems of the light-dependent reactions (photosystem II and photosystem I) are found in the thylakoids of a chloroplast and are numbered based upon the order they were discovered, not by the order in which they proceed. There are two regions found within a photosystem, a light-harvesting pigment complex and reaction center. Both regions contain chlorophylls, which are responsible for exciting electrons.

In 1932, two American scientists, Robert Emerson and William Arnold, were the first to suggest the presence of photosynthetic reaction centers within chloroplasts. Additional observations published in 1936 by two German scientists, Hans Gaffron and Kurt Wohl, suggested several chlorophylls were responsible for exciting electrons, which then passed to a specific area where important chemical reactions began. By the 1950s, key observations were being made on photosystems, primarily PSI, including the existence of chlorophyll *a* and the ability of the system to reduce $NADP^+$ to NADPH. However, none of these observations were coherently integrated. By 1960, two British biochemists, Robert Hill and Fay Bendall, proposed a hypothesis for the existence of two photosystem structures, but the results of their experiments provided little support. However, in 1961, two teams of scientists provided the first clear experimental evidence supporting the existence of the two photosystem structures.

Electrons in the light-dependent reactions flow along a linear pathway, beginning in PSII (Figure 5.12). Solar energy is absorbed by the photosynthetic pigments located within the light-harvesting pigment complex, and this energy passes from one pigment to another until it reaches the specialized pair of chlorophyll *a* molecules located in the reaction center. Once chlorophyll *a* receives the energy, this energy is used to excite electrons in the reaction center that are donated by the splitting of water molecules. When water molecules split, electrons are released into the reaction center, while the remaining hydrogen protons will be used to make ATP and NADPH, and the oxygen atoms combine to form oxygen gas, which was first demonstrated by C. B. van Niel in 1931.

Next to the reaction center is an electron transport chain (Figure 5.12), where high-energy electrons travel down once released from the reaction center of PSII. As the high-energy electrons move along the electron transport chain, they lose energy, which is used to pump hydrogen protons from the stroma of the chloroplast into the thylakoids, creating a proton gradient. The increase in hydrogen protons within the thylakoids travel through an enzyme called ATP synthase, where ADP is phosphorylated to produce ATP used during the Calvin cycle.

The low-energy electrons from PSII travel to the reaction center of PSI (Figure 5.12), where they are once again excited by the absorption of solar energy by photosynthetic pigments. The high-energy electrons leave the reaction center of PSI and pass down another electron transport chain, where they are transferred to $NADP^+$ to form the high-energy carrier molecule, NADPH, used during the Calvin cycle.

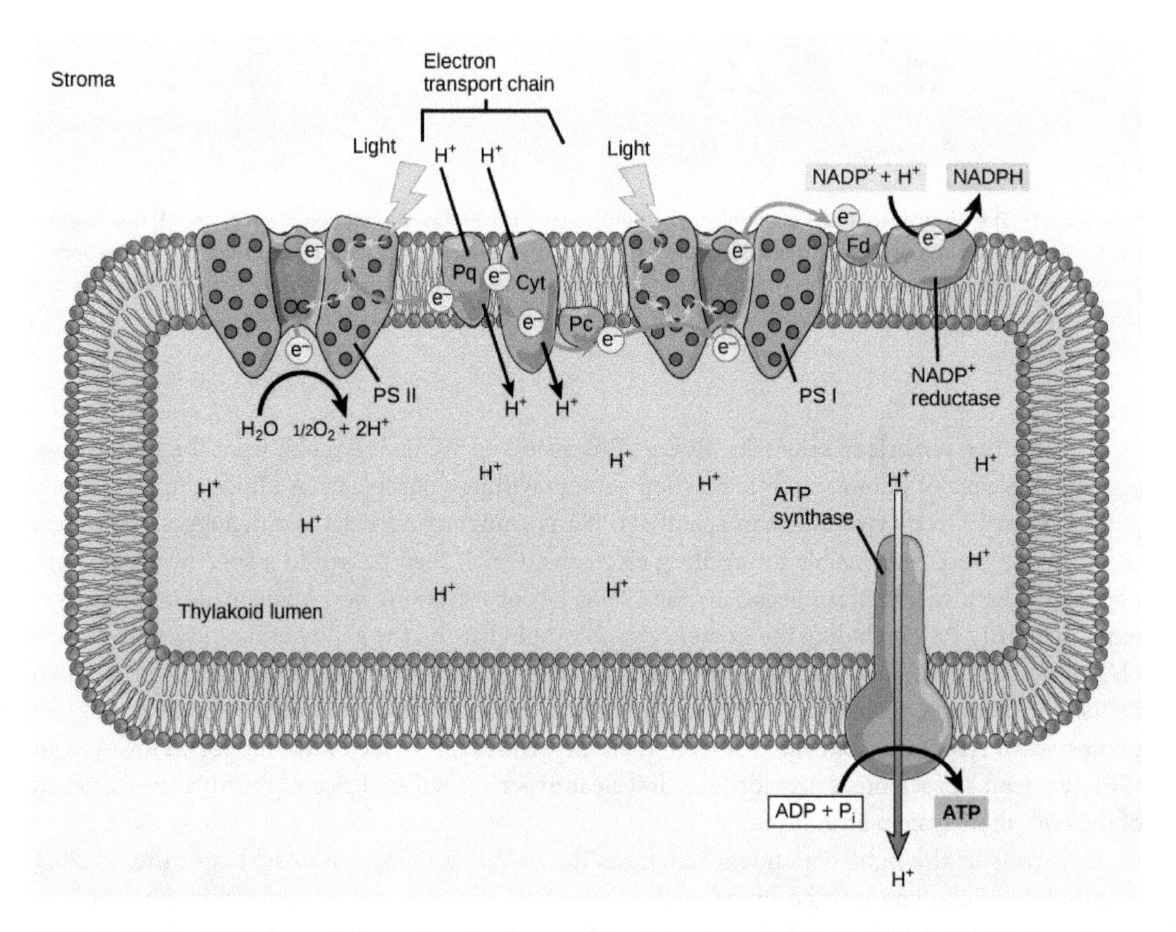

FIGURE 5.12 The light-dependent reactions involve a series of reactions responsible for exciting electrons to produce ATP and NADPH for the Calvin Cycle. The light-dependent reactions occur along the thylakoids of the chloroplasts.

Light-Independent Reactions (Synthesis Reactions)

The **light-independent reactions**, also known as the synthesis reactions, are a group of reactions in which energy from ATP and NADPH generated during the light-dependent reactions is used by enzymes in the stroma of the chloroplast to convert carbon dioxide (CO_2) into organic molecules like glucose.

FIGURE 5.13 Melvin Calvin.

The light-independent reactions involve a series of enzyme-catalyzed reactions that occur in a cyclic pathway known as the **Calvin cycle**, in which carbohydrates are produced using carbon dioxide from the atmosphere and energy formed during the light-dependent reactions. The Calvin cycle is named for American biochemist Melvin Calvin (Figure 5.13), who led a team of scientists to the discovery of these reactive processes during the 1950s.

The experiments performed by the scientists involved exposing a culture of green algae to carbon dioxide (CO_2) containing the radioactive isotope Carbon-14 (^{14}C). The team traced the path of the ^{14}C and discovered that small amounts of products were produced as the algae converted carbon dioxide (CO_2) to sugar. When Calvin and the team exposed the green algae to the radioactive carbon dioxide for 60 seconds, many different compounds with ^{14}C were discovered; however, after running the experiment again for only seven seconds, ^{14}C was primarily discovered in 3-phosphoglycerate. This suggested that carbon fixation must involve carbon dioxide attaching to a two-carbon molecule, since 3-phosphoglycerate is a three-carbon molecule. Although Calvin and his team could not immediately identify the molecule, the group knew they were looking for a molecule that, when combined with carbon dioxide, would produce a three-carbon product. After subsequent experiments, Calvin and team discovered that ribulose biphosphate (RuBP) was the molecule carbon dioxide would combine with to form the three-carbon product.

Figure 5.14 illustrates the Calvin cycle, which is a three-step process that takes place in the stroma of the chloroplast:

1. **Carbon fixation.** Carbon fixation is an important process in which carbon dioxide (CO_2) gas is taken in through the stomata of a plant and converted into organic molecules readily utilized by cells. In this particular step of the Calvin cycle, the carbon atom from carbon dioxide attaches to a five-carbon molecule called ribulose biphosphate (RuBP). The attachment of the carbon to ribulose biphosphate (RuBP) is catalyzed by an enzyme known as rubisco, considered one of the most abundant and important proteins in the world. The result of this enzyme-catalyzed reaction is an unstable six-carbon molecule, which splits in half, forming two three-carbon molecules called 3-phosphoglycerate (3PG).
2. **Synthesis of glyceraldehyde-3-phosphate (G3P).** The synthesis of glyceraldehyde-3-phosphate (G3P) occurs as a result of carbon reduction. In this step, 3PG is converted in two enzyme-catalyzed reactions, using ATP and NADPH from the light-dependent reactions. In the first enzyme-catalyzed reaction, a phosphate from ATP is added to 3PG to form 1,3-biphosphoglycerate, while 1,3-biphosphoglycerate is reduced to G3P by NADPH. The resulting product is G3P, a precursor for the synthesis of important carbohydrates.
3. **Regeneration of ribulose biphosphate (RuBP).** While some of the G3P produced during the Calvin cycle will be used as a building block for glucose molecules and other important carbohydrates such as starch and cellulose, as well as for the synthesis of fatty acids and amino acids, a series of enzyme-catalyzed reactions is responsible for converting G3P back to additional ribulose biphosphate (RuBP) in order for the cycle to repeat. Energy for these enzyme-catalyzed reactions is provided by ATP generated during the light-dependent reactions.

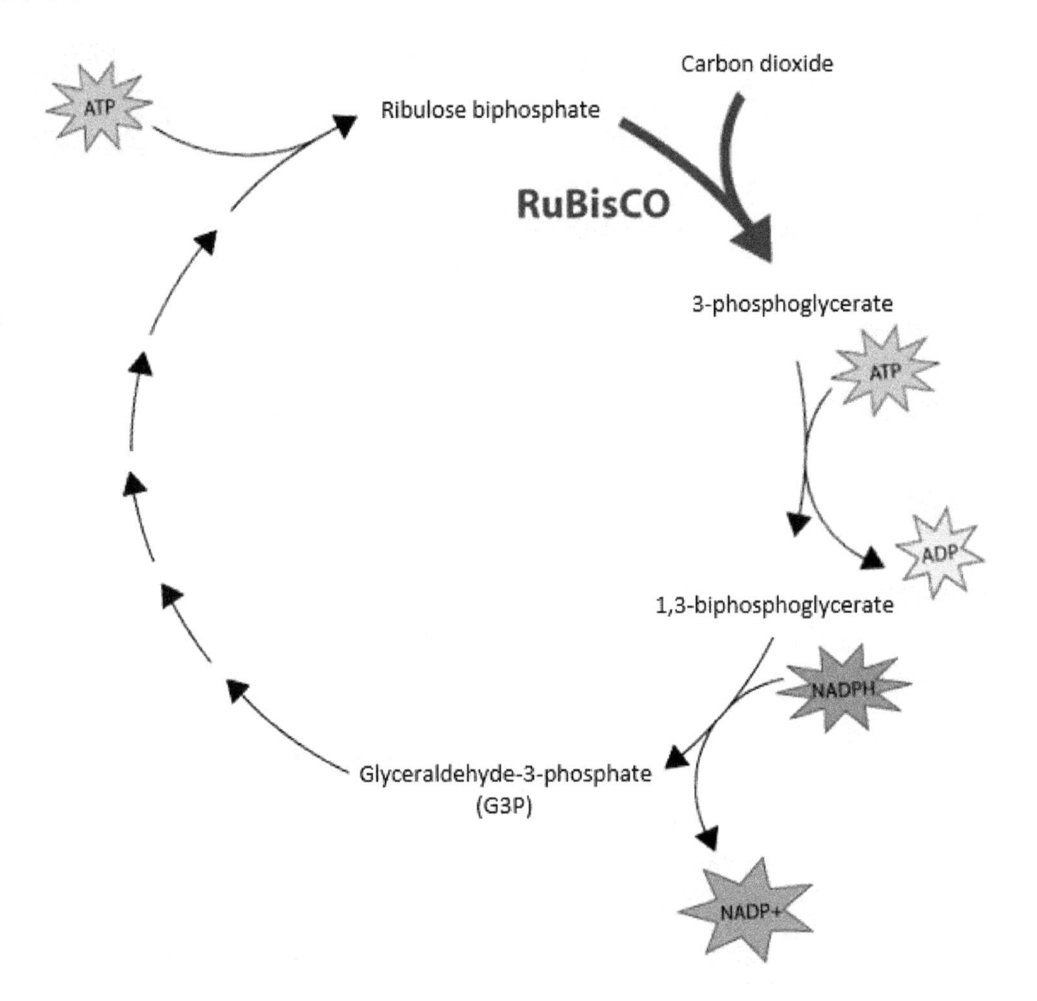

FIGURE 5.14 The Calvin Cycle is a photosynthetic process that takes place in the stroma of the chloroplast that uses ATP and NADPH from the light-dependent reactions to synthesize the necessary components required to build important carbohydrates, lipids, and proteins.

Alternative Photosynthetic Pathways

There are various different types of alternative photosynthetic pathways, as a result of evolutionary adaptations made by plants to help conserve water and improve photosynthetic efficiency in hot, arid climates. Loss of water can be detrimental or fatal to land plants.

C_3 Plants (About 85 Percent of All Plant Species)

C_3 plants include the majority of recognizable plants, like azaleas and maples, as well as common agricultural plants, such as rice and wheat (Figure 5.15, left). These particular types of plants carry out photosynthesis as outlined, utilizing the enzyme rubisco to fix carbon dioxide (CO_2) during the carbon fixation step of the Calvin cycle. However, during hot and dry weather, C_3 plants will use a variant of photosynthesis in order to conserve water. This process is called **photorespiration**, aptly named because it involves light and the consumption of oxygen (O_2), while releasing CO_2. During hot and dry days, C_3 plants close their stomata to prevent water loss due to evaporation. This phenomenon prevents CO_2 from entering the leaf, thereby decreasing its concentration, while at the same time preventing O_2 from leaving, increasing its concentration. As a result, rubisco begins to fix O_2 during the Calvin cycle, resulting in the release of CO_2 as a byproduct. Photorespiration is not ideal for plants, because the process does not produce ATP or glucose and decreases photosynthetic output by releasing CO_2, which would normally be fixed during photosynthesis.

C_4 Plants (About 0.4 Percent of All Plant Species)

Photorespiration is considered a wasteful process because it does not play a role in the Calvin cycle; therefore, some plants have evolved an adaptation in order to reduce photorespiration. This alternative photosynthetic process, occurring in plants such as corn (Figure 5.15, middle), sugarcane, and crabgrass, is called **C_4 photosynthesis**. Reports of this new alternative photosynthetic pathway began to appear as early as 1954, when studies of sugarcane revealed additional compounds other than phosphoglyceric acid (PGA) produced when exposed to radioactive carbon dioxide ($^{14}CO_2$). During subsequent studies and reports in the 1960s, scientists identified these compounds in both sugarcane and maize as the four-carbon molecules malate and aspartate. A detailed study and description of work performed on sugarcane was published in 1965 confirming the presence of malate and aspartate when exposed to $^{14}CO_2$ and concluding that carbon assimilation was different than that found in other types of plants. The process was introduced as the **C_4 dicarboxylic acid pathway** (later named the C_4 pathway) and extensively studied by Australian scientists Hal Hatch and Roger Slack in the mid-to-late 1960s.

C_4 photosynthesis is a two-stage carbon-fixation process in which plants contain an enzyme, similar to that of rubisco, called **PEP carboxylase**. Unlike rubisco, PEP carboxylase has a strong affinity for CO_2, especially when concentrations are low, and is not influenced by high levels of O_2. During hot and dry weather conditions, C_4 plants will partially close their stomata, reducing water loss but also allowing small amounts of CO_2 to enter the leaf. PEP carboxylase binds to CO_2, forming four-carbon molecules (oxaloacetate, malate, or aspartate), which operate like shuttles, moving the CO_2 to areas of the leaf where the Calvin cycle can occur as normal. Although this process provides a significant advantage to plants in hot and dry environments, the extra step of having to fix CO_2 with PEP carboxylase requires additional energy, a disadvantage.

CAM Plants (About 10 Percent of All Plant Species)

Crassulacean acid metabolism, or **CAM photosynthesis**, is another alternative photosynthetic process occurring in plants such as pineapple and many different cacti (Figure 5.15, right). This type of photosynthetic process has been traced back to the time of the Romans, but it was first detailed in writing in the 1800s. Some of the first scientists to research and observe CAM plants were Swiss botanist Nicolas-Théodore de Saussure (*Chemical Research on Plant Growth* (1804)), French botanist Ephrem Aubert (*Physiological Research on Succulent Plants* (1892)), and American botanist Herbert M. Richards (*Acidity and Gas Interchange in Cacti* (1915)). More detailed studies of the sequence of biochemical reactions that CAM plants undergo were performed by Stanley L. Ranson and Meirion Thomas during the 1960s. The term *crassulacean acid metabolism* originated from the initial work and discovery of the process, performed with a group of succulent plants belonging to the family *Crassulacea*, but it was first used in public to describe the process by Thomas in 1947.

FIGURE 5.15 Various different plants, such as wheat (left), corn (middle), and cacti (right), utilize different photosynthetic pathways, which have resulted from years of evolutionary adaptation, in order to maximize photosynthetic efficiency, while conserving water during hot and dry weather.

CAM photosynthesis is a photosynthetic adaptation primarily used by desert plants in response to the differing temperatures and humidities they experience during the day and night. Desert nights are characterized by lower temperatures and higher humidities, and at this time, the plants will open their stomata in order to allow CO_2 to enter. During the day, which is generally hot and dry, the plants will close their stomata to conserve water. Despite stomata being closed during the day, high concentrations of CO_2 were collected at night, reducing the likelihood of photorespiration. With the stomata closed, PEP carboxylase, the same enzyme found in C_4 plants, will bind to CO_2 to form the four-carbon molecule malate. Malate is stored within vacuoles, which are transported to the chloroplasts. Once at the chloroplasts, malate enters the organelle, releasing CO_2, which is then fixed during the Calvin cycle of photosynthesis. Due to closing their stomata during the day, the amount of CO_2 entering the plants is significantly reduced, thereby hindering overall plant growth. In addition, because CAM plants require an additional step of binding CO_2 to PEP carboxylate, similar to that of C_4 plants, more energy is required.

Note—the remaining percentage of plants (about 4.6 percent) use a combination of these alternative photosynthetic pathways (e.g., C_3 plants may switch to CAM during times of drought).

5-4. Cellular Respiration

Cellular respiration is a metabolic process in which cells extract energy stored within the bonds of nutrients and convert this energy into a usable form, known as ATP. Occurring within the mitochondria of eukaryotic cells, this catabolic pathway involves the complete breakdown of glucose, in the presence of oxygen, into carbon dioxide and water. As can be seen in the following equation, cellular respiration is virtually the reverse of photosynthesis.

$$C_6H_{12}O_6 + O_2 \rightarrow CO_2 + H_2O + \text{energy}$$

Because this process occurs in the presence of oxygen, it is called an aerobic process. Cellular respiration breaks down glucose slowly in order to ensure that the ATP that is produced occurs at a gradual pace. As the chemical bonds of the glucose molecule are broken, the molecule is rearranged into various different types of intermediate molecules, releasing energy. This energy is carried by high-energy electron-carrying molecules to an electron transport chain, where the energy is transferred to 32 to 36 molecules of ATP per one glucose, depending on the tissue and organism. Upon leaving the electron transport chain, the electrons combine together with oxygen, the final electron acceptor, and hydrogen ions to form water molecules. Cellular respiration is one of the most efficient energy conversion processes in nature, capturing 35 to 40 percent of the energy available in glucose to produce large quantities of ATP. The remaining energy is lost as heat. As a comparison, a gasoline engine only converts about 25 percent of the energy available in gasoline to motion.

Cellular respiration is a four-step process: glycolysis, preparatory reaction, the Krebs cycle, and the electron transport chain.

Glycolysis

Glycolysis is a catabolic process that occurs within the cytoplasm of a cell and does not require the addition of oxygen. Therefore, the process is anaerobic. Glycolysis involves a series of ten enzyme-catalyzed chemical reactions, which are responsible for breaking down, converting, and rearranging a glucose molecule into two three-carbon molecules called **pyruvate** (Figure 5.17)

The full discovery of the process as we know it today took about 100 years to compile, starting with work performed by French microbiologist Louis Pasteur on fermentation in 1860, described in section 5-5. Additional research on fermentation performed by other scientists, including German chemist Eduard Buchner in 1897 and English biochemists Arthur Harden and William Young in 1905, provided the links necessary to outline our current understanding of this catabolic pathway. The most common type of glycolysis occurring in organisms was first described by two German biochemists, Otto Meyerhof (Figure 5.16, left) and Gustav Embden, and Austro-Hungarian biochemist Jakub Karol Parnas (Figure 5.16, right). In 1913, Meyerhof gave a lecture that suggested energy within living cells begins with the intake of food, then is converted through a series of transformative steps, finally being released as heat. In addition, in 1918, Meyerhof began to pursue research with oxygen respiration in muscle and alcoholic fermentation by yeast and shortly thereafter discovered the same coenzymes involved in lactic acid fermentation that had been discovered by Harden and Young in 1905. After the start of World War I, Meyerhof began research on heat production in muscle tissue and detailed the lactic acid cycle, in which glycogen is converted to lactic acid during anaerobic conditions. His research on the lactic acid cycle confirmed the work performed by Pasteur in the 1860s, and

FIGURE 5.16 Gustav Embden (not pictured), Otto Meyerhof (left), and Jakub Karol Parnas (right) have been credited with our current understanding of how glycolysis occurs within living organisms. This is the reason why the process is often called the Embden-Meyerhof-Parnas (EMP) pathway.

in 1922, Meyerhof won the Nobel Prize in Physiology or Medicine. During the 1930s, the combined efforts of Meyerhof and additional research performed on carbohydrate metabolism in muscle tissue by Embden and Parnas led to a more detailed description of the glycolytic pathway by 1940. As a result, glycolysis is oftentimes referred to as the Embden-Meyerhof-Parnas (EMP) pathway.

Glycolysis begins when the cell utilizes two molecules of ATP, which activates glucose, a six-carbon molecule. Upon being activated with the addition of the ATP molecules, the glucose molecule becomes unstable, allowing the enzymes of glycolysis to rearrange its energy. The six-carbon glucose molecule is first converted to fructose diphosphate, then split into two three-carbon molecules, glyceraldehyde-3-phosphate (G3P). Each G3P molecule proceeds through a series of chemical reactions, in which high-energy electrons are removed and energy is released,

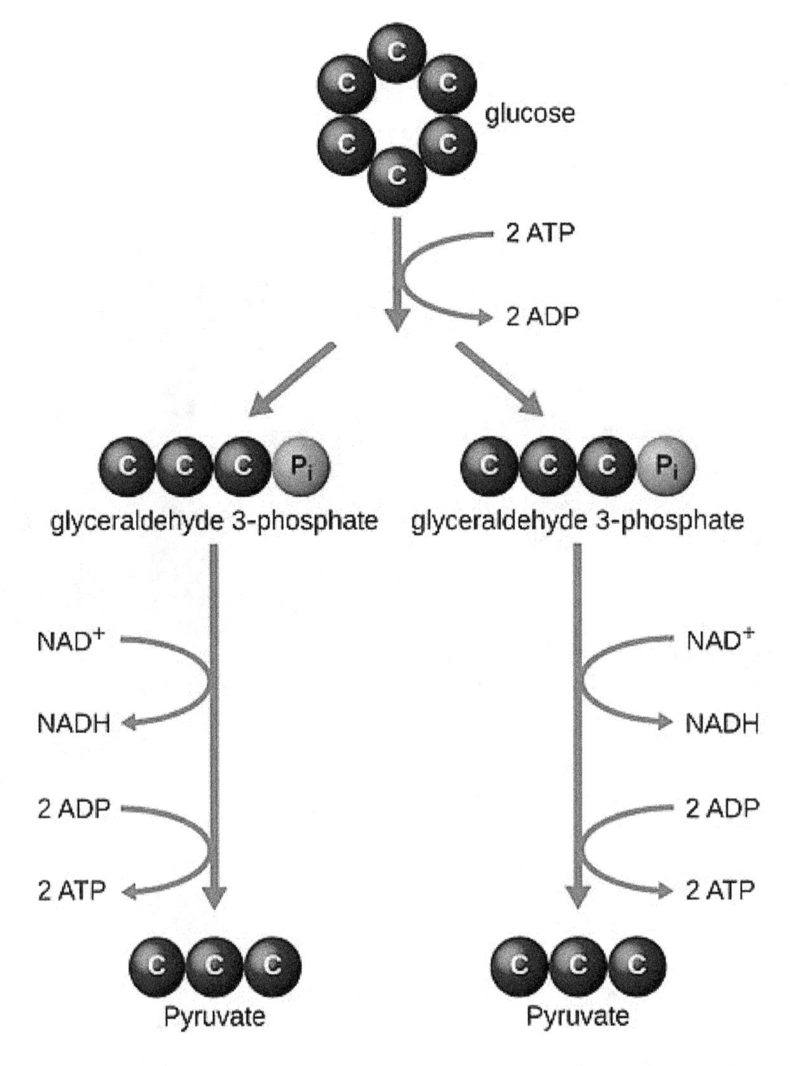

FIGURE 5.17 Glycolysis is an anaerobic process that occurs within the cytoplasm of a cell. The process involves ten enzyme-catalyzed reactions, which are responsible for converting glucose into two pyruvate molecules.

until both molecules are converted to two molecules of pyruvate (a three-carbon molecule). The high-energy electrons, along with hydrogen ions, are added to a coenzyme, NAD^+, to form two high-energy electron-carrying molecules, NADH, responsible for carrying two high-energy electrons to the electron transport energy. Energy released when the glucose is broken down is used to attach phosphate groups to ADP, producing four ATP molecules. However, since the cell had to use two ATP molecules at the beginning of glycolysis, the net gain of ATP is two. If oxygen is available, the pyruvate molecules will enter the mitochondria and proceed through the rest of the steps of cellular respiration. However, if oxygen is not available, the pyruvate molecules will be broken down further in the cytoplasm during the process of fermentation. The final products of glycolysis include four ATP (two net); two pyruvate molecules; and two NADH molecules.

Preparatory Reactions

If oxygen is available, pyruvate molecules will enter the mitochondrial matrix, where each will undergo a series of enzyme-catalyzed reactions (Figure 5.18) that modify the molecules in order for them to enter the Krebs cycle. During the reactions, the pyruvate molecules react with a molecule called *coenzyme* A (CoA), which removes one carbon and two oxygen atoms to produce carbon dioxide that diffuses out of the cell into the bloodstream and is transported to the lungs to be exhaled. In addition, as the pyruvate molecules undergo modifications, high-energy electrons are removed from the molecules and are added to coenzyme NAD^+, along with a hydrogen ion, to form NADH. The remaining two-carbon molecule attaches to CoA, which is responsible for carrying the molecule to the Krebs cycle, producing the molecule *acetyl-CoA*. Since there are two pyruvate molecules entering the mitochondria after glycolysis, this group of reactions occurs twice, producing two molecules of acetyl-CoA, two molecules of carbon dioxide, and two molecules of NADH.

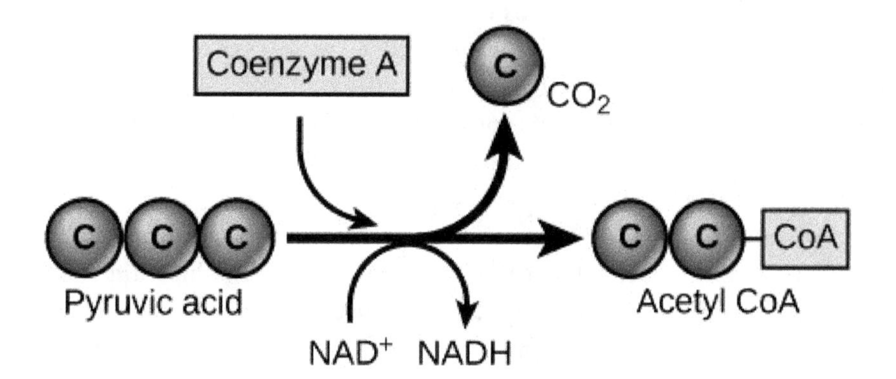

FIGURE 5.18 In order for pyruvate to enter the Krebs cycle, it must be converted to an acetyl-CoA molecule during a group of enzyme-catalyzed reactions known as the preparatory reactions. During the course of these reactions, high-energy electrons are removed from pyruvate, forming NADH, carbon dioxide is produced when a carbon atom is removed from the molecule, and coenzyme A (CoA) attaches to the remaining carbon atoms to produce acetyl-CoA.

Krebs Cycle

The Krebs cycle is a cyclic metabolic pathway involving a series of eight enzyme-catalyzed chemical reactions occurring within the mitochondrial matrix.

The Krebs cycle is named after German biochemist Hans Krebs; however, key components and reactions involved in the cycle were discovered by Hungarian biochemist Albert Szent-Györgyi (Figure 5.19, left) in 1937, for which he received the Nobel Prize in Physiology or Medicine the same year. Hans Krebs (Figure 5.19, right), along with a research student, William A. Johnson, were responsible for establishing the sequence of the cyclic pathway. In their paper entitled "The Role of Citric Acid in Intermediate Metabolism in Animal Tissues" (1937), Krebs and Johnson outlined the cycle while studying the effects of succinate, fumarate, and malate on oxygen consumption, all of which resulted in an increase. Also in 1937, another group of German biochemists, Franz Koop and Carl Martinus, demonstrated that reactions using citrate produced a four-carbon molecule called oxaloacetate. As a result, Krebs and Johnson suggested that citrate and oxaloacetate were the missing intermediates and filled in the gaps that remained in their own research. Krebs and Johnson called this cycle the citric acid cycle. Because of this discovery, Krebs received the Nobel Prize in Physiology or Medicine in 1953.

FIGURE 5.19 The primary components of the Krebs cycle were discovered by Albert Szent-Györgyi (left) in 1937; however, in the same year, Hans Krebs (right), outlined the sequence of events in the cyclic pathway, calling the process the citric acid cycle.

When each acetyl-CoA enters the Krebs cycle (Figure 5.20) from the preparatory reaction, they combine with a four-carbon molecule, *oxaloacetate*, to form a six-carbon molecule, releasing coenzyme A (CoA). The six-carbon molecule proceeds through a series of enzymatic reactions in which it is rearranged into several different types of intermediates, releasing high-energy electrons transferred to NAD^+ and FAD to form three molecules of NADH and one molecule of $FADH_2$ and one molecule of ATP. NADH and $FADH_2$ are responsible for carrying high-energy electrons to the electron transport chain. In addition, two carbon atoms and four oxygen atoms are released during the rearrangement, producing two carbon dioxide molecules, which diffuse out of the cell into the bloodstream and are sent to the lungs to be exhaled. After the removal of the two carbon atoms, the four-carbon molecule that remains goes through a series of enzyme-driven reactions and is rearranged into the original four-carbon starting molecule, oxaloacetate, and the cycle repeats. Since there are two acetyl-CoA molecules produced during the preparatory reaction, the Krebs cycle turns twice for each original glucose molecule. Therefore, the final products of two turns of the Krebs cycle include two ATP molecules, six NADH molecules, two $FADH_2$ molecules, and four carbon dioxide molecules.

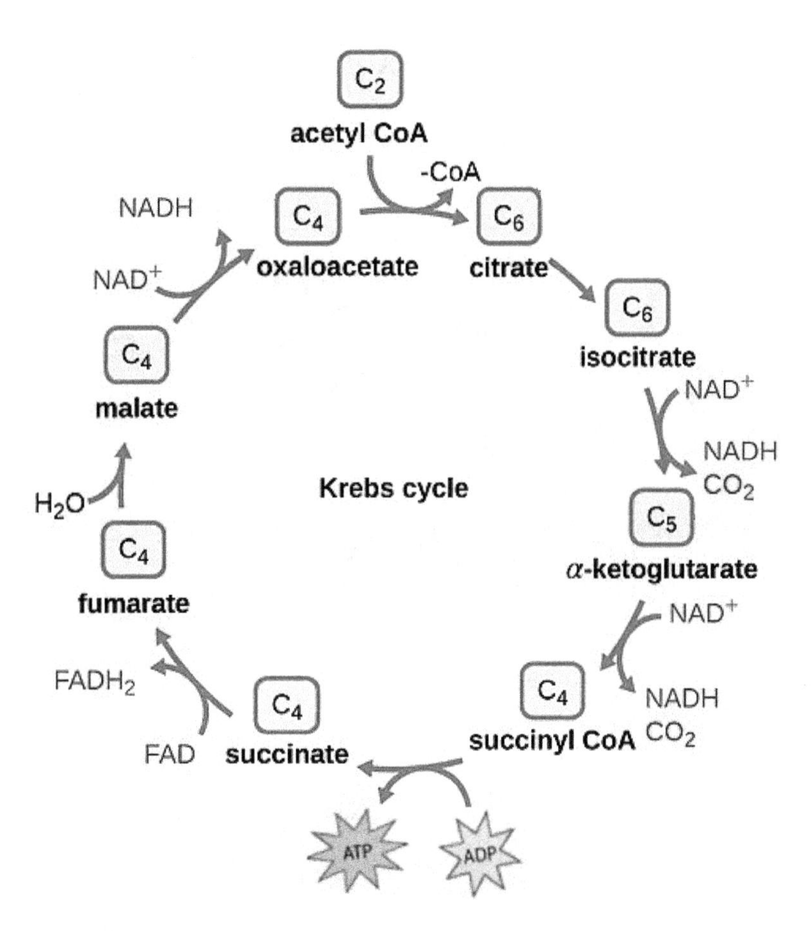

FIGURE 5.20 The Krebs cycle is a cyclic metabolic pathway that involves eight enzyme-catalyzed reactions that occurs within the mitochondrial matrix. During the process, acetyl-CoA attaches to a four-carbon molecule (oxaloacetate) then proceeds through several reactions, producing intermediate molecules that release high-energy electrons. These high-energy electrons are transferred to three NAD+ molecules to form NADH and one FAD molecule to form FADH2. In addition, the Krebs cycle also produces two carbon dioxide molecules and an ATP molecule. Due to the presence of two acetyl-CoA molecules resulting from the preparatory reactions, the Krebs cycle must turn twice for every glucose molecule.

Electron Transport Chain

The electron transport chain is the final step in the process of cellular respiration and occurs within the cristae of the mitochondria (Figure 5.21). The electron transport chain consists of membrane-bound proteins embedded within the mitochondrial cristae. The membrane-bound proteins act as electron carriers as they move high-energy electrons donated by NADH and FADH2 "down the chain," releasing energy as they pass from one set of molecules to the next. When NADH and $FADH_2$ release the high-energy electrons, they become NAD^+ and FAD, respectively, and return to the Krebs cycle to pick up additional electrons. Some of the energy released as the electrons pass down the electron transport chain is used to pump hydrogen ions across the mitochondrial membrane, creating an ion gradient. As the hydrogen ions pass back across the mitochondrial membrane, they flow through a membrane-bound protein called ATP synthase, which uses energy provided by the ion gradient to attach a phosphate group to ADP, producing ATP. This process is known as **chemiosmosis** (Figure 5.21).

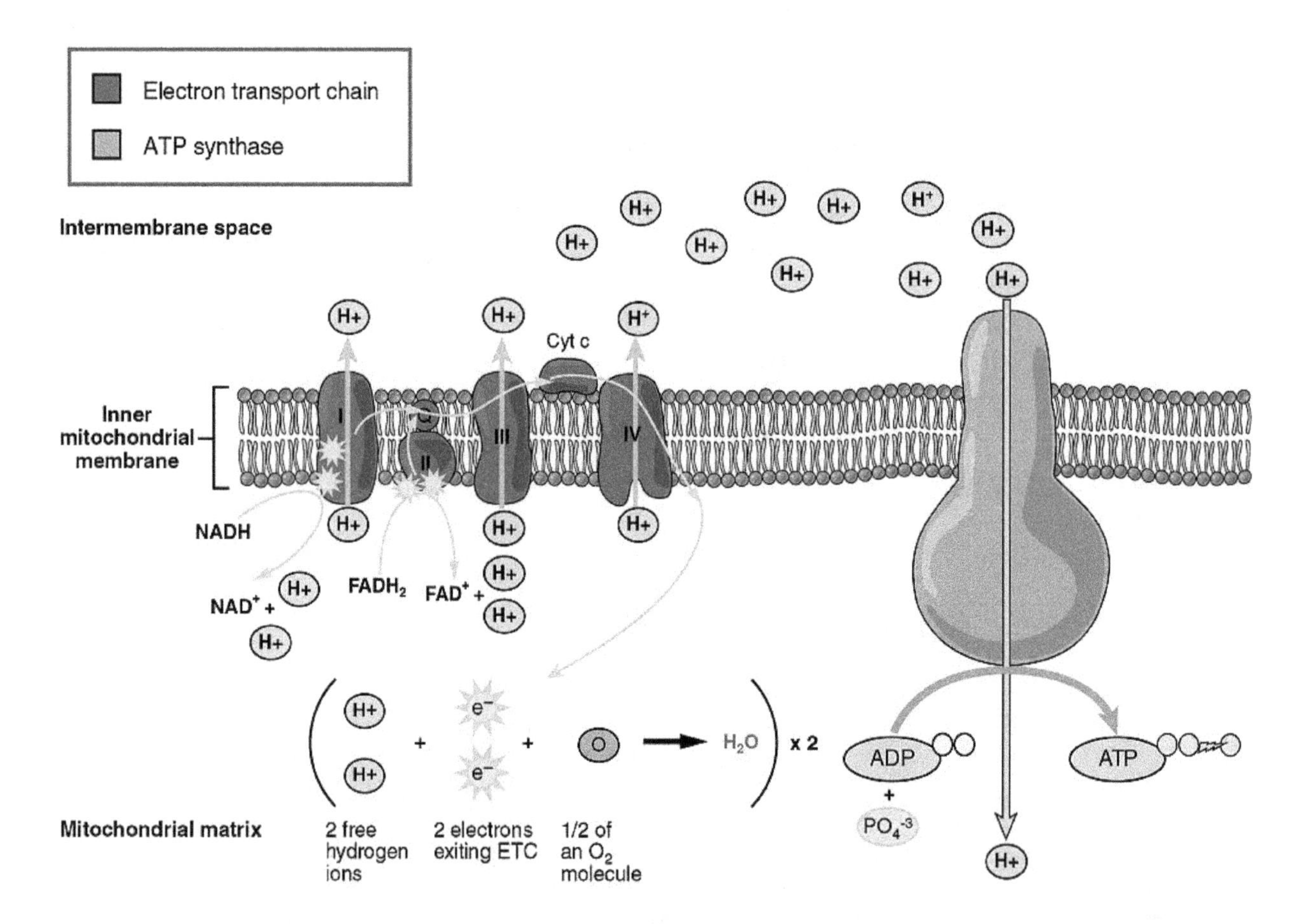

FIGURE 5.21 The final step of cellular respiration is the electron transport chain, which occurs within the mitochondrial cristae. High-energy electrons are carried into the electron transport by NADH and $FADH_2$, molecules produced throughout the reactions of cellular respiration. These high-energy electrons are passed from proteins inset within the mitochondrial cristae and the energy released is used to pump hydrogen across the mitochondrial membrane to produce an ion gradient. As hydrogen ions pass across the membrane, they pass through ATP synthase, producing ATP in a process known as chemiosmosis. Upon reaching the end of the electron transport chain, the electrons are met by oxygen and hydrogen atoms to form water molecules.

Chemiosmosis was first proposed in 1961 by British biochemist Peter D. Mitchell, for which he received a Nobel Prize in Chemistry in 1978. Many scientists prior to Mitchell had concentrated their research on what powered ATP synthesis. Mitchell, conversely, focused his research on where ATP synthesis occurred within the mitochondria, and his experiments involved isolating the inner membranes by making them into vesicles. He then exposed the vesicles to oxygen and NADH and observed the vesicles functioning as they do within the cell. However, he made another observation: the pH of the vesicles began to decrease as a result of an increase in hydrogen ions. As a result, he developed a new hypothesis suggesting that an ion gradient must form as the hydrogen ions are pumped out of the vesicles, providing an energy source used to synthesize ATP. What Mitchell did not observe at the time was the existence of "knobby" structures protruding from the vesicles, which were later discovered to be ATP synthase complexes, revealed through the use of electron microscopy.

Once the electrons reach the end of the electron transport chain, they are met by oxygen, the final electron acceptor. The electrons and oxygen atoms combine with hydrogen ions to form water molecules.

The theoretical yield of ATP produced from one glucose molecule at the end of cellular respiration ranges from 32 to 36 molecules. The actual yield is hard to predict for various reasons. Oftentimes, NADH and $FADH_2$ molecules may not always be used to generate ATP. However, it has been noted in scientific studies that about 2.5 molecules of ATP are generated for each NADH molecule, while 1.5 ATP molecules are produced from $FADH_2$, a lower yield, because it enters the electron transport chain after NADH. Cellular respiration produces a total of ten NADH molecules, which equates to 25 ATP molecules, while a total of two $FADH_2$ molecules are produced, generating three ATP molecules. The remaining ATP molecules, four total from glycolysis and the Krebs cycle, results in an estimated 32 total molecules. Another reason for the variable number of ATP molecules produced is due to varied respiration rates in active cells versus inactive cells. For example, muscle cells are highly active and therefore require more ATP, but adipose tissue, composed of relatively inactive cells, requires less energy. It should also be noted that not all hydrogen ions are transported through the ATP synthase complex. The number of theoretical ATP produced is based on the assumption that all hydrogen ions are transported through. Finally, some of the intermediates formed during the Krebs cycle are used to make other types of molecules, such as proteins and fats.

5-5. Fermentation

Fermentation is a metabolic process that has been used in the production of alcoholic beverages and foods, dating back to China in 7,000 BCE. Many scientists believed that air caused fermentation to occur; however, a group of scientists in the 1830s, including Theodor Schwann (Figure 5.22, left), published several papers concluding that fermentation was the result of living organisms called yeasts. Schwann demonstrated this assertion by boiling grape juice, killing all the organisms within, including the yeasts, discovering that fermentation did not occur again until yeasts were reintroduced. At the same time, other scientists did not follow this logic and still believed that yeasts were nonliving, composed of chemicals and gases; others thought yeasts were alive and generated spontaneously, while others felt that yeasts were responsible for fermentation.

The question of the cause of fermentation began to change in 1856 when French winemakers discovered that the wine they were producing from sugar beets was turning sour and the alcohol content was low. The winemakers asked Louis Pasteur (Figure 5.22, middle) to help. Therefore, Pasteur designed and carried out several different experiments to help determine what was causing the winemakers' issues. He first demonstrated that yeast cells gave rise to other yeast cells. During his next set of experiments, Pasteur sealed sterile flasks containing grape juice and yeast and left other sterile flasks exposed to the air. Upon observation, he noticed that the yeasts were growing with or without oxygen. His final group of experiments involved placing bacteria and yeast in different sterile flasks and demonstrated that bacteria produce acids during fermentation, while yeasts produce alcohol. This was an important discovery, because it led to the process of pasteurization, which was a commonly used procedure in winemaking in the 19th century and is used today to eliminate pathogens from milk.

Although Pasteur demonstrated that fermentation was caused by living organisms, he was not able to explain the chemical reactions associated with the process. In 1897, the idea that fermentation involved chemical reactions was introduced by German scientist Eduard Buchner (Figure 5.22, right). Buchner performed an experiment in which he ground up yeast cells, extracted a liquid from the cells, and observed that the liquid fermented a sugar solution, producing alcohol and carbon dioxide gas much like yeast cells. He concluded that enzymes, proteins produced by the cells, were responsible for the chemical reactions of fermentation.

FIGURE 5.22 The study of fermentation began in the 1830s with Theodor Schwann (left), who concluded that fermentation was the result of reactions performed by yeast. In 1856, Louis Pasteur (middle) demonstrated that different types of fermentation occurred based upon which microorganism was responsible for the process. He discovered that bacteria produced acids, while yeast produced alcohol. In the late 1890s, Eduard Buchner (right) concluded that yeast produced enzymes, which they utilize in order to carry out the chemical reactions of fermentation.

Fermentation is defined as the process by which carbohydrates are broken down during anaerobic (lacking oxygen) conditions, using different molecules as the final electron acceptor. The process of fermentation occurs within the cytoplasm and includes glycolysis, where a glucose molecule is split into two pyruvate molecules. However, without the presence of oxygen, electrons from NADH are transferred to the pyruvate molecules during fermentation. Depending on the final electron acceptor, the pyruvate molecules are either converted into lactic acid or ethanol. As a result, there are two different types of fermentation.

Lactic Acid Fermentation

During lactic acid fermentation (Figure 5.23), pyruvate molecules are the final electron acceptors and are converted to lactic acid, a common phenomenon occurring in muscle cells. Muscle cells use fermentation when they are vigorously worked and the amount of pyruvate exceeds oxygen delivery, leading to "oxygen-debt." Muscle cells will continually use fermentation in order to provide small quantities of ATP (only two ATP are produced) to sustain activity over short periods of time; however, as the amount of lactic acid increases, the muscle cells become more acidic, resulting in muscle fatigue. When oxygen delivery increases, the lactic acid is transported in the bloodstream to the liver, where it is converted back to pyruvate and broken down through cellular respiration. Lactic acid fermentation is an important process in the food industry, especially in making different types of cheese and yogurt, and is an important process used when making chocolate from cacao beans.

Alcohol Fermentation

Alcohol fermentation (Figure 5.23) occurs in a two-step process: pyruvate is converted to acetaldehyde upon the release of carbon dioxide, and acetaldehyde is the final electron acceptor, resulting in the production of ethanol. This type of fermentation is a commonly used process in the production of wine, beer, other alcoholic spirits, and baked goods. For example, wine is produced from fermenting sugar in grapes; beer is produced from fermenting grains, like barley; and vodka is produced from fermenting sugar in potatoes. It is important to note that ethanol is also an important fuel source generated from the fermentation of sugar, found in sugarcane, and starch, found in corn.

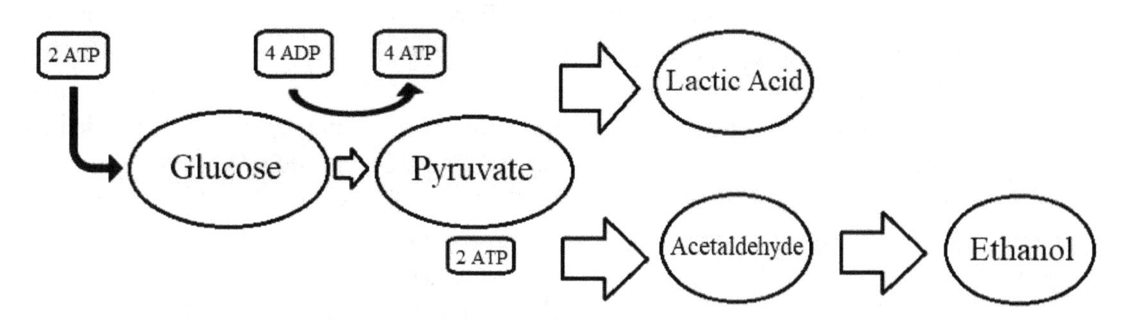

FIGURE 5.23 Two types of fermentation occur during anaerobic conditions upon the completion of glycolysis. Lactic acid fermentation utilizes pyruvate as the final electron acceptor and commonly occurs within muscle cells under extensive use. Alcohol fermentation is a two-step process that uses acetaldehyde as its final electron acceptor, resulting in the production of ethanol.

It has been outlined that fermentation can be advantageous to humans, especially in the production of food and alcoholic beverages. However, fermentation is an important process for other types of organisms, like bacteria and algae, in that it enables them to produce ATP under anaerobic conditions. Despite these advantages, fermentation does have its disadvantages as well. Fermentation is considerably inefficient, because only 2 percent of the energy available in glucose is captured during fermentation, compared to 35 to 40 percent captured during cellular respiration. In addition, fermentation produces toxins that are potentially harmful to a cell. The acids produced during lactic acid fermentation can lower the pH of the cell's environment and also affect enzymatic activity. Alcohol waste is of great concern to cellular proteins, as it can cause these proteins to denature, affecting their overall function.

Chapter Summary

Below is a summary of the main ideas from each section of the chapter.

5-1. Metabolism

- Metabolism is the sum of all chemical reactions that occur in a cell that are responsible for transforming energy. Metabolic reactions within a cell occur in a sequential manner, known as metabolic pathways. There are two different types of metabolic pathways. Anabolism is the metabolic pathway responsible for synthesizing new, complex molecules. Catabolism is the metabolic pathway responsible for breaking down large, complex molecules. Metabolic pathways occur independently and simultaneously, producing energy and molecules that each can utilize during specific reactions.

5-2. Energy

- Energy is defined as the ability to do work. Energy is measured in calories, which is the amount of energy needed to raise one gram of water one degree Celsius. A kilocalorie is the measure of the amount of energy in food and heat released by an organism.
- There are two primary sources of energy found on Earth. Solar energy is the energy provided by the sun. Photoautotrophs are organisms, such as plants, algae, and cyanobacteria, that capture solar energy and use the energy to convert carbon dioxide and water into organic molecules in a process known as photosynthesis. Geothermal energy is energy provided by inorganic chemicals, such as iron, sulfur, and/or ammonia. Chemoautotrophs are organisms, such as bacteria and archaea, that obtain their energy from inorganic chemicals to synthesize organic molecules. Heterotrophs are organisms that consume other organisms and rely on photoautotrophs and chemoautotrophs for their energy needs. Energy obtained from both the sun and inorganic chemicals is important for development, growth, repair, and reproduction.
- There are two different forms of energy. Potential energy is stored energy, or energy available to do work. Chemical energy, the energy stored in the chemical bonds of molecules, is an example of potential energy. Kinetic energy is the energy used to move an object. Various examples of kinetic energy include thermal, solar, and sound energy.
- The flow and transformation of energy between and within organisms can be explained by the laws of thermodynamics. The first law of thermodynamics states that energy cannot be created nor destroyed, but can be converted from one form to another, as in the conversion of solar energy to chemical energy. The second law of thermodynamics states that the conversion of energy from one form to another is quite inefficient and results in the loss of usable energy in the form of heat. Entropy is defined as disorder in the universe, which occurs as the amount of heat increases, decreasing the amount of energy available to do work.
- Adenosine triphosphate (ATP) is the main energy source found within a cell. ATP is responsible for providing the energy required for different types of chemical reactions. ATP is composed of adenine, ribose, and three phosphate groups. The primary source of energy found

within an ATP molecule is found in the covalent bonds of the three phosphate groups. As the covalent bond holding the last phosphate group is broken during hydrolysis, it produces about 7.3 kilocalories of energy, an amount sufficient for biological reactions. When the last phosphate group is removed from ATP, it comes adenosine diphosphate (ADP), which is quickly resynthesized when energy is released during glucose catabolism. ATP is an effective source of energy for three reasons. The ATP molecules are constantly resynthesized. The covalent bond holding the last phosphate group is unstable and easy to break, releasing energy. The energy released when the covalent bond holding the last phosphate group is broken amounts to twice the amount of energy required for biological reactions. ATP provides sufficient energy for the cell to perform three main functions. Chemical work—energy is used to synthesize important biological molecules, such as proteins. Transport work—energy is used to help transport substances across the plasma. Mechanical work—energy is used for mechanical functions, such as muscle contraction.

5-3. Photosynthesis

- Photoautotrophs are a group of organisms that include plants, algae, and cyanobacteria that convert solar energy into chemical energy in order to synthesize organic molecules, in a process known as photosynthesis. Photosynthesis involves the conversion of carbon dioxide and water into glucose and oxygen with the addition of solar energy. Photosynthesis occurs in specialized organelles called chloroplasts. Chloroplasts are found within mesophyll cells, cells located between the upper and lower epidermis of a leaf. The upper epidermis of the leaf contains a waxy substance called the cuticle. The lower epidermis of the leaf contains the stomata, small pore openings that allow carbon dioxide to enter and oxygen to exit the leaf. Chloroplasts consist of a double-membrane system surrounding a gelatinous fluid called the stroma. The stroma contains membranous structures called thylakoids. Within the thylakoids are important photosynthetic pigments called chlorophylls, which are responsible for capturing solar energy.
- Energy produced by the sun is called electromagnetic radiation, which travels in waves toward Earth. Only 1 percent of the sun's electromagnetic radiation reaches Earth's surface, as a majority is trapped by the ozone layer or water vapor and carbon dioxide. The radiation that reaches Earth's surface provides a sufficient amount of energy to power photosynthesis and is called visible light. Visible light consists of various different colors (violet to red), makes up a small portion of the electromagnetic spectrum (380 to 750 nm), and is composed of small packets of energy called photons. Photons help determine the brightness of light, contain energy that is directly correlated to how fast they move, and travel in wavelengths, which is inversely related to a photon's energy level. Visible light reaching Earth has three fates upon striking an object. Reflected: Light bounces off an object and can determine the color of that object. Transmitted: Light passes through an object and can determine the color of that object. Absorbed: Light is captured by the object and is important in driving photosynthesis.
- Photosynthetic pigments are found within the thylakoids of chloroplasts and are responsible for absorbing various different wavelengths of light. There are three different types of photosynthetic pigments. Chlorophyll *a* is the most abundant and important, functioning as

the main light-capturing pigment, which absorbs violet, blue, red, and orange wavelengths of light, but reflects and transmits green wavelengths, giving leaves their green color. Chlorophyll *b* is a much smaller and structurally different photosynthetic pigment and absorbs different blue and red-orange wavelengths, while reflecting or transmitting green and yellow wavelengths, which also contributes to a leaf's color. Carotenoids are composed of carotenes and xanthophylls that are responsible for absorbing violet and blue-green wavelengths of light and reflecting yellow, orange, and red wavelengths. Carotenoids are masked by chlorophyll *a* and chlorophyll *b* during the summer, but as the weather changes, the chlorophyll pigments break down much sooner than carotenoids, giving rise to the colors of autumn.

- Light-dependent reactions, or photo reactions, occur within the thylakoids of the chloroplasts and are responsible for absorbing solar energy and converting it to chemical energy. Chemical energy is stored in two high-energy carrier molecules, ATP and NADPH, which provides the energy necessary to synthesize glucose. Oxygen gas is also produced during the light-dependent reactions and is either released through the stomata or used by the plant during cellular respiration. The light-dependent reactions occur along a group of specialized units called photosystems located within the thylakoids. The photosystems consist of two main regions. The light-harvesting pigment complexes, containing up to 400 photosynthetic pigments bound to proteins. The reaction center, which is surrounded by the light-harvesting pigment complexes and contains special chlorophyll *a* molecules and proteins. There are two different types of photosystems, photosystem I (PSI) and photosystem II (PSII), but their names do not reflect the order in which they occur. Several scientists were responsible for the discovery and outlining the functions of the two photosystems, including Robert Emerson, William Arnold, Hans Gaffron, Kurt Wohl, Robert Hill, and Fay Bendall. The light-dependent reactions begin in photosystem II (PSII) and end in photosystem I (PSI). In PSII, solar energy is absorbed by photosynthetic pigments in the light-harvesting pigment complexes and passes from one pigment to another until the energy reaches the reaction center. Within the reaction center, specialized chlorophyll *a* molecules receive the energy and excite electrons donated by the splitting of water molecules. The remaining hydrogen protons resulting from the splitting of the water molecules is used to make ATP and NADPH, while oxygen atoms combine to form oxygen gas. Once released from PSII, the high-energy electrons travel down an electron transport chain, where energy is lost in order to pump hydrogen protons into the thylakoids, resulting in a proton gradient. The increased number of hydrogen protons in the thylakoids travel through ATP synthase to make ATP. The reaction center of PSI receives the low-energy electrons, and the absorption of solar energy by photosynthetic pigments excites the electrons again. The high-energy electrons leave the reaction center of PSI and travel down another electron transport chain to be transferred to $NADP^+$ to form NADPH.
- Light-independent reactions, or synthesis reactions, occur within the stroma of the chloroplasts and use the energy in ATP and NADPH formed during the light-dependent reactions to convert carbon dioxide into glucose. The light-independent reactions involve a series of enzyme-catalyzed reactions occurring in a cyclic pathway known as the Calvin cycle. The Calvin cycle was named for Melvin Calvin, whose team discovered the series of reactions involved in the process. The Calvin cycle is a three-step process. Carbon fixation involves attaching carbon dioxide to ribulose biphosphate (RuBP) with the help of an enzyme called

rubisco to produce an unstable six-carbon molecule, which splits to form two three-carbon molecules known as 3-phosphoglycerate (3PG). The synthesis of glyceraldehyde-3-phosphate begins when 3-phosphoglycerate (3PG) is converted through a series of enzyme-catalyzed reactions using energy from ATP and NADPH produced during the light-dependent reactions. The original starting material, ribulose biphosphate (RuBP), is regenerated as some of the remaining glyceraldehyde-3-phosphate not used to make glucose and other molecules is converted back through a series of enzyme-catalyzed reactions using energy from ATP.

- There are various different types of alternative photosynthetic pathways that have developed as a result of evolutionary adaptations that plants have made in order to conserve water. C_3 plants are a group of plants that utilize a process known as photorespiration, in which they close their stomata during hot and dry weather to reduce water loss due to evaporation. The closing of the stomata prevents carbon dioxide from entering the leaf and prevents oxygen from exiting. The increased concentration of oxygen within the leaf begins to become fixed by rubisco, producing carbon dioxide as a byproduct. Photorespiration does not produce ATP or glucose and decreases photosynthetic output. C_4 plants are a group of plants that utilize a two-stage carbon-fixation process with the help of an enzyme known as PEP carboxylase. PEP carboxylase has a strong affinity for carbon dioxide when concentrations are low. During hot and dry weather, C_4 plants will partially close their stomata, allowing some carbon dioxide to enter. Carbon dioxide entering the stomata binds to PEP carboxylase to form a four-carbon molecule, such as malate, which is then moved to parts of the leaf where the Calvin cycle can occur. The additional step of fixing carbon dioxide using PEP carboxylase requires more energy, a disadvantage to the C_4 plant. CAM (crassulacean acid metabolism) plants are a group of plants that conserve water by closing their stomata during the day when the weather is hot and dry and opening them at night to allow carbon dioxide to enter. CAM plants reduce the likelihood of photorespiration, because high concentrations of carbon dioxide are collected at night, which binds to PEP carboxylase during the day to form malate. Malate enters the chloroplasts, releasing carbon dioxide, which is fixed in the Calvin cycle. Both the insufficient amount of carbon dioxide collected during the day resulting in reduced plant growth and the additional energy required to use PEP carboxylase are considered disadvantages to CAM plants.

5-4. Cellular Respiration

- Cellular respiration is an aerobic metabolic process in which cells extract energy from the complete breakdown of glucose to form ATP, carbon dioxide, and water. Cellular respiration occurs within the mitochondria of the cell. As the chemical bonds in glucose are broken, the molecule is rearranged into intermediate forms, releasing energy. The energy is transferred by high-energy electron-carrying molecules along an electron transport chain, where 32 to 36 molecules of ATP are produced per glucose molecule. The remaining electrons combine with oxygen and hydrogen atoms to form water molecules. The process of cellular respiration is an efficient process, extracting 35 to 40 percent of available energy in a glucose molecule, with the remaining energy lost as heat.

- Glycolysis is an anaerobic process occurring in the cytoplasm of the cell. Glycolysis requires ten enzyme-catalyzed reactions to break down, convert, and rearrange a glucose molecule into two pyruvate molecules. A hundred years of experimental research compiled by Louis Pasteur, Eduard Buchner, Arthur Harden, William Young, Otto Meyerhof, Gustav Embden, and Jakub Karol Parnas outlined our current understanding of the process. Glycolysis begins when two ATP molecules are used to activate the glucose molecule. Activation of glucose results in an unstable molecule, which allows enzymes to begin rearranging the molecule's energy, producing two three-carbon molecules called glyceraldehyde-3-phosphate (G3P). The two glyceraldehyde-3-phosphate (G3P) molecules begin losing high-energy electrons and releasing energy as they proceed through the different chemical reactions, as the molecules are converted to two three-carbon molecules of pyruvate. The high-energy electrons, along with hydrogen atoms, are added to two NAD^+ coenzymes to form two NADH molecules, which are responsible for transporting the high-energy electrons to the electron transport chain. Energy released during the breakdown of glucose is also used to attach phosphate to ADP to produce four ATP molecules. The overall net gain of ATP during glycolysis is two ATP, since two were used to activate the glucose molecule at the start of the process. If oxygen is available, the two pyruvate molecules will enter the remaining steps of cellular respiration, but if oxygen is not available, the pyruvate molecules will be broken down within the cytoplasm during fermentation. The final products of glycolysis are four ATP (two net), two pyruvate molecules, and two NADH molecules.
- The preparatory reactions occur when oxygen is available and the two pyruvate molecules enter the mitochondrial matrix. While in the mitochondrial matrix, a series of enzyme-catalyzed reactions convert the pyruvate molecules into molecules that can enter the Krebs cycle. The pyruvate molecules react with coenzyme A (CoA), which removes carbon and oxygen atoms to form carbon dioxide. Additional modifications to the pyruvate molecules remove high-energy electrons, which are added, along with hydrogen atoms, to NAD^+ to form NADH molecules. The remaining two-carbon molecule attaches to CoA to form acetyl-CoA, and the molecule is carried to the Krebs cycle. Since two pyruvate molecules enter the mitochondria, the final products of the preparatory reactions are two molecules of acetyl-CoA, two molecules of carbon dioxide, and two molecules of NADH.
- The Krebs cycle is a metabolic process named for Hans Krebs, who was responsible for establishing the sequence of the eight enzyme-catalyzed reactions in the cyclic pathway, although key components and reactions of the process were discovered by Albert Szent-Györgyi, Franz Koop, and Carl Martinus. The Krebs cycle occurs within the mitochondrial matrix and begins when each acetyl-CoA molecule from the preparatory reaction combines with oxaloacetate to form a six-carbon molecule. The six-carbon molecule is rearranged into different types of intermediate molecules. During the rearrangement process, high-energy electrons are transferred to NAD^+, FAD, and ADP to form NADH, $FADH_2$, and ATP. NADH and $FADH_2$ carry high-energy electrons to the electron transport chain. In addition, carbon dioxide is produced when carbon and oxygen atoms are removed as the six-carbon molecule is rearranged. The remaining four-carbon molecule proceeds through a series of enzyme-catalyzed reactions to produce the original four-carbon molecule, oxaloacetate, causing the cycle to repeat. Since two acetyl-CoA molecules are produced during the

preparatory reactions, the Krebs cycle turns twice. The final products of the Krebs cycle are two ATP molecules, six NADH molecules, two $FADH_2$ molecules, and four carbon dioxide molecules.

- The electron transport chain occurs along membrane-bound proteins located within the mitochondrial cristae. The membrane-bound proteins act as high-energy electron carriers as NADH and $FADH_2$ pass along the chain, releasing energy and becoming NAD^+ and FAD, which return to the Krebs cycle. The energy released from NADH and $FADH_2$ pumps hydrogen ions across the mitochondrial membrane, producing an ion gradient. As hydrogen ions pass back across the mitochondrial membrane, they flow through ATP synthase, an enzyme that uses energy from the ion gradient to produce ATP. This process is known as chemiosmosis and was first proposed by Peter D. Mitchell in 1961. Upon reaching the end of the electron transport chain, electrons combine with oxygen, the final electron acceptor, and hydrogen atoms to form water molecules.
- Upon completion of cellular respiration, a theoretical yield of 32 to 36 molecules of ATP is produced for every glucose molecule. The actual yield of ATP molecules produced during cellular respiration varies for different reasons. NADH and $FADH_2$ molecules may not always be used to generate ATP. Studies have shown, however, that 2.5 molecules of ATP are generated for each NADH molecule, and 1.5 ATP molecules are produced from one $FADH_2$. This would equate to 25 molecules of ATP from NADH and three molecules of ATP from $FADH_2$, equaling a total of 28 ATP molecules. With the addition of four ATP molecules from glycolysis and the Krebs cycle combined, 32 ATP molecules are produced. Respiration rates in cells may produce a variable number of ATP molecules. Muscle cells are highly active and require more ATP, while adipose tissue is less active and requires less ATP. Not all hydrogen ions are transported across the mitochondrial membrane. The theoretical number of ATP molecules produced is based on the assumption that all hydrogen ions are transported across, when some are not. Intermediates formed during the Krebs cycle are also used to make proteins and fats.

5-5. Fermentation

- Fermentation is a metabolic process that has been used for thousands of years to produce alcoholic beverages and food. Beginning in the 1830s, scientists began to rethink the age-long thought that air caused fermentation to occur. Theodor Schwann concluded that fermentation was the result of yeast, a living organism. Louis Pasteur demonstrated that yeasts were reproducing and growing with or without the presence of oxygen and produced alcohol during fermentation, while bacteria produced acids, leading to the process we now know as pasteurization. Eduard Buchner concluded that enzymes produced by yeast cells were responsible for the chemical reactions associated with fermentation.
- Fermentation is an anaerobic process in which carbohydrates are broken down using different molecules as the final electron acceptor. The process occurs within the cytoplasm. During glycolysis, the glucose molecule is split into two pyruvate molecules, but without the presence of oxygen, electrons are transferred from NADH to pyruvate molecules.

- There are two different types of fermentation that can occur, depending on which molecule is the final electron acceptor. Lactic acid fermentation occurs when pyruvate molecules, the final electron acceptors, are converted to lactic acid. This is a common process in muscle cells when oxygen levels are low and pyruvate molecules exceed oxygen delivery while the cells are being vigorously worked. The pyruvate molecules are converted to lactic acid, which builds in the muscles, resulting in muscle fatigue. When oxygen levels return to normal, the lactic acid is delivered to the liver, where it is converted back to pyruvate and enters cellular respiration. Lactic acid fermentation is an important process in making foods such as cheeses, yogurt, and chocolate. Alcohol fermentation is a two-step process in which pyruvate is converted to acetaldehyde, which acts as the final electron acceptor, resulting in the production of ethanol. Alcohol fermentation is an important process in making alcoholic beverages, such as beer and wine, and food products like bread.
- Fermentation is an advantageous process not only for humans, in the production of food and alcoholic beverages, but for bacteria and archaea as well, enabling them to produce ATP under anaerobic conditions.
- Fermentation also has its disadvantages, including being inefficient in energy extraction (about 2 percent) and producing toxins that can be harmful to a cell. Acids produced during lactic acid fermentation lower the pH of a cell's environment, affecting enzymatic activity. Alcohols produced during alcohol fermentation cause proteins to denature, affecting their overall function

End-of-Chapter Activities and Questions

Directions: Please refer back to what you learned in this chapter to complete the following activities.

Define Each Term in Your Own Words

1. Metabolic Pathway
2. Adenosine Triphosphate (ATP)
3. Chlorophyll
4. Pyruvate
5. Fermentation

Chapter Review

1. Explain the following statement: "Metabolic pathways are both functionally opposite and simultaneous."
2. What are the two forms of energy? Give an example of each form and indicate which law of thermodynamics matches best with your example.
3. Summarize the reactions involved in the process of photosynthesis. Make sure to include in your summary where these reactions occur within a chloroplast and the products of each.
4. Summarize the reactions involved in the process of cellular respiration. Make sure to include in your summary where these reactions occur within a mitochondria and the products of each.
5. Define fermentation and explain the differences between lactic acid and alcohol fermentation.

Multiple Choice

1. The metabolic pathways of both anabolism and catabolism require which of the following?
   - **a.** small, simple molecules
   - **b.** enzymes
   - **c.** energy
   - **d.** all of the choices are required

2. The majority of the energy located within an ATP molecule is found between the ______.
   - **a.** three phosphate groups
   - **b.** adenine and ribose molecules
   - **c.** ribose and three phosphate groups
   - **d.** none of the above

3. Oxygen production is derived from the splitting of ________ molecules during photosynthesis.
   - **a.** glucose
   - **b.** light
   - **c.** carbon dioxide
   - **d.** water

4. Which of the following molecules is not matched with the correct metabolic reaction of cellular respiration?
   - **a.** pyruvate—glycolysis
   - **b.** ATP—electron transport chain
   - **c.** oxygen—Krebs cycle
   - **d.** NADH—preparatory reaction

5. Who was the scientist credited with introducing the idea that chemical reactions are involved in the process of fermentation?
   - **a.** Eduard Buchner
   - **b.** Louis Pasteur
   - **c.** Hans Krebs
   - **d.** C. B. van Niel

Image Credits

CHAPTER

6

DNA and DNA Technology

PROFILES IN SCIENCE

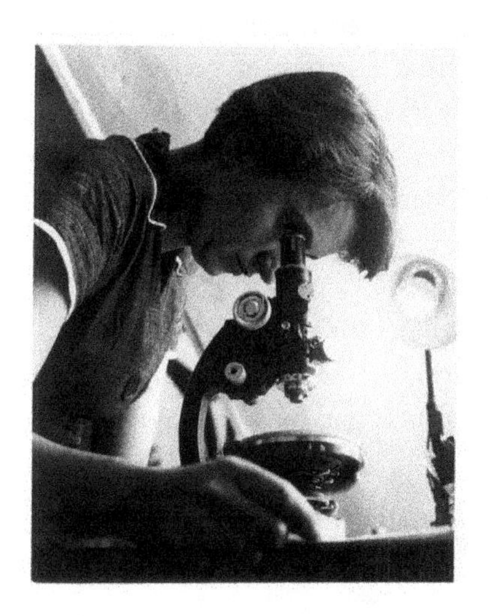

FIGURE 6.1 Rosalind Franklin.

Rosalind Franklin was born July 25, 1920, in London, England, and died in London in 1958. Franklin demonstrated a remarkable scholastic aptitude at an early age, playing memory games and "doing arithmetic for pleasure." By age 11, she was enrolled at St. Paul's Girls' School, where she excelled in multiple subjects, including science. After finishing a year early in 1938, she received a scholarship to attend Newnham College (Cambridge) to study chemistry and physics, graduating in 1941. In 1942, Franklin was commissioned by the British Coal Utilization Research Association to work on the permeability of coal, in addition to classifying it and predicting its performance for the purpose of fuel. Her work led to the development of a new gas mask, the basis of her thesis for PhD, which she received in 1945. In 1947, Franklin moved to Paris and began learning more about X-ray crystallography, a technology introduced to her at Newnham College. Franklin left Paris in 1951 to accept a position at King's College, where she was hired to perform X-ray diffraction on proteins and lipids; however, the focus of her work was redirected to the DNA molecule instead. While at King's College and working alongside her graduate student, Raymond Gosling, Franklin's X-ray diffraction photos suggested DNA consisted of a double-helical backbone, evident in her most famous image, Photo 51, taken in 1952. Upon mathematical analysis of the image, the size and structure of the double helix was calculated, providing the essential information required for the development of the DNA model and discovery of the DNA structure in 1953. Shortly thereafter, Franklin accepted a position as a senior scientist at Birkbeck College, using X-ray crystallography to study the genetic material and structure of viruses, leading to further studies abroad, including the University of California-Berkley in 1956. Unfortunately, the discovery of abdominal tumors and an eventual ovarian cancer diagnosis shortened her visit to the United States. Despite her diagnosis and frequent treatments, Franklin continued her work until her death, publishing multiple papers on the genetic makeup and structure of viruses. Franklin's name appears on multiple college buildings, awards, and an asteroid in recognition of her scientific accomplishments.

Introduction to the Chapter

In the late 1860s, Swedish physician Friedrich Miescher began isolating and categorizing various lipids and proteins in white blood cells he obtained from pus in surgical bandages. During the course of his experiments, Miescher had successfully isolated and discovered an unknown substance from the nuclei of the white blood cells. This unknown substance, which he called *nuclein*, had a unique chemical structure containing high amounts of phosphorous, but it lacked sulfur. Consequently, Miescher refocused his research on characterizing the chemical nature of *nuclein* and demonstrating that it was a primary component of the nucleus that may influence the inheritance of traits. Miescher's discovery initiated the multitude of studies performed over the next 100 years to determine the structure and function of *nuclein*, a molecule we know now as deoxyribonucleic acid (DNA). It was not until 1953 that three scientists discovered the double-helix structure they exclaimed was "the secret of life." Many scientists whose tireless research on the structure and function of DNA have not been universally recognized for their contribution to its final discovery. This chapter will introduce the scientists responsible for laying the foundation for the discovery of the structure and function of DNA. The chapter also provides a detailed description of the structure of DNA, how the molecule replicates, and how it acts as a template for protein synthesis. The chapter concludes with information on different types of mutations and biotechnology, a process that has benefited the agricultural and medical fields in recent years.

Chapter Objectives

In this chapter, students will learn the following:

6-1. Several early scientists are credited with discovering and isolating DNA, identifying its primary components, and proposing its structure and genetic function.

6-2. Deoxyribonucleic acid (DNA) is a nucleic acid composed of three main components twisted in a unique double helix structure whose main function is carrying the genetic information for all living organisms on genes located on chromosomes.

6-3. DNA replication is a carefully regulated process in which a variety of different molecules and enzymes are utilized in order to synthesize a new DNA strand.

6-4. Protein synthesis is a two-step process, involving transcription and translation, where the DNA base sequence of a gene is transcribed into an mRNA strand and then translated into an amino acid sequence specifying a particular protein.

6-5. Mutations are permanent alterations in the base pair sequences of DNA and RNA nucleotides, resulting in nonfunctional proteins, and can occur in a variety of ways.

6-6. Biotechnology is a field of biology that includes genetic engineering, a process that involves DNA manipulation to produce beneficial genetically modified organisms.

6-7. Biotechnology is a beneficial process used in agriculture, medicine, and *CRISPR-Cas9*, but the general public has raised concerns about its practice and overall benefit.

6-1. Discovery of the Structure of DNA

When Friedrich Miescher (Figure 6.2a) discovered *nuclein* at the end of the 1860s, it laid the foundation for what would become the "race for the double helix," a dramatic interaction among scientists to discover the structure of deoxyribonucleic acid (DNA) in the 1950s. However, prior to the discovery of the structure of DNA, a multitude of scientists studied the molecule and made defining discoveries that would pave the way for James Watson, Francis Crick, and Maurice Wilkins in 1953. Unfortunately, some of these scientists have been obscured by history; therefore, they deserve recognition for their contributions to the overall understanding of the structure of DNA.

Albert Kossel (Figure 6.2b) was a German biochemist who was greatly influenced by Miescher's discovery of *nuclein* in 1869. Beginning in 1878, Kossel began investigating the *nuclein* of yeast cells and successfully isolated the nucleic acid portion of the molecule. Between 1885 and 1901, Kossel discovered and isolated the five primary nitrogenous bases: adenine, guanine, cytosine, thymine, and uracil. In 1910, Kossel received the Nobel Prize in Physiology or Medicine for his work on *nuclein*.

Phoebus Levene was an American biochemist who was the first to isolate and identify the basic components of DNA in 1919, including its bases, adenine, guanine, cytosine, and thymine; its sugar, deoxyribose; and the phosphate group. Levene suggested that these components were linked together through the phosphate groups to form a string of units he called **nucleotides**.

Nikolai Koltsov was a Russian biologist, and in 1927, he was the first to propose the idea that a large hereditary molecule was responsible for the inheritance of traits. This molecule, Koltsov suggested, was composed of "two mirror strands that would replicate in a semi-conservative fashion using each strand as a template" (Soyfer 2011, 726).

Frederick Griffith was a British bacteriologist who, in 1928, was studying two forms of the bacterium *Pneumococcus*—the "smooth" or virulent form and the "rough" or nonvirulent form. Upon injecting mice with a heat-killed version of the "smooth" form of bacteria, he observed that the mice did not develop disease. However, after mixing a heat-killed version of the "smooth" form with a live version of the "rough" form and injecting this into mice, disease quickly developed and the mice died. Griffith suggested that an unidentified transforming factor was responsible for the transfer of traits that occurred between the two forms of bacteria.

Jean Brachet was a Belgian biochemist who in 1933 discovered that DNA was located within the chromosomes of cells.

William Astbury was a British molecular biologist who studied biological molecules utilizing a technique known as X-ray diffraction. In 1937, Astbury produced the first X-ray diffraction pattern of DNA, suggesting it had a regular structure that could be easily deduced.

Oswald Avery (Figure 6.2c), **Colin MacLeod**, and **Maclyn McCarty** were a group of American and Canadian scientists who, in 1943, discovered that the unidentified transforming factor that had caused trait transfer in Griffith's 1928 experiment was DNA.

Alfred Hershey and **Martha Chase** (Figure 6.2d) were American scientists who conducted a series of experiments using bacteriophages, a group of viruses composed primarily of protein and DNA that can infect and replicate within bacterial cells, to determine whether DNA or proteins were the genetic material. In 1952, their experiments revealed that viral DNA enters the bacterial host cell, not the protein, leading them to conclude that DNA was the genetic material. Despite working together as a team to make this discovery, Hershey was awarded the 1969 Nobel Prize in Physiology or Medicine for his work; however, Chase was excluded.

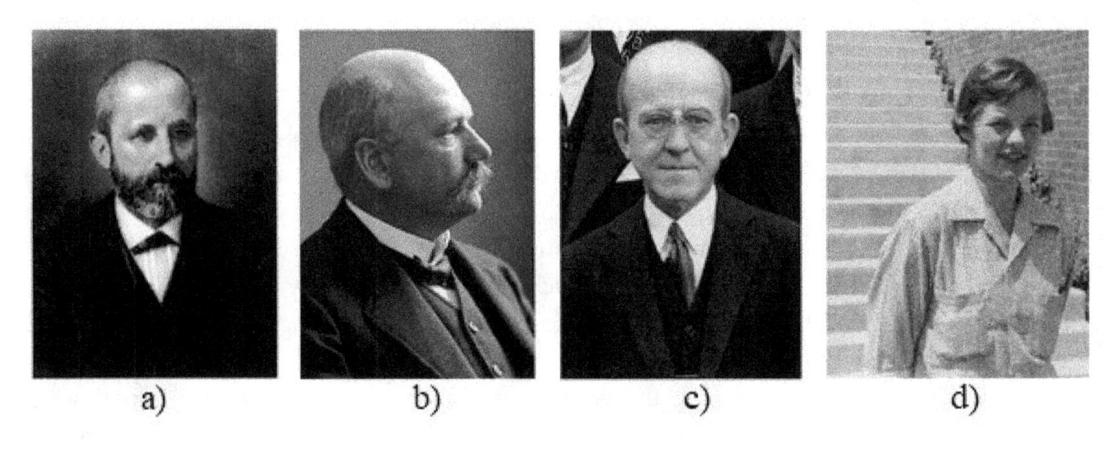

FIGURES 6.2A–D Many scientists contributed to the eventual discovery of the structure of DNA, including a) Friedrich Miescher, b) Albert Kossel, c) Oswald Avery, and d) Martha Chase.

6-2. DNA Structure and Function

Deoxyribonucleic acid (DNA) is a nucleic acid composed of a group of molecules known as nucleotides. Nucleotides of DNA consist of three main components (Figure 6.3):

1. five-carbon sugar molecule called deoxyribose
2. phosphate group
3. nitrogenous bases

DNA contains two strands of alternating, covalently bonded sugar and phosphate molecules, twisted into a double-helix structure, similar to a spiral staircase. These covalently bonded molecules make up the sugar-phosphate backbone of DNA and were first deduced by Watson and Crick in 1953, who both won the 1962 Nobel Prize in Physiology or Medicine for this discovery. However, it has been alleged that this deduction was based on mathematical data compiled by Rosalind Franklin during her many years of X-ray diffraction experiments on the molecule. Each strand differs from the other, with one strand ending in a nonbonded sugar and the other ending

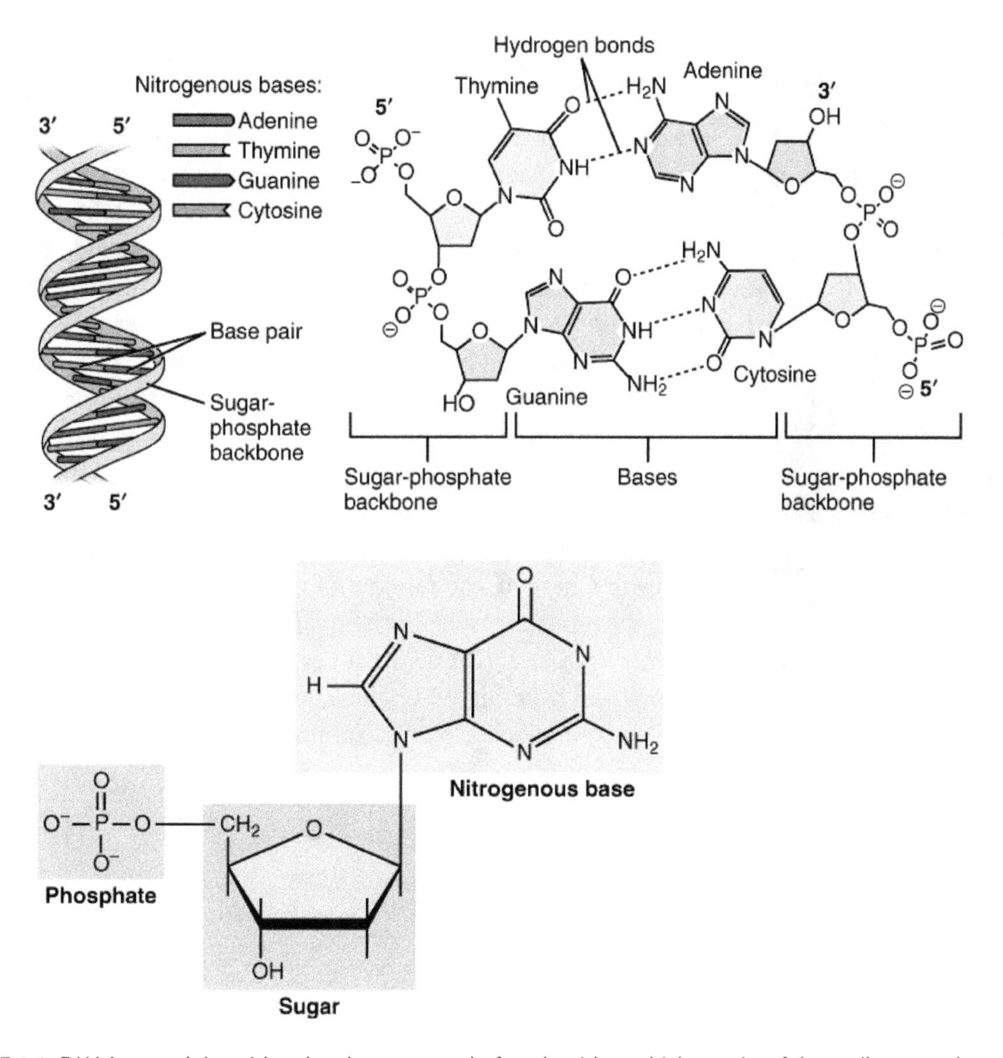

FIGURE 6.3 DNA is a nucleic acid molecule composed of nucleotides, which consist of deoxyribose, a phosphate group, and four nitrogenous bases: adenine, guanine, thymine, and cytosine. The phosphate groups and deoxyribose sugars are covalently bonded together forming two strands of nucleotides that make up the DNA backbone. The two strands run in opposite directions (5' to 3' and 3' to 5') and are twisted together similar to a spiral staircase. On the inside of the molecule are the nitrogenous bases that practice complemenary base pairing with adenine bonding with thymine and guanine bonding with cytosine by means of hydrogen bonding.

in a nonbonded phosphate group. This results in the strands being arranged in opposite directions, for which DNA is oftentimes described as exhibiting an **antiparallel arrangement**. Because of this arrangement, the end with the nonbonded phosphate group is called 5' and the nonbonded sugar is called 3'. These numbers refer to the carbons to which the phosphate and sugar molecules are attached, respectively. Since the arrangement of DNA is in an antiparallel state, one strand of DNA runs in a 5' to 3' direction, while the other runs in a 3' to 5' direction. The antiparallel arrangement of DNA is important in ensuring that the nitrogenous bases (discussed shortly) are properly oriented, as suggested by Rosalind Franklin when deducing the width of the DNA molecule (Figure 6.4). The DNA sugar-phosphate backbone helps provide support for the DNA molecule.

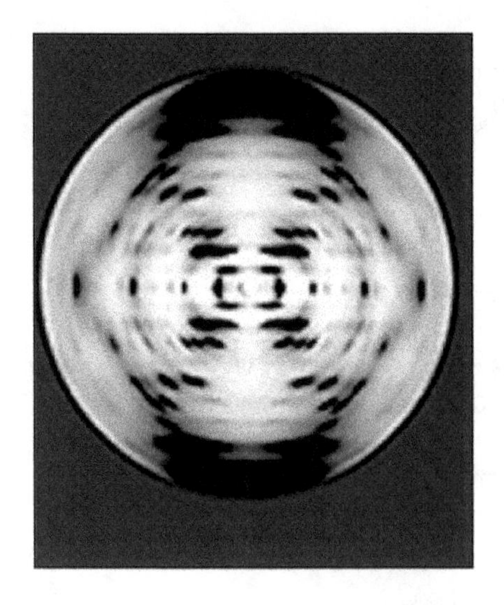

FIGURE 6.4 Photo 51, an X-ray diffraction photo taken by Rosalind Franklin and Raymond Gosling in 1952, helped provide the information needed to help deduce the structure of DNA.

Attached to the sugar molecules, extending into the middle of the spiral staircase, are the nitrogenous bases, whose sequences carry the genetic information and instructions for building all aspects of a living organism, from the cell to complex structures. There are two groups of nitrogenous bases that make up a DNA nucleotide: double-ring structures known as *purines,* which include two types of bases—adenine and guanine—and single-ring structures known as *pyrimidines,* which include two additional bases—thymine and cytosine. Hydrogen bonds form between a purine and pyrimidine, pairing the nitrogenous bases in what is known as complementary base pairing. Three hydrogen bonds occur between guanine and cytosine, while two hydrogen bonds form between adenine and thymine.

Many scientists once thought the DNA of every species on Earth contained 25 percent of each nitrogenous base; however, according to Erwin Chargaff, the number of guanine equals that of cytosine, while the number of adenine is equal to the number of thymine. This became known as Chargaff's rule and is evident in the table below (Figure 6.5) from his 1951 publication entitled "The Composition of the Deoxyribonucleic Acid of Salmon Sperm."

TABLE II

Purine and Pyrimidine Contents of Salmon Sperm DNA

The results are expressed in moles per mole of P in the hydrolysate.

Experiment No.*	Preparation No.	Hydrolysis procedure	Nitrogenous constituent				Recovery of nitrogenous constituents		
			Adenine	Guanine	Cytosine	Thymine	Purines	Pyrimidines	Total
1	1	1	0.27	0.18			0.45		
2		1	0.26	0.19			0.45		
3		1			0.17	0.28		0.45	
4		1			0.18	0.28		0.46	
5		2	0.28	0.20	0.21	0.27	0.48	0.48	0.96
6		2	0.30	0.22	0.20	0.29	0.52	0.49	1.01
7		2	0.27	0.18	0.19	0.25	0.45	0.44	0.89
8		2	0.28	0.21	0.20	0.27	0.49	0.47	0.96
9	2	1	0.25	0.18			0.43		
10		1	0.29	0.20			0.49		
11		2	0.29	0.18	0.20	0.27	0.47	0.47	0.94
12		2	0.28	0.21	0.19	0.26	0.49	0.45	0.94
13		2	0.30	0.21	0.20	0.30	0.51	0.50	1.01

* In each experiment between twelve and twenty-four determinations of individual purines and pyrimidines were performed.

FIGURE 6.5 Erwin Chargaff's 1951 data on the DNA composition of salmon sperm demonstrates Chargaff's Rule, which states that the number of adenine equals the number of thymine, while the number of guanine equals that of cytosine.

In understanding how complementary base pairing occurs, if one knows the base sequence of one DNA strand, then one should be able to determine the base sequence of the second strand. Complementary base pairing is important because it helps determine the function of a gene.

A **gene** (Figure 6.6) is a unique sequence of DNA bases, up to 3,000 in total, that spell out codes responsible for detailing an organism's characteristics and for building proteins. Genes are located on **chromosomes** (Figure 6.6), first outlined by American biologist Thomas Hunt Morgan in his 1910 paper entitled "Sex Limited Inheritance in Drosophila." Morgan's work on fruit flies was a catalyst for understanding where the genes were located and their importance in an offspring's inheritance of parental traits. Chromosomes make up a **genome**, which consists of an organism's entire set of DNA. The genome of a prokaryotic cell exists as a single chromosome in the form of a circular structure called a nucleoid, found directly attached to the plasma membrane of the cell. The chromosomes within the genome of eukaryotic cells are located within a cell's nucleus, wrapped around a group of proteins called **histones** (Figure 6.6). As mentioned in chapter 4, the histone proteins are responsible for keeping the DNA from tangling and enabling the DNA to be tightly packed within the nucleus. In the largest of all known genomes, the human genome, there are 23 pairs of chromosomes, containing about three billion base pairs. The smallest cellular genome known to date is about 112,091 base pairs and is found in an insect-dwelling bacteria, *Nasuia deltocephalinicola*. Other notable organism and corresponding chromosome numbers include dogs, with 39 pairs, and fruit flies, with four pairs. The number of base pairs in an organism's genome can be used to help determine the length of a DNA molecule.

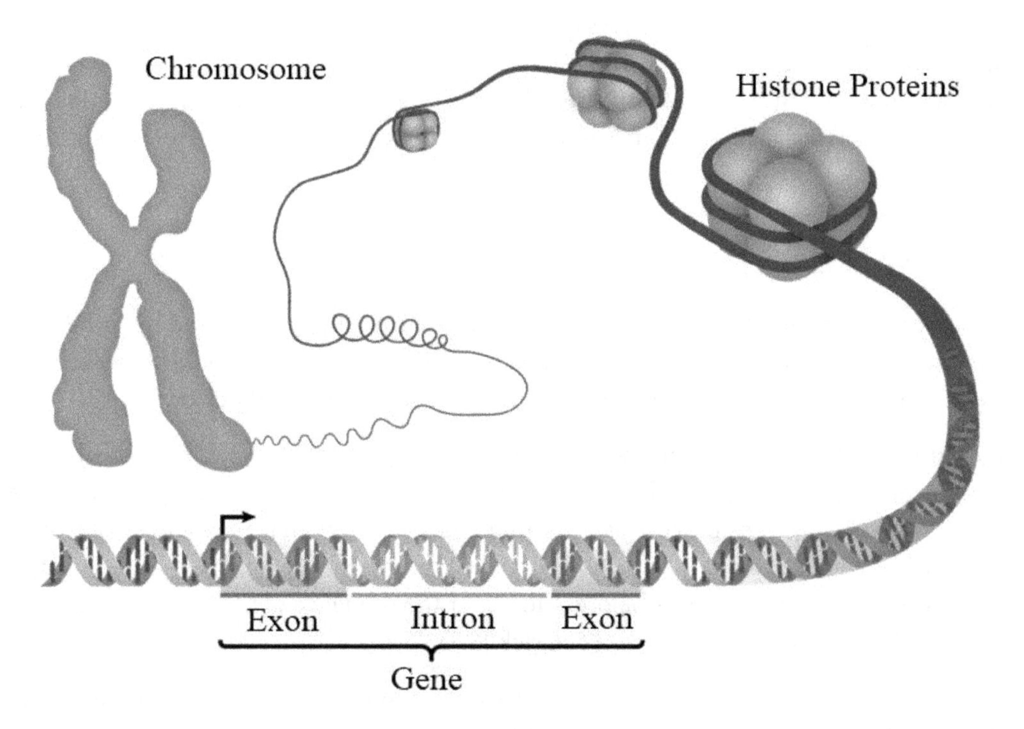

FIGURE 6.6 An organism's genome is composed of chromosomes, which are wrapped around a group of proteins called histones. Within chromosomes are genes, which are responsible for providing the code necessary to build proteins. Genes are composed of non-coding DNA called introns, while the DNA that codes for proteins is called exons.

Many organisms have more DNA than the necessary amount to make proteins. For example, only about 2 percent of the DNA in humans exists as genes carrying codes for protein-making. This suggests that a huge amount of DNA is considered noncoding DNA. Noncoding DNA is found primarily in eukaryotic cells and exists as thousands of repeating base sequences. This type of DNA may exist in one of three forms—gene fragments, duplicate genes, or mutated genes that have lost their ability to code for proteins. Roughly 25 percent of noncoding DNA occurs within genes known as **introns** (Figure 6.6). Introns were first discovered in the protein-making genes of adenovirus in 1977 by two molecular biologists, American Phillip Sharp and Englishman Richard J. Roberts. Although their discoveries were made independently of each other, they shared the Nobel Prize in Physiology or Medicine in 1993. The term *intron*, however, was not introduced until 1978, coined by American biochemist Walter Gilbert in his paper entitled "Why Genes in Pieces?" At the time of their discovery, introns consisted of ten to 10,000 bases in length, but Gilbert suggested that as more information became available about introns, the base length would be considerably greater. Today, an intron may range from 65 to 100,000 bases in length. The remaining 75 percent of noncoding DNA is found between genes.

Although scientists do not fully understand the function of noncoding DNA, there are some who propose that introns may be responsible for regulating the formation of messenger RNA (mRNA) strands during the process of translation. In other words, introns can pick and choose which coding pieces of DNA are incorporated into the mRNA strand. Other scientists suggest that introns may be involved in a process known as crossing over during meiosis, which will be discussed in chapter 7. Finally, introns may also be involved in switching genes on or off or influencing the rate at which proteins are synthesized.

DNA located within the gene that is responsible for coding proteins is called **exons** (Figure 6.6.) and may average between 100 to 300 bases in length. Exons are discussed further in section 6-4.

6-3. DNA Replication

In 1953, James Watson and Francis Crick discovered the structure of DNA and hypothesized that its specific structural arrangement suggested a mechanism by which DNA could copy itself. Shortly thereafter, Watson and Crick hypothesized how DNA replication would occur:

> Now our model for deoxyribonucleic acid is, in effect, a *pair* of templates, each of which is complementary to the other. We imagine that prior to duplication the hydrogen bonds are broken and the two chains unwind and separate. Each chain then acts as a template for the formation onto itself of a new companion chain so that eventually we shall have two pairs of chains, where we only had one before. Moreover the sequence of the pairs of bases will have been duplicated exactly. (Watson & Crick 1953, 966)

Cell division is one of the most important processes a cell will undergo. Prior to dividing, a cell must generate additional cellular materials, a large number of nucleotides, and energy. These processes must occur within the cell prior to the replication of DNA. DNA replication is a carefully regulated process, and once a cell commits to DNA replication, the cell must divide. After DNA replication occurs, each daughter cell produced during division receives the same amount of genetic material.

The process of DNA replication is best described using the **semiconservative model** (Figure 6.7), proposed by Watson and Crick. The model suggests that upon replication, the resulting DNA molecules in each daughter cell contain one parental strand and one newly formed strand. Prior to this model, other models (Figure 6.7) were introduced, including the **conservative model**, stating that parental DNA molecule was preserved and came back together after replication, and the **dispersive model**, where all four strands of DNA, the parental strands and the new strands, consisted of a mixture of both new and old DNA.

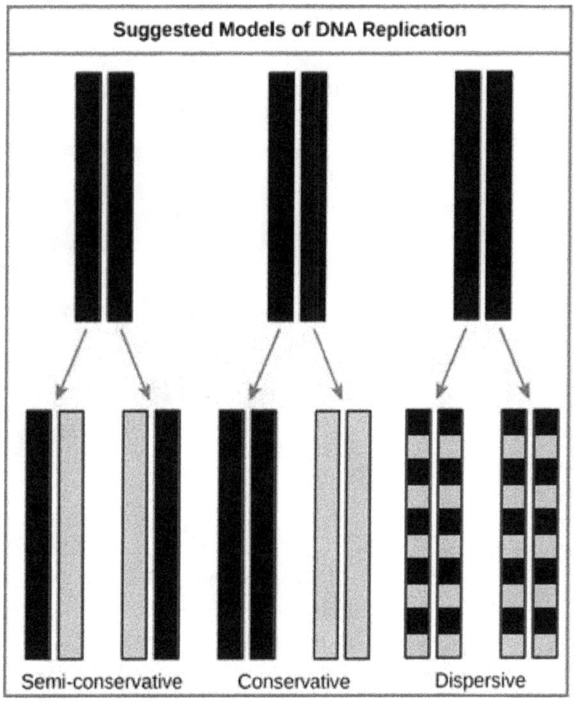

In 1958, the results of experiments performed by two American molecular biologists Matthew Meselson and Franklin Stahl demonstrated and supported the semiconservative model, as suggested by Watson and Crick, while in addition disproving the other two models.

DNA replication involves a large number of enzymes, which the cell will use to ensure accurate replication of the molecule. Once the cell has the replication enzymes it requires, it proceeds with DNA replication, which is detailed in Figure 6.8.

Suggested Models of DNA Replication

Semi-conservative Conservative Dispersive

FIGURE 6.7 Three models were introduced to describe what happens to DNA strands during the replication process. The best model is the semiconservative model, proposed by Watson and Crick, which states that each new DNA molecule contains one parental strand and one newly formed strand. The other two models, conservative and dispersive, were both disproved by experiments performed by Meselson and Stahl in 1958.

DNA replication begins at a region known as the **origin of replication**, a short, specific sequence of nucleotides within the DNA molecule. Here, a specialized enzyme called **helicase** recognizes and attaches to the origin of replication and begins to unwind the helix. As helicase unwinds the helix, the enzyme separates the two parent strands by breaking the hydrogen bonds connecting the nitrogenous bases, making each parental strand available as a template for replication, producing a Y-shaped region known as the **replication fork**. As each strand begins to separate, a group of proteins called **single-strand binding proteins** attaches to each parental strand, keeping them separated from each other during replication.

The next step in DNA replication is the addition of a molecule consisting of a short chain of five to ten nucleotides called **RNA primer**, which is synthesized by an enzyme called **DNA primase**. RNA primer is responsible for attracting **DNA polymerase**, an enzyme that initiates complementary base pairing to the template strand according to the following base-pairing rules: A with T and C with G. Because of DNA's antiparallel arrangement, DNA replication must also occur antiparallel to the parental strands. The complementary bases are added and joined by hydrogen bonds of DNA polymerase to the 3' end of the template strand, elongating a new DNA strand in a 5' to 3' direction along the replication fork. The new strand created in this direction is called the **leading strand**. Since DNA can only be replicated in the 5' to 3' direction by DNA polymerase, the synthesis of the **lagging strand** along the the other template strand is created discontinuously in a series of fragments. These fragments are called **Okazaki fragments** and can be between 100 and

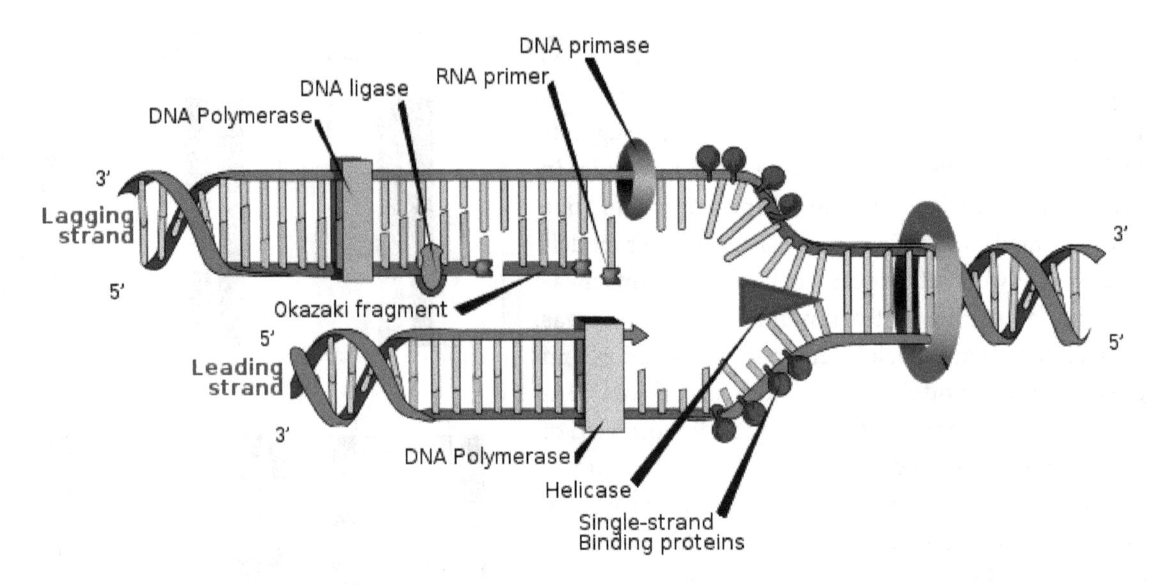

FIGURE 6.8 DNA replication begins along a specified sequence of nucleotides known as the origin of replication. Helicase begins unwinding the DNA molecule, exposing each strand to act as a template, producing a region known as the replication fork. Here, each strand remains separated by single-strand binding proteins. DNA primase synthesizes a short sequence of nucleotides called an RNA primer, which is responsible for attracting DNA polymerase, an enzyme that begins complementary base pairing to the template strands. Along the leading strand, DNA is continuously synthesized, while the lagging strand DNA synthesis occurs discontinuously with Okazaki fragments, joined together by DNA ligase. Errors during the replication process are corrected by DNA nuclease, which cuts out the error and replaces it with the correct base sequence.

2,000 nucleotides long, depending on the type of cell. The Okazaki fragments are joined together by an enzyme called **DNA ligase** to form a continuous strand of DNA.

As each parental strand is being replicated, various other types of DNA polymerases are responsible for proofreading the complementary base pairings. Upon finding an incorrect pairing due to replication errors or DNA damage, an enzyme called **nuclease** will cut out the fragment of DNA with the incorrectly paired bases. The resulting gap will be replaced with the correct sequence of bases by DNA polymerase and DNA ligase, using an undamaged template strand. The elongation rate of DNA replication is quite fast, with about 50 nucleotides added per second in human cells. Replication of both the leading and lagging strands will continue in both directions and at the same rate until the original DNA molecule has been replicated. Replication ends once the RNA primer has been removed by DNA polymerase and the correct sequence of DNA nucleotides have been added to the primer region.

Repeated DNA replications in a cell produce short, uneven DNA molecules. However, in eukaryotic cells, the ends of chromosomes have short sequences of repeated nucleotides called **telomeres**. Telomeres do not contain genes; therefore, they do not code for proteins but provide two important functions: protecting the ends of genes from the loss of DNA during each replication and limiting the number of divisions a cell can undergo (about 50 in humans). Portions of telomeres are lost per division until there is a loss of functional DNA, resulting in cell death. However, some cells, such as cancer cells, express an enzyme known as **telomerase**, which counteracts the shortening of telomeres by rebuilding them regardless of the number of cell divisions. Cancer is a concept that will be discussed further in chapter 7.

6-4. Gene Expression and Protein Synthesis

The sequence of nitrogenous bases in DNA is responsible for providing the genetic information and instructions for building an organism. However, the same sequence of nitrogenous bases of DNA is also responsible for directing the synthesis of proteins, known as **gene expression**.

Protein synthesis is a process that takes place within a group of structures known as ribosomes, found within the cytoplasm of the cell. The sequence of DNA bases only provides the code necessary for protein synthesis but cannot direct the process. Therefore, protein synthesis requires an intermediary molecule in order to decipher the code outlined by the DNA sequences. This intermediary molecule is called **ribonucleic acid,** or **RNA**. The idea of DNA providing the code necessary for protein synthesis was first proposed by Francis Crick in 1958 in his **sequence hypothesis**. The sequence hypothesis "assumes that the specificity of a piece of nucleic acid is expressed solely by the sequence of its bases, and that this sequence is a (simple) code for the amino acid sequence of a particular protein" (Crick 1985, 152). The relationship of DNA making RNA and RNA making proteins was described by Crick as the central dogma of molecular biology (Figure 6.9).

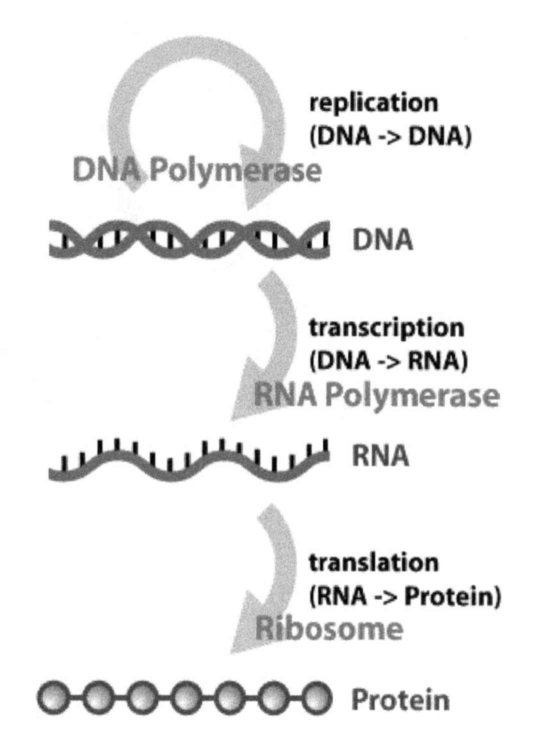

FIGURE 6.9 The central dogma of molecular biology was introduced by Francis Crick in 1958 to describe the relationship between DNA and RNA to make proteins.

RNA was described above as the intermediary molecule between DNA and proteins. However, what differentiates RNA from DNA is several structural differences (Figure 6.10). RNA is a single-stranded molecule composed of the sugar ribose. In addition, the nitrogenous base, thymine, is absent in the RNA molecule and replaced with another type of nitrogenous base called uracil. Another difference between DNA and RNA is a functional difference—RNA can also function as an enzyme. DNA codes for three different types of RNA molecules, all responsible for the conversion of DNA base sequences into proteins.

1. **Messenger RNA (mRNA).** mRNA is the intermediary molecule responsible for deciphering the DNA code used to build a protein. mRNA carries the message from the nucleus to the ribosomes, where the information is used to synthesize a protein.
2. **Ribosomal RNA (rRNA).** rRNA is the RNA component of the ribosome, composing about 60 percent of a ribosome by weight.
3. **Transfer RNA (tRNA).** tRNA is responsible for translating the code on the mRNA strand and transferring a specific amino acid to the ribosome to help build the protein.

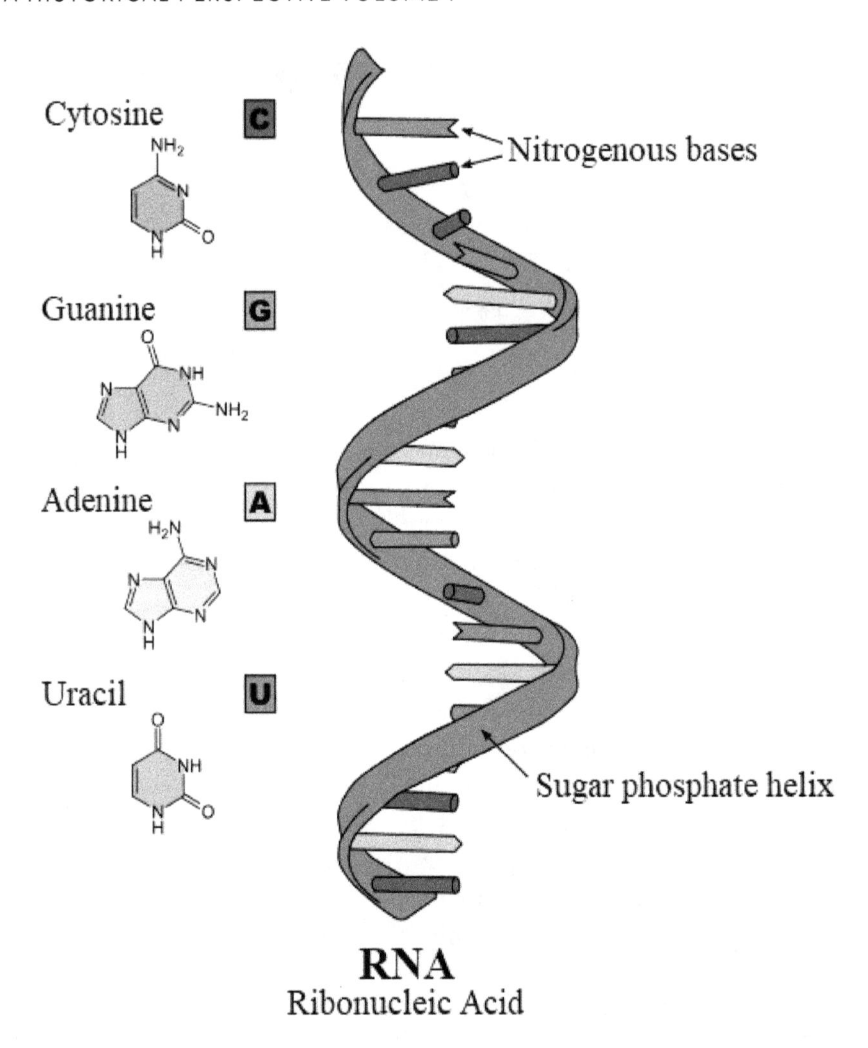

FIGURE 6.10 Ribonucleic acid (RNA) is a nucleic acid that plays an integral role in the formation of proteins. The molecule consists of a single strand of nucleotides with thymine replaced with uracil.

Recall from chapter 3 that proteins are composed of amino acids and there are 20 different amino acids that can be used to build proteins. But how does the cell know which of the 20 different amino acids to use when building the protein? The DNA base sequences are translated into amino acid sequences by using an important tool known as the **genetic code**. The genetic code consists of 64 different combinations of three-base sequences, or triplets, called **codons**, each of which correspond to a specific amino acid (Figure 6.11). The idea that the genetic code consisted of three base sequences occurring as triplets, with each triplet coding for an individual amino acid, was first proposed by a group of scientists led by Francis Crick in 1961. During the experiment, the researchers isolated mutations within a T4 bacteriophage gene, *rIIB*, and demonstrated that upon inserting or deleting single nucleotides into the gene, the result was a nonfunctional *rIIB* gene. However, upon the insertion or deletion of three base pairs into the gene, the researchers discovered that the gene would remain functional. Also in 1961, two biochemists, American Marshall

Nirenberg and German J. Heinrich Matthaei, from the National Institutes of Health, were the first to experimentally demonstrate that codons specify particular amino acids, in what was known as the "poly-U" experiment. During this experiment, Nirenberg and Matthaei discovered that the codon UUU coded for the amino acid phenylalanine. Throughout the remainder of the early 1960s, different research groups continued working on cracking the genetic code, until 1966, when a team led by Nirenberg revealed all 64 codons responsible for the 20 amino acids. Of the 64 total codons discovered, 61 specify for a particular amino acid, including the start codon—AUG, which codes for the amino acid methionine. The remaining codons are called stop codons, UAG, UAA, and UGA, and are responsible for stopping the process of protein synthesis when the ribosome encounter these codons.

Second letter

First letter	U	C	A	G	Third letter
U	UUU Phe	UCU Ser	UAU Tyr	UGU Cys	U
	UUC Phe	UCC Ser	UAC Tyr	UGC Cys	C
	UUA Leu	UCA Ser	UAA Stop	UGA Stop	A
	UUG Leu	UCG Ser	UAG Stop	UGG Trp	G
C	CUU Leu	CCU Pro	CAU His	CGU Arg	U
	CUC Leu	CCC Pro	CAC His	CGC Arg	C
	CUA Leu	CCA Pro	CAA Gln	CGA Arg	A
	CUG Leu	CCG Pro	CAG Gln	CGG Arg	G
A	AUU Ile	ACU Thr	AAU Asn	AGU Ser	U
	AUC Ile	ACC Thr	AAC Asn	AGC Ser	C
	AUA Ile	ACA Thr	AAA Lys	AGA Arg	A
	AUG Met	ACG Thr	AAG Lys	AGG Arg	G
G	GUU Val	GCU Ala	GAU Asp	GGU Gly	U
	GUC Val	GCC Ala	GAC Asp	GGC Gly	C
	GUA Val	GCA Ala	GAA Glu	GGA Gly	A
	GUG Val	GCG Ala	GAG Glu	GGG Gly	G

FIGURE 6.11 The codon table lists all 64 codons associated with protein synthesis, including the 61 responsible for specifying a particular amino acid, and the three stop codons.

Finally, there are several descriptive words used to describe the genetic code. The genetic code is said to be **degenerate**, which means that many amino acids are specified by more than one codon, helping protect an organism against harmful mutations. Another term used to describe the genetic code is **universal**. All living organisms are composed of DNA; therefore, they use the same mRNA codons to specify amino acids. This suggests evidence that all living organisms share a common ancestor and that genes can be exchanged from one organism to another. Finally, the term

unambiguous is oftentimes used to describe the genetic code, because each codon specifies only one amino acid.

Transcription

Transcription is the first step in the process of protein synthesis and occurs in the nucleus of eukaryotic cells. Transcription involves transcribing the information contained within the DNA base sequences of a gene into an mRNA strand in three different steps: **initiation**, **elongation**, and **termination**. Each of these steps are specifically associated with a distinct region of a gene:

1. **Promoter.** The promoter is a nontranscribed region of DNA consisting of two short sequences of DNA bases—TATAAA and transcription factor binding sites—which initiate the binding of the enzyme **RNA polymerase**. The promoter is responsible for initiating the start of transcription and determining the direction along which DNA strand transcription will occur.
2. **Gene body.** The region of the gene where elongation occurs.
3. **End of gene.** The end of the gene consists of a specific sequence of nucleotides responsible for terminating mRNA synthesis.

Initiation

The initiation of transcription is catalyzed by the enzyme RNA polymerase, which recognizes and binds to the promoter site of the gene. However, before RNA polymerase can bind to the promoter site, proteins called **transcription factors** bind to the promoter (Figure 6.12). The formation of this protein complex enables RNA polymerase to properly bind to the promoter to initiate transcription.

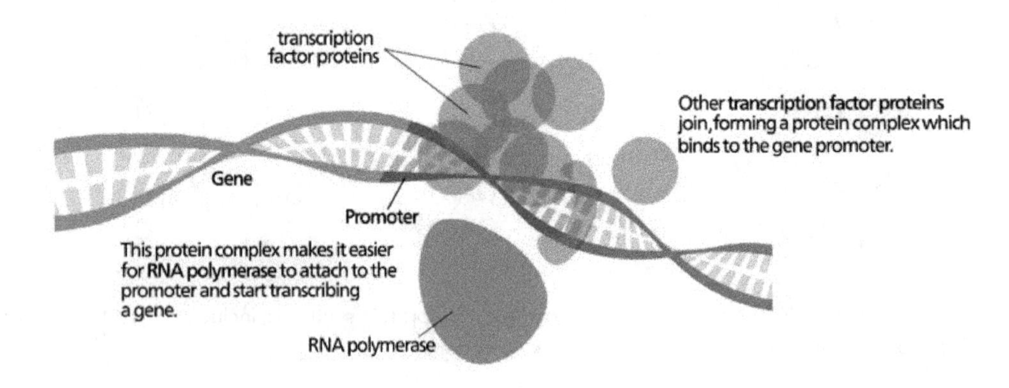

FIGURE 6.12 Transcription factors bind to the promoter of a gene to form a protein complex that initiates transcription along the DNA strand.

Upon binding to the promoter, RNA polymerase begins to unwind the DNA double helix and break the hydrogen bonds between the nitrogenous bases, separating the two strands. For every gene, only one of the exposed strands will be transcribed, and this strand is called the **template strand** because it contains the information needed in order to produce the required mRNA strand.

The template strand can be used to make hundreds of mRNA molecules, but the number varies depending on the amount of protein needed by the cell. The other exposed strand, not being transcribed, is called the **coding strand**; however, it may serve as a template for another gene (Figure 6.13)

Elongation

Elongation involves RNA polymerase moving along the body of the template strand of the gene, synthesizing a single strand of mRNA (Figure 6.13). As RNA polymerase moves along the template strand in the 5' to 3' direction, it unwinds the DNA molecule, exposing up to 20 DNA nucleotides at a time, pairing each exposed DNA nucleotide with its complementary RNA nucleotide at the 3' end of the growing strand. In the case of an exposed adenine nucleotide, RNA polymerase base pairs the nitrogenous base uracil. Once the mRNA strand is about ten nucleotides long, the strand begins to separate from the DNA template strand, allowing hydrogen bonds to reform and the DNA molecule to rewind. In eukaryotic cells, transcription along the template strand can occur about 40 nucleotides per second.

Termination

At the end of the gene, there is a short sequence of DNA bases called the **termination signal**. When RNA polymerase reaches the termination signal, transcription stops. Once transcription stops, RNA polymerase releases the newly formed mRNA strand, which can be up to 3,000 bases long, and separates from the template strand (Figure 6.13).

After mRNA is released from the template strand, it undergoes an important modification—**capping** (Figure 6.13). After several nucleotides along the 5' end of the mRNA strand are transcribed, a modified guanine nucleotide cap forms. On the 3' end of the mRNA stand, a sequence of 100 to 200 adenine nucleotides are added, forming a tail. Capping is an important step for three reasons: it helps protect the mRNA strand from degradation by cellular enzymes, enables it to be transported from the nucleus to the cytoplasm, and helps the mRNA strand align and bind correctly within a ribosome.

The final modification of the mRNA strand is the editing process. This process involves cutting and removing the noncoding portions of the strand (introns), while cutting and combining the protein-coding portions of the mRNA strand (exons) (Figure 6.13). Once the edited mRNA strand is ready, it leaves the nucleus through the nuclear pores and travels out into the cytoplasm, where it may remain for a few minutes to a few days until it binds within ribosomes, initiating the translation process.

Translation

Translation is the final process in the synthesis of proteins and involves decoding the mRNA nucleotide sequence into an amino acid sequence of a protein. Translation occurs at the ribosomes within the cytoplasm. The process involves transfer RNA (tRNA), which is responsible for delivering the appropriate amino acid, as coded by the mRNA strand, to the ribosome. The tRNA molecule is three-dimensional and shaped much like a clover leaf. It consists of a sequence of three exposed bases, which complement the mRNA codon found on the mRNA strand, called an **anticodon**. The anticodon recognizes and reads the mRNA codon sequence, delivers the correct amino acid

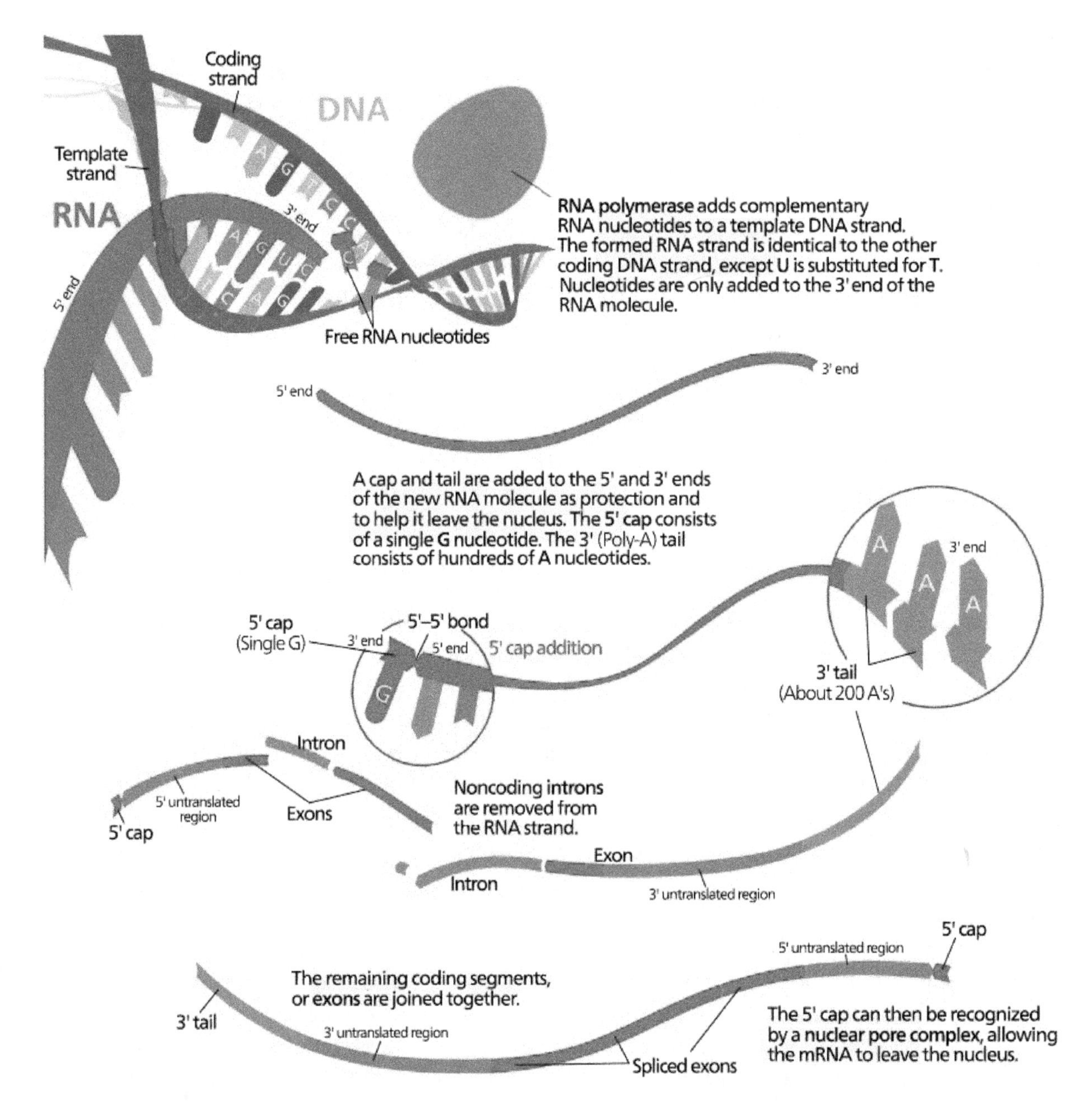

FIGURE 6.13 Transcription begins when RNA polymerase attaches to the promoter site with the help of transcription factors. As the enzyme moves along the DNA strand, the double helix unwinds and separates, exposing the template strand, the strand containing the gene to be transcribed. As RNA polymerase moves along the template strand, it begins synthesizing an mRNA strand in the 5' to 3' direction, complementary base pairing the appropriate RNA nucleotides with the exposed DNA nucleotide. Upon reaching a DNA nucleotide with the base adenine, RNA polymerase base pairs the RNA nucleotide containing uracil. Finally, when RNA polymerase reaches a termination signal on the DNA strand, transcription stops and the newly synthesized mRNA strand is released. Prior to its use in making proteins, the mRNA strand receives additional modifications, including a cap and tail and any non-coding introns are edited out.

to the ribosomes, and uses ATP to attach an amino acid to the overall sequence. About 40 different molecules of tRNA exist in a cell, which is considerably less than the number of codons coding for amino acids. Since at least one tRNA molecule exists for each of the 20 amino acids, this suggests that some tRNA molecules are capable of recognizing multiple codons. This phenomenon was first observed by Francis Crick in 1966, who called it the **wobble effect**, a characteristic that is important in ensuring that a functional protein is synthesized, even if changes occur within the DNA base sequences.

Like transcription, translation is a three-step process: **initiation, elongation**, and **termination**.

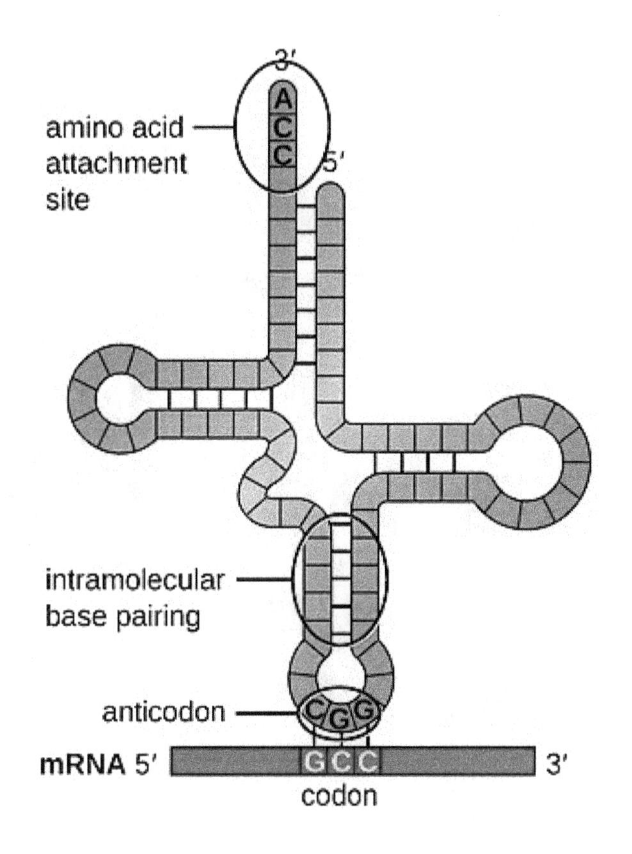

FIGURE 6.14 Transfer RNA (tRNA) is an RNA molecule responsible for carrying the appropriate amino acid to the ribosome as coded by the mRNA strand. The tRNA molecule consists of a three-base sequence called an anticodon that recognizes and binds to the mRNA codon.

Initiation

Initiation begins with the first codon in the amino acid sequence—the start codon AUG—which codes for the amino acid methionine. Methionine is found at the beginning of all amino acid sequences of proteins. Once translation is initiated, an initiation complex forms, consisting of a small ribosomal subunit containing mRNA binding sites and an initiator tRNA molecule that binds to the start codon (Figure 6.15). The formation of an initiation complex is important in properly aligning the ribosome to the mRNA strand prior to the start of translation.

Elongation

Elongation refers to the building of the amino acid sequence that will become the eventual protein. Elongation occurs as the ribosome moves down the mRNA strand from the 5' end to the 3' end of the strand, translating each codon as it passes (Figure 6.15). When the ribosome reaches a codon, a tRNA molecule releases the specified amino acid, then moves to the next codon, then the next, continually building the amino acid sequence. As the amino acids are released, a peptide bond forms (a type of covalent bond) between them, which is responsible for helping maintain the long chain of amino acids (polypeptide).

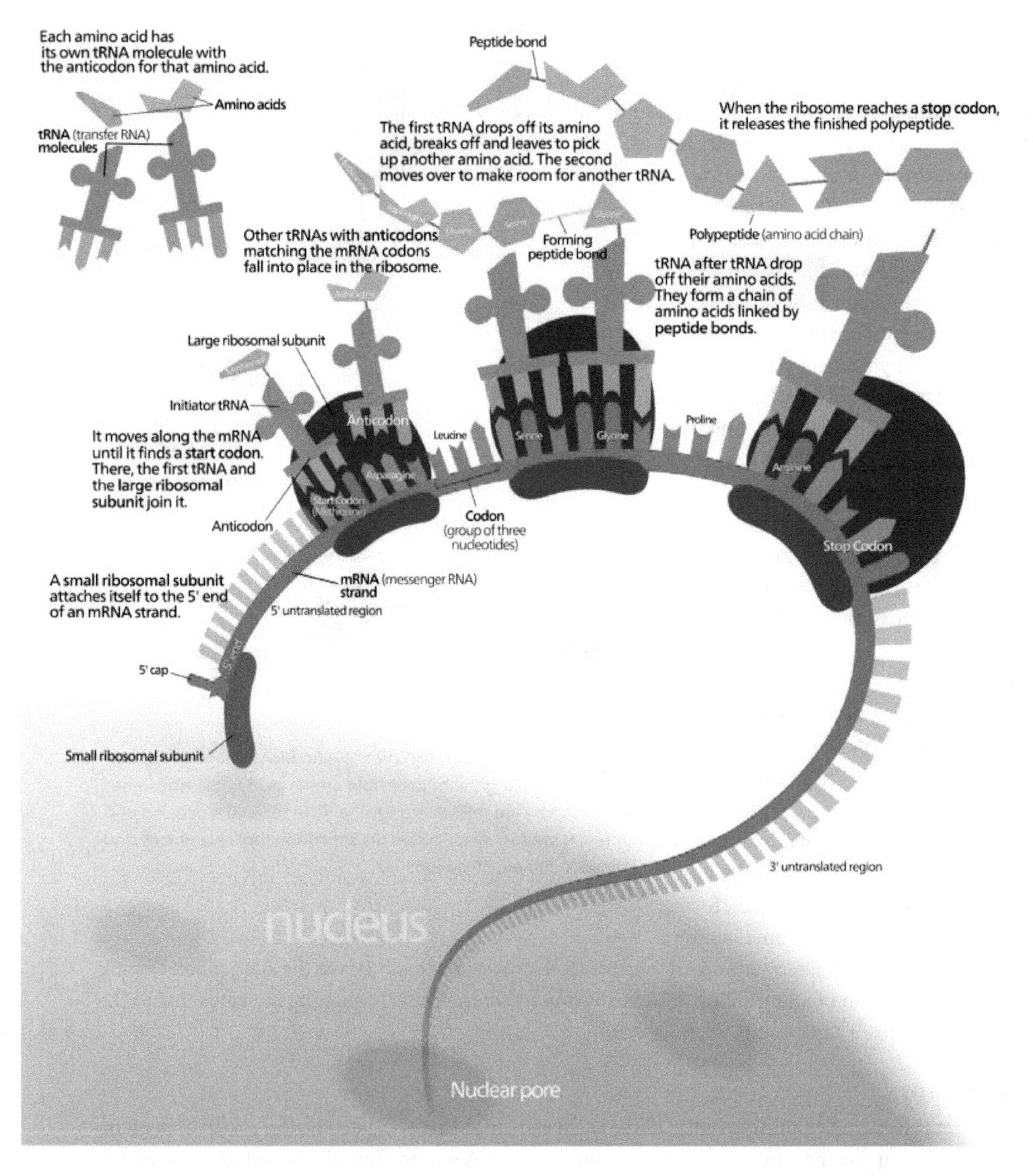

FIGURE 6.15 Translation begins when an initiator tRNA molecule recognizes the start codon, AUG, on the mRNA strand, which codes for the amino acid methionine (Met). Once initiated, the ribosome moves down the mRNA strand from the 5' to 3' end, translating each codon on the strand and tRNA brings in the appropriate amino acid. The ribosome continues along each subsequent codon and as amino acids are released from the tRNA molecule, a peptide bond forms between them forming a long amino acid sequence. Translation terminates when the ribosome reaches one of three stop codons—UGA, UAG, or UAA—and the ribosome releases the mRNA strand and amino acid sequence, which then folds into a functional protein with the help of chaperone proteins.

Termination

Once the ribosome reaches a stop codon—UGA, UAG, or UAA—elongation stops, terminating protein synthesis. A group of proteins called releasing factors bind to the ribosome, causing the ribosome to release the amino acid sequence and the mRNA strand (Figure 6.15).

After translation has terminated, the amino acid sequence begins to fold into its functional three-dimensional structure as soon as it is released from the ribosomes. A group of proteins called **chaperone proteins** are responsible for folding the amino acid sequence by influencing correct molecular interactions and stabilizing important folds influential to the protein's final structure. Finally, the protein may undergo additional structural modifications, including the cutting and rearranging of amino acids to the attachments of other types of molecules, like lipids or sugars.

6-5. Mutations

A **mutation** is a permanent alteration in the base pair sequences of DNA or RNA nucleotides, resulting in a change in the structure and function of proteins. Mutations can either occur spontaneously or can be induced by **mutagens**, environmental agents that result in a mutation.

Spontaneous mutations are the most common type of mutation, occurring when nucleotide bases are substituted, inserted, or deleted during DNA replication. Since DNA replication is quite rapid (more than 1,000 bases are replicated per minute in humans), accidental errors do occur, such as incorrectly paired bases, which occur about one in every 1,000 to 100,000 base pairs. However, spontaneous mutations are quite rare and are minimized due to the "proofreading" ability of DNA polymerase during and after replication. If DNA polymerase finds a mismatched pair of base sequences, DNA repair enzymes will cut out the incorrect sequence of nucleotides and replace it with the correct, complementary nucleotide sequence. This process is quite efficient! A newly replicated DNA strand may only contain one mistake in every one billion nucleotides. Although most errors are corrected by DNA polymerase and DNA repair enzymes, there are some errors that are not discovered, leading to mutations.

Induced mutations result from exposure to toxic environmental factors called mutagens. Examples of mutagens include chemical and radiation mutagens, and many of these mutagens are also carcinogenic (cancer-causing). **Chemical mutagens** are mutagens that alter the base pair sequencing by reacting with the atoms found within the DNA molecule (carbon, hydrogen, oxygen, nitrogen, and phosphorus) and interfering with DNA replication. Chemical mutagens can be found in a variety of places, such as the food we eat and industrial chemicals. However, the most common type of chemical mutagen is cigarette smoke. Cigarette smoke contains several different types of carcinogenic chemicals, contributing to about 33 percent of all cancer deaths. The most common and most lethal in the United States is lung cancer; however, other types of cancer can be attributed to cigarette smoke, including mouth, larynx, and pancreas. The risk of cancer from cigarette smoke increases when combined with drinking alcohol.

Radiation mutagens include two types of radiation: ionizing or nonionizing radiation. **Ionizing radiation** is a form of radiation that produces ionized atoms with unpaired electrons, called free radicals and includes X-rays and gamma rays. In the 1920s, American geneticist Hermann Muller suggested a connection between the use of X-rays and lethal mutations when genetic changes were discovered during experiments performed on *Drosophila*.

Other scientists replicated these results using other types of organisms, such as wasps and corn. As a result of these discoveries, Muller began publicizing the dangers of using high-energy forms of radiation, such as X-rays and gamma rays, due to their potential genetic risk to humans and other types of organisms. Exposure to X-rays and gamma rays can result in alterations in the atomic structure of DNA or the breaking apart of chromosomes as tightly bound electrons are removed. **Nonionizing radiation**, such as ultraviolet (UV) radiation, the type of radiation given off by the sun, can also affect the structure of the DNA molecule. Upon exposure to UV radiation, the pyrimidine bases of DNA, thymine and cytosine, absorb the radiation, resulting in bond rearrangement. This rearrangement prevents complementary base pairing within the DNA strand. In addition, kinks can form within the DNA molecule as a result of the formation of thymine dimers, which affect DNA replication. Thymine dimers form when adjacent thymine bases within the DNA strand are covalently linked when exposed to UV radiation but are removed by DNA repair enzymes.

Exposure to mutagens can result in large-scale mutations, in which the overall organization of chromosomal genes is affected, or small-scale mutations, where one or more nucleotide base pairs are affected. Large-scale mutations are called **chromosomal aberrations** and can affect the organization of chromosomal genes in four ways: inversion, translocation, deletion, or duplication (Figure 6.16).

Inversion

Inversion occurs when a piece of DNA is removed, reoriented 180°, then reinserted back into the original chromosome. An example is hemophilia, the disease in which blood is unable to clot.

Translocation

Translocation is the removal and relocation of a DNA segment onto the same chromosome or onto another chromosome. An example is translocation Down syndrome, a form of Down syndrome that results from the translocation of DNA from one of the three 21 chromosomes to chromosome 14.

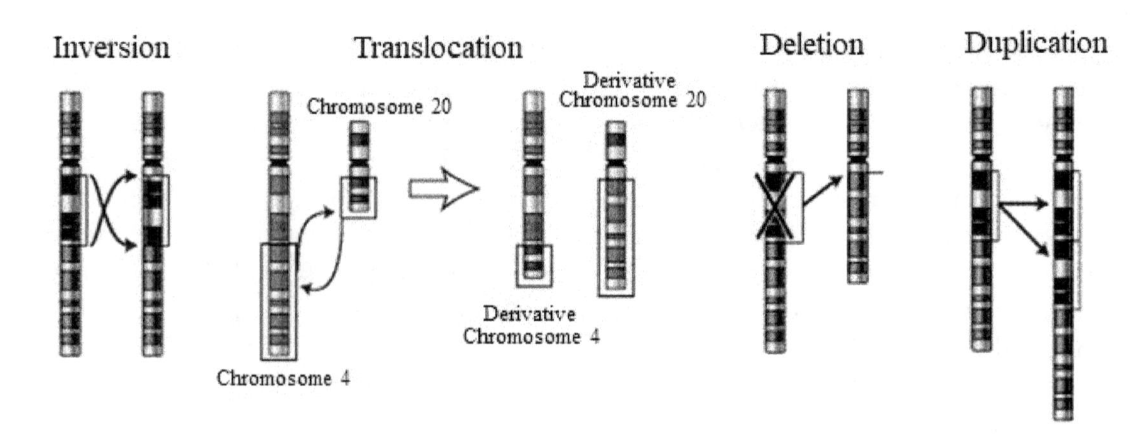

FIGURE 6.16 Chromosomal aberrations occur when the organization of a gene on a chromosome is affected by a mutation. Examples of these types of mutations include inversion, translocation, deletion, or duplication. These types of mutations can give rise to a variety of different types of genetic disorders.

Deletion

Deletion occurs when an entire portion of DNA is removed or lost from a chromosome. An example is Williams syndrome, a disease characterized by mild to moderate intellectual disability or learning disorders, unique personality traits, distinct facial features, and cardiovascular issues resulting from the deletion of DNA from chromosome 7.

Duplication

Duplication is the presence of an additional copy of a DNA segment on the same chromosome or on a different chromosome. An example is Charcot-Marie-Tooth (CMT) disorder, the most common inherited neurological disorder, characterized by the progressive loss of muscle tissue and touch sensation resulting from a duplication of DNA on chromosome 17.

Small-scale mutations are called **point mutations** and result when one or more nucleotide base sequences are changed, potentially affecting the overall amino acid sequence. There are two types of point mutations: substitution mutations and frameshift mutations (Figure 6.17).

Substitution Mutations

Substitution mutations are a group of mutations that occur when a single nucleotide has been substituted for another. A substitution mutation can give rise to three different varieties of mutations:

1. **Silent mutation.** A silent mutation is a mutation that occurs when the nucleotide substitution does not change the amino acid sequence.
2. **Missense mutation.** Most substitution mutations are missense mutations, which are mutations that occur when the nucleotide substitution changes one of the amino acids in the overall amino acid sequence. The resulting change in amino acid sequence may have little effect; however, if the change is within a crucial area of the protein, such as the active site, it can greatly alter the activity of the protein. Although missense mutations may lead to an improved protein, the majority of proteins produced are useless or vary in function. An example is sickle-cell anemia, a blood disorder characterized by abnormally shaped red blood cells due to a mutation in the gene that produces hemoglobin.
3. **Nonsense mutation.** A nonsense mutation is a mutation that occurs when the nucleotide substitution results in a STOP codon, UAG, UAA, or UGA. The resulting STOP codon terminates translation, producing a short, nonfunctional protein. Although rare, a nonsense mutation is known to cause cystic fibrosis, a disease that produces thick and sticky mucus, clogging the lungs, making breathing difficult, and obstructing the pancreas, affecting the digestive system.

Frameshift Mutations

Frameshift mutations are mutations that occur from the insertion or deletion of DNA nucleotides, which shifts the codons into a new sequence, resulting in a nonfunctional protein. This type of mutation is much more harmful than a substitution mutation, because the entire amino acid sequence has been altered, having a detrimental effect on the resulting protein. An example is Tay-Sachs, a disease characterized by the destruction of nerve cells in the brain and spinal cord.

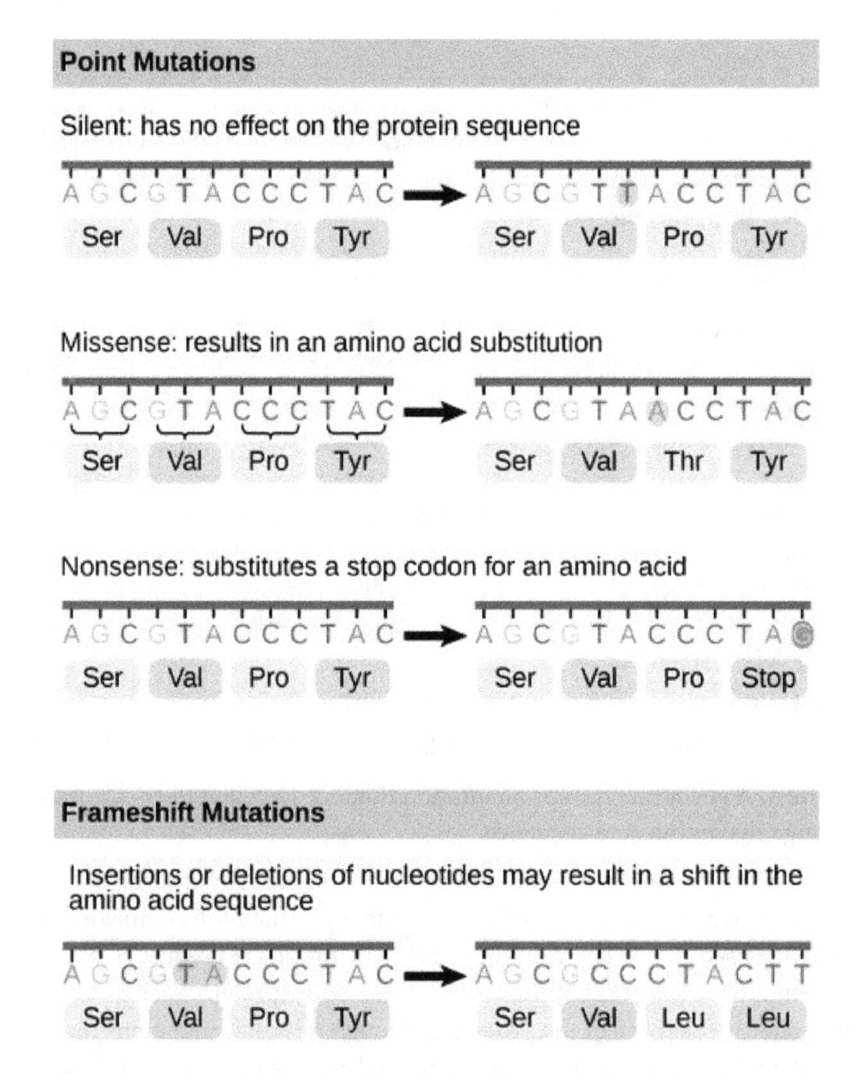

FIGURE 6.17 Point mutations are mutations that arise when a nucleotide base sequence in DNA has been affected and include substitution and frameshift mutations. Substitution mutations occur when a single nucleotide has been substituted for another producing three different types of mutations: silent mutation (does not affect amino acid sequence), missense mutation (affects amino acid sequence), or nonsense mutation (results in a STOP codon). Frameshift mutations usually occur when a nucleotide base sequence has either been inserted or deleted from the overall genetic sequence.

Most mutations are considered harmful, because they generally produce nonfunctional proteins, leading to serious diseases or even death. The nonfunctional protein, more commonly an enzyme, cannot catalyze important reactions, resulting in the buildup of molecules that would have reacted with the enzyme. This accumulation of molecules leads to a specific illness or death. There are, however, some mutations that have little to no detrimental effect. These types of mutations are usually neutral in nature, producing neither a positive nor negative effect. Finally, although rare, some mutations can be beneficial to an organism. If the mutation increases the survival and reproductive output of an organism, the mutation will be favored by natural selection, which is essential to evolution.

6-6. Biotechnology and Genetic Engineering

Our current understanding of DNA structure and how the molecule replicates is important in the field of **biotechnology**. Biotechnology is the process by which biological molecules, cells, and/or organisms are modified in order to achieve a practical benefit. A commonly practiced aspect of biotechnology is **genetic engineering**, a process in which genetic material is directly manipulated by means of adding, deleting, or transplanting genes from one organism to another. This produces genes of interest, which are used for practical benefit. Genetic engineering gives rise to **transgenic organisms**, often called **genetically modified organisms (GMOs)**, such as bacteria, plants and animals, and a variety of biotechnology products. Genetic engineering is important for a variety of reasons, including helping researchers understand how genes work, improving agriculture, and developing better treatments for disease.

In order to produce certain organisms of choice, researchers determine the desirable trait, then manipulate the DNA using five main steps.

1. **Cut.** This process involves a group of enzymes called **restriction enzymes** (Figure 6.18), first discovered by researchers studying bacterial enzymes in the late 1960s, most notably American biochemist Paul Berg. Berg discovered that the enzymes were responsible for protecting bacteria from invasive organisms and viruses by cutting and destroying the DNA of the invaders, thereby restricting their growth. Berg not only observed that the bacterial enzymes could recognize a specific nucleotide sequence but could also cut the DNA straight across the helix at this precise location and remove it without damaging the gene.

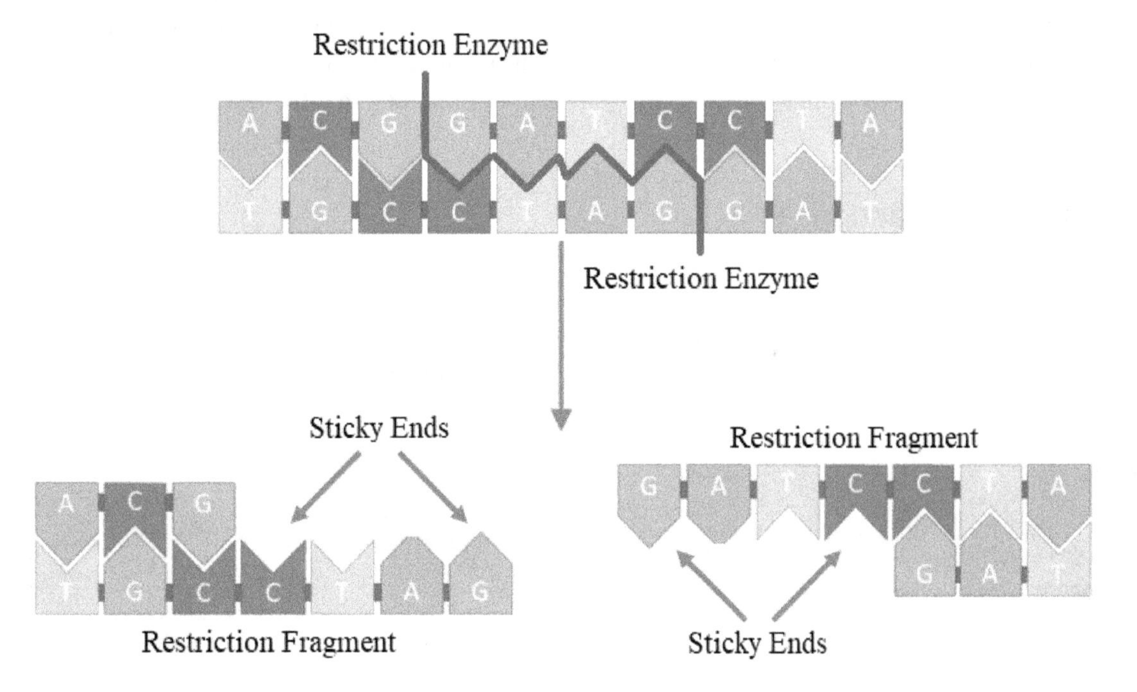

FIGURE 6.18 Restriction enzymes are used to cut out a specific gene of interest along an area known as the restriction site. This results in DNA fragments with "sticky ends," which can recognize each other and complementary base-pair to form double-stranded DNA molecules.

Since this time, hundreds of restriction enzymes have been identified, with each recognizing a short, specific DNA nucleotide sequence, consisting of no more than eight nucleotides, called a **restriction site**. Upon recognizing the restriction site, the enzymes cut across both DNA strands along the sugar-phosphate backbones to produce **restriction fragments** (Figure 6.18). The resulting double-stranded restriction fragments have at least one single-stranded end, known as a "sticky end." Since each restriction enzyme exposes a specific set of restriction fragments, the fragments will recognize each other, then complementary base-pair along the "sticky ends" to form double-stranded DNA molecules.

2. **Amplify.** The next step is amplification, using a process known as the **polymerase chain reaction (PCR)**, developed in 1985 by American biochemist Kary Mullis, who shared a Nobel Prize in Chemistry for this technique in 1993. The polymerase chain reaction is a very specific process in which an unlimited number of copies of a desirable DNA sequence are produced quickly (Figure 6.19). The process begins in a test tube, in which a small amount of DNA is mixed with primers, a supply of free nucleotides and DNA polymerase, the enzyme responsible for carrying out DNA replication. Once loaded into an automated PCR machine, the DNA is run through a series of heating and cooling cycles, which only takes a few minutes. Heating the DNA causes the DNA strands to separate, while cooling enables DNA polymerase to base-pair free nucleotides, forming the new double-stranded DNA molecule of interest. After about 30 or so cycles, the PCR machine can produce billions of copies of the desired DNA sequence from a single molecule of DNA. The available DNA from the PCR is then used for a variety of purposes, including forensics, identifying genes associated with genetic disorders, and making transgenic organisms. The greatest strength of PCR is its ability to amplify specific DNA sequences from the smallest of samples, such as blood droplets. However, it is quite sensitive and can yield false positives resulting from contamination. Contamination can occur from either leftover DNA from a previous analysis, hair, or breathing into the test tube.

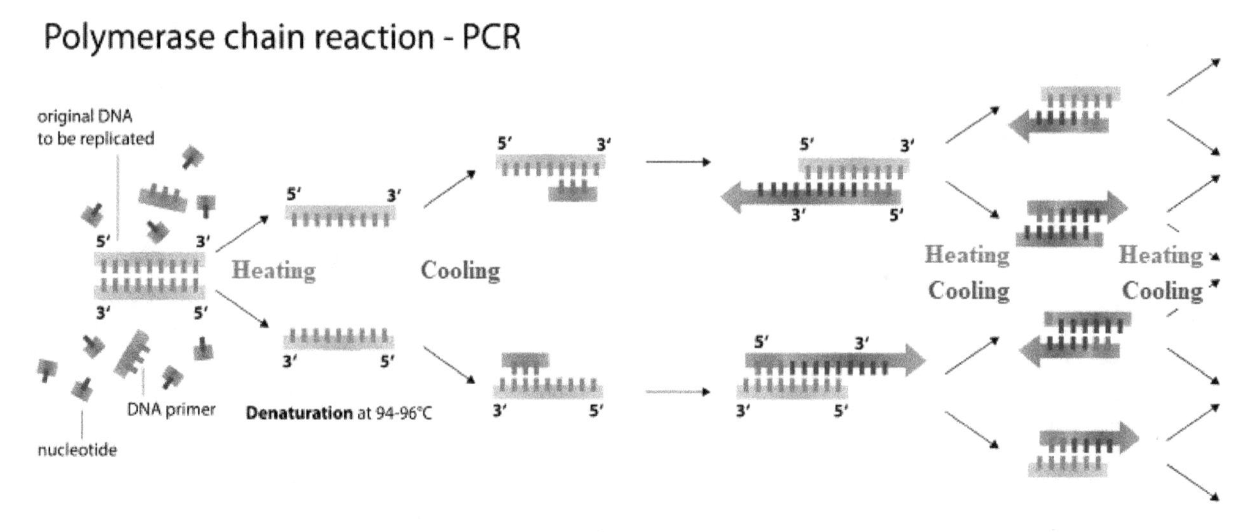

FIGURE 6.19 The polymerase chain reaction is a useful biological tool in genetic engineering because it can produce an unlimited number of identical and desirable DNA sequences in a short period of time. The process requires a combination of DNA primers, DNA polymerase, and a series of heating and cooling stages in order to produce copies of the DNA molecule.

Once the DNA has been amplified, it must be analyzed. The most common analysis technique at one time was exposing an entire genome to restriction enzymes, which produced a large collection of different-sized fragments. The DNA fragments were then separated and could be visualized using a technique called **gel electrophoresis** (Figure 6.20). Gel electrophoresis involves loading DNA fragments into shallow wells within an agarose gel made of carbohydrates from seaweed. The gel is then placed within a chamber containing two electrodes—a positive and a negative—which provides electrical current to help separate the DNA fragments. Since the sugar-phosphate groups are negatively charged molecules, DNA will move toward the positive electrode, with smaller DNA fragments moving much faster than the larger. Upon completion, the DNA fragments have separated by size and appear as distinct bands within the gel.

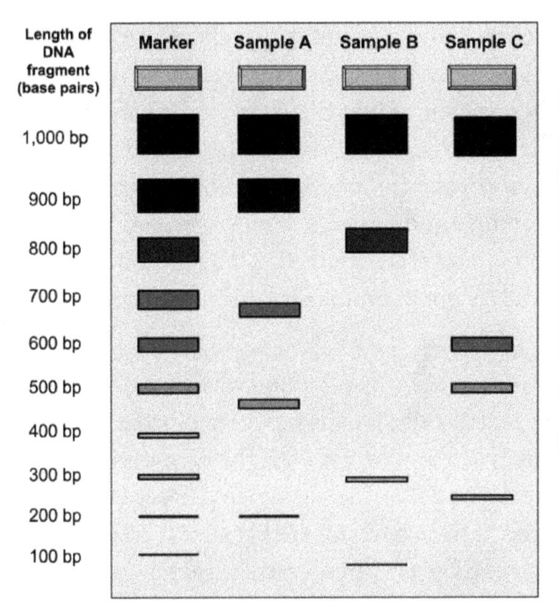

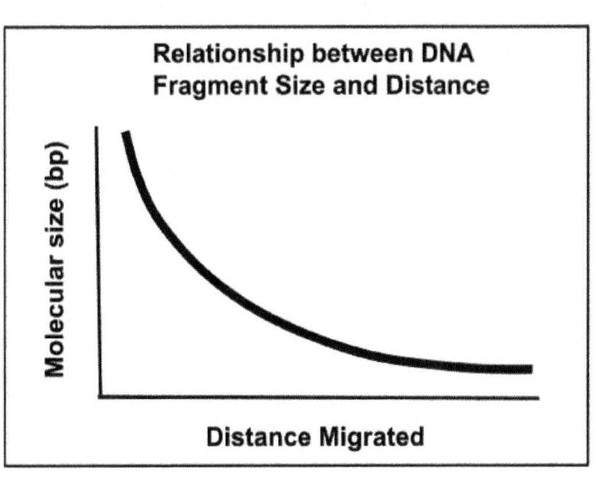

FIGURE 6.20 Gel electrophoresis is a tool used to separate DNA fragments using an electrical current. Since DNA is a negatively charged molecule, the DNA fragments will migrate towards the positively charged electrode. In addition, large DNA fragments move much slower than smaller DNA fragments.

However, the preferred method used today does not involve the use of restriction enzymes, but rather short sequences of DNA bases repeated over and over again, called **short tandem repeats (STRs)**. STRs consist of short sequences of nucleotides, no longer than five total, repeated ten to 15 times, located adjacent to one another. The use of STRs in analyzing DNA has become increasingly useful in helping identify people, especially at crime scenes or after a natural disaster, since each person has different STRs. Since each individual has a different number of repeats, STRs can be used to identify an individual's unique set of genetic markers, called a genetic profile, or DNA fingerprint.

3. **Inserting.** Once the multiple copies of the desired DNA sequence have been produced, the next step is to insert the sequences into a host cell that is able to maintain the DNA as its own. This combination of DNA from two or more sources is known as **recombinant DNA technology (rDNA)** (Figure 6.21), which results in transgenic or genetically modified organisms. In order to introduce the desired DNA sequence into the host cell, the researcher will use a structure called a **vector**. The vector of choice for many researchers is the small circular pieces of DNA found within bacterial cells, known as **plasmids,** structures first discovered by researchers in the 1970s while studying *Escherichia coli* (*E. coli*) and antibiotic resistance. Plasmids contain only a small number of genes, which are not part of the bacterial chromosome but can replicate as separate entities. Since these genes are not part of the bacterial chromosome, some scientists suggest that plasmids may not be necessary for survival or reproduction. The most commonly used plasmids are those found in *E. coli,* because the bacteria grows relatively easily and there is much published about its genetics.

   In order to combine the desired DNA sequence into the plasmid, the plasmid is cut using the same restriction enzyme used to cut the desired DNA sequence, producing sticky ends. This enables the researcher to combine the DNA sequence into the plasmid, because they were cut with the same restriction enzyme. The researcher then uses DNA ligase to seal the DNA sequence into the plasmid, which works by covalently bonding the sugar-phosphate backbones of each DNA strand together. This combination results in a new plasmid with recombinant DNA that can then be inserted into a bacterial host cell. There are times when researchers may use other means to introduce DNA into a multicellular host, including viruses, gametes, or a fertilized ovum.

4. **Grow.** Once the plasmid with recombinant DNA has been inserted into the host bacterial cell, the cell will replicate through a series of repeated cell divisions to form **clones**, cells that are genetically identical. As the bacterial cell divides, the plasmid with the recombinant DNA also replicates, producing multiple copies of the desired DNA sequence in a process known as **gene cloning** (Figure 6.21). Gene cloning is important to researchers because it enables them to make multiple copies of a desired gene, which can be used in medical research or used to genetically modify an organism to produce beneficial proteins.

5. **Identify.** The last step requires identifying the bacterial cells with the desired DNA sequence. One of the ways to locate the DNA sequence involves the addition of a chemical that would break down the double strands of DNA into single-stranded molecules within the bacteria cell. Next, a **DNA probe** is applied, consisting of a short, single-stranded sequence of synthetically produced radioactive-DNA with the same nucleotide sequence as the desired DNA sequence. The DNA probe would radioactively identify the desired DNA sequence within the bacterial cell, causing the sequence to glow.

   DNA probes have also been used to identify specific STRs after gel electrophoresis. These particular probes can be radioactive or have colored molecules attached. Once the gel electrophoresis is finished, the separated DNA fragments are transferred to a piece of nylon paper. The paper is soaked with a solution containing the specific DNA probe, which binds to the STR, making it visible within the gel. This method is common in most forensic applications; however, the use of DNA probes is not necessary, because the STRs are visible after gel electrophoresis, because they were labeled during the PCR reaction.

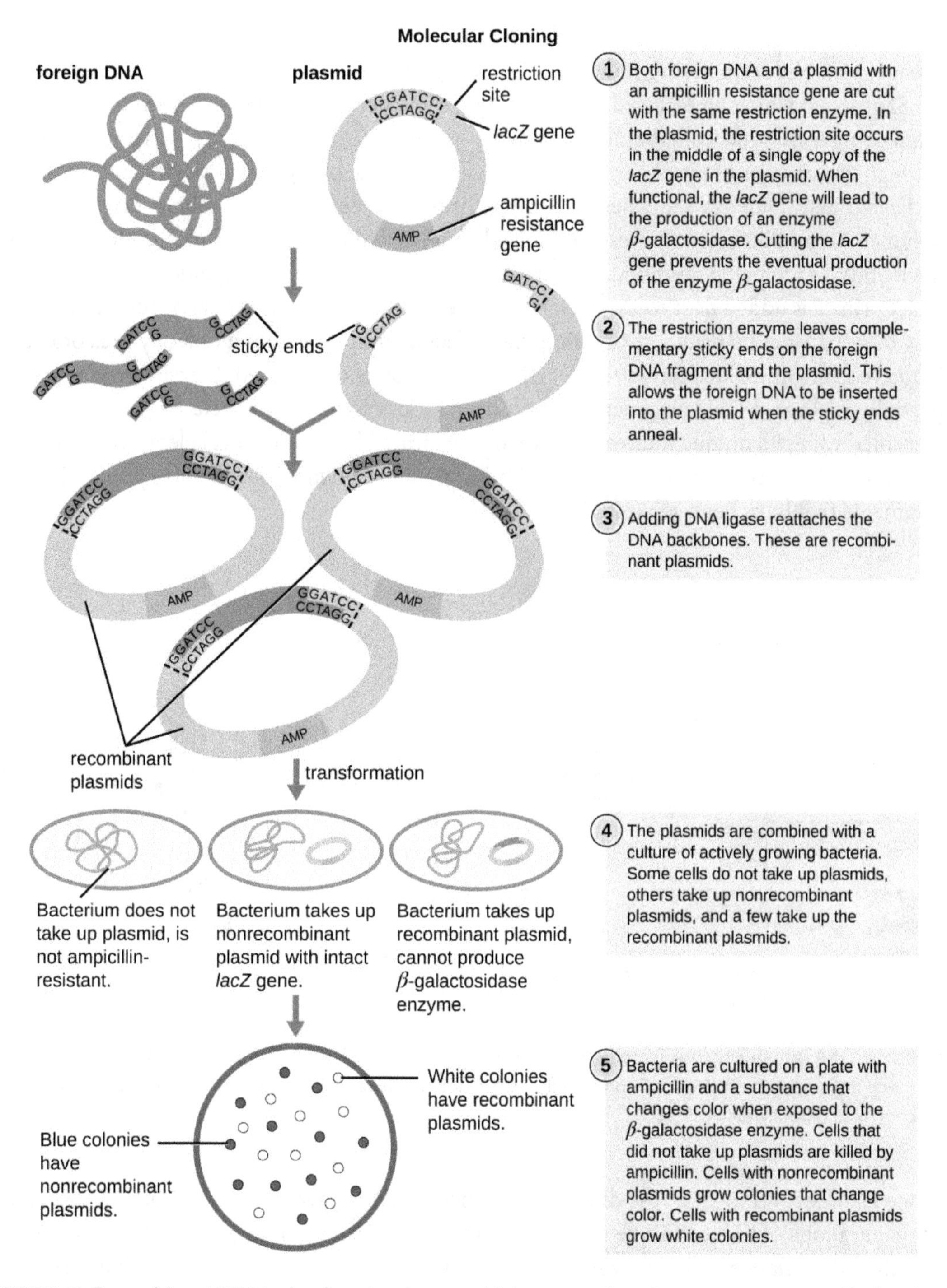

FIGURE 6.21 Recombinant DNA technology involves combining genes from two or more sources. During the insertion step, the desired gene is inserted into a small, circular piece of DNA found in bacterial cells called plasmids. A section of the plasmid is cut out and replaced with the desired gene using DNA ligase, which helps seal the DNA sequence into place. The recombinant plasmid is then reinserted back into the bacterial cell and begins undergoing repeated cell divisions to produce genetically identical cells called clones. This series of cell divisions produces multiple copies of the recombinant plasmid in a process known as gene cloning. Gene cloning is an important step because it helps produce multiple copies of the desired gene.

A more common approach to identifying a cloned gene, proven to be much faster and cheaper, is **DNA sequencing** (Figure 6.22), introduced in 1977 by British biochemist Frederick Sanger, for which he received the Nobel Prize in Chemistry in 1980. The initial DNA sequencing technique, called dideoxy sequencing (or the Sanger method), was a major breakthrough, because its use in determining the gene sequence of interest occurred more rapidly and accurately. Dideoxy sequencing requires three main components: a DNA primer, DNA polymerase, and different fluorescent-labeled chain-termination nucleotides, called dideoxynucleotides, each being modified versions of the original bases. During dideoxy sequencing, DNA primer is added next to the region of interest and DNA polymerase begins complementary base-pairing the primer, extending the molecule until it reaches a dideoxynucleotide. Upon reaching the dideoxynucleotide, DNA elongation is terminated, thereby producing a partially replicated DNA fragment. After termination, each dideoxynucleotide fragment is identified using its unique fluorescent dye and separated by means of gel electrophoresis. The results of gel electrophoresis are analyzed by computer software, producing a pool of DNA fragments that have been separated and measured base by base.

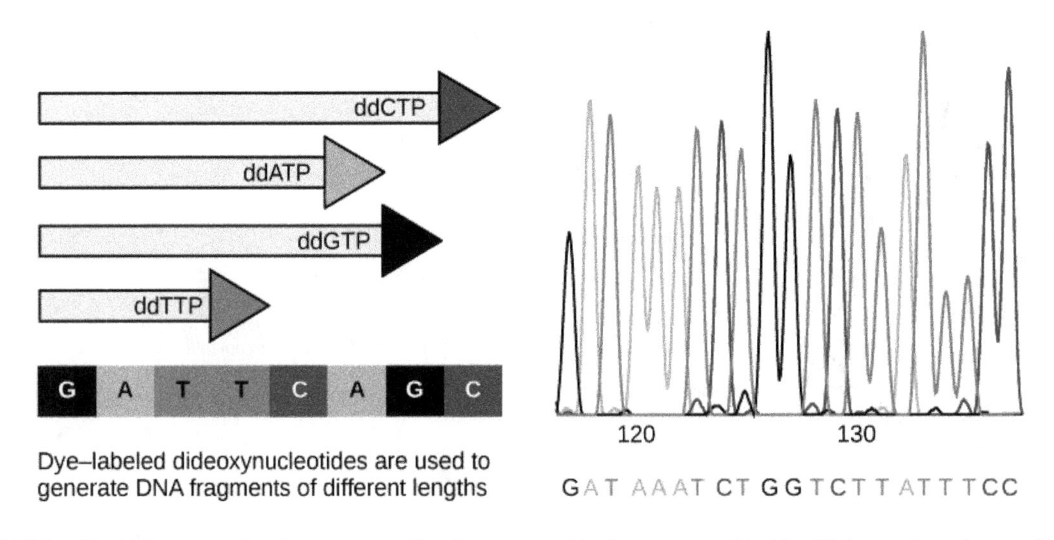

FIGURE 6.22 DNA sequencing has become the cheaper and faster approach to identifying a cloned gene. This process was developed by Frederick Sanger in the 1970s and involves producing partially replicated DNA fragments known as dideoxynucleotides.

Since its induction in 1977, the dideoxy sequencing method was used for about 40 years, until newer generations of the technique were introduced. Ever since, the practice of DNA sequencing has continued to change, drastically improving both the efficiency and cost at which desired genes are identified. The "next generation" technique, a method known as **sequencing by synthesis**, identifies a DNA sequence by synthesizing the complementary strand and electronically monitoring the individual nucleotides used to construct the strand. This technique is slowly being replaced by a "third generation" sequencing technique, a process that requires slowly moving the DNA strand through a small membranous pore to identify each individual base as it passes through an electrical current.

6-7. Modern Uses of Biotechnology

Agricultural Biotechnology

Agricultural biotechnology is the use of genetic engineering to modify plants and animals to increase agricultural productivity. With almost all genetically modified organisms being plants, this is an especially important practice for farmers, because it provides them with the tools they need to increase crop production, while lowering overall cost. The cultivation of genetically engineered crops in the United States began in the 1990s. Currently, there are ten approved GMO crops, including corn, cotton, and soybeans. According to the USDA, approximately 92 percent of corn, 94 percent of cotton, and 94 percent of soybean acres are planted using genetically engineered seeds.

Corn, cotton, and soybeans are some of the most commonly modified crops, in order to improve their resistance to herbicides and insects (Figure 6.23). Herbicides are a group of substances that are used to kill unwanted plants and work by inhibiting enzymes important in synthesizing amino acids. If a plant is unable to synthesize important amino acids, it cannot produce functional proteins, resulting in death. Herbicide-tolerant crops were created by means of recombinant DNA technology in order to tolerate potent herbicides such as glyphosate. Within the genome of the herbicide-tolerant crops, a bacterial gene was inserted that codes for an enzyme capable of functioning in the presence of an herbicide, without affecting the ability to synthesize amino acids and proteins. This enables farmers to kill weeds, the crop's main competitors for important resources, such as water, light, and nutrients, without harming their crops. As of 2019, 94 percent of domestic soybean acres, 98 percent of cotton acres, and 90 percent of corn acres planted were produced using herbicide-tolerant seeds.

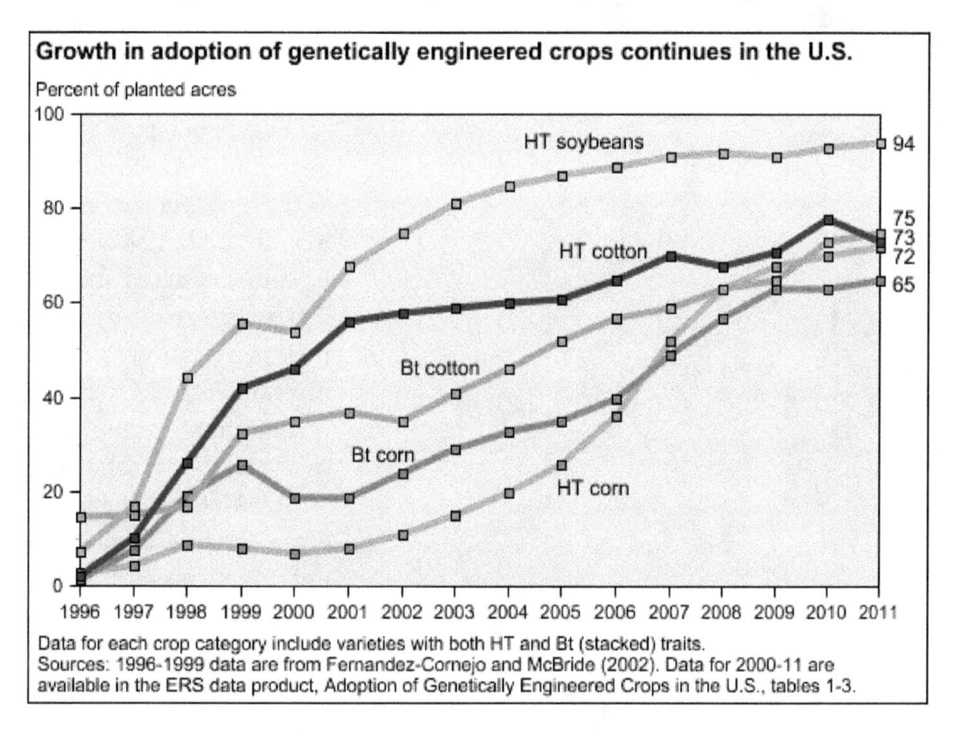

FIGURE 6.23 The graph shows the steady increase in the percent of herbicide-tolerant (HT) and insectide-resistant (*Bt*) crops grown in the United States from 1996 to 2011. As of 2019, the number of HT and *Bt* corn and cotton crops have grown dramatically.

Besides being herbicide-tolerant, many crops are also insect-resistant as well. Many insect-resistant crops have been genetically modified with the *Bt* gene, a gene found in the soil bacterium *Bacillus thuringiensis*. The *Bt* gene encodes for a protein that produces crystal-containing spores, which are poisonous to insects but not plants or humans. The spores damage the digestive system of the insect by binding to the gut wall and creating pores within. This causes paralysis of the gut wall, and the insect stops feeding, resulting in death due to tissue damage and starvation. Genetically engineered *Bt* corn and cotton have been available since 1996, and in 2019, about 83 percent of domestic corn acreage and 92 percent of domestic cotton acreage were planted with genetically engineered, insect-resistant seeds.

Many crops are genetically modified to be both herbicide-tolerant and insect-resistant, known as "stacked" varieties. As of 2019, about 80 percent of corn and 89 percent of cotton acres were planted using "stacked" seeds.

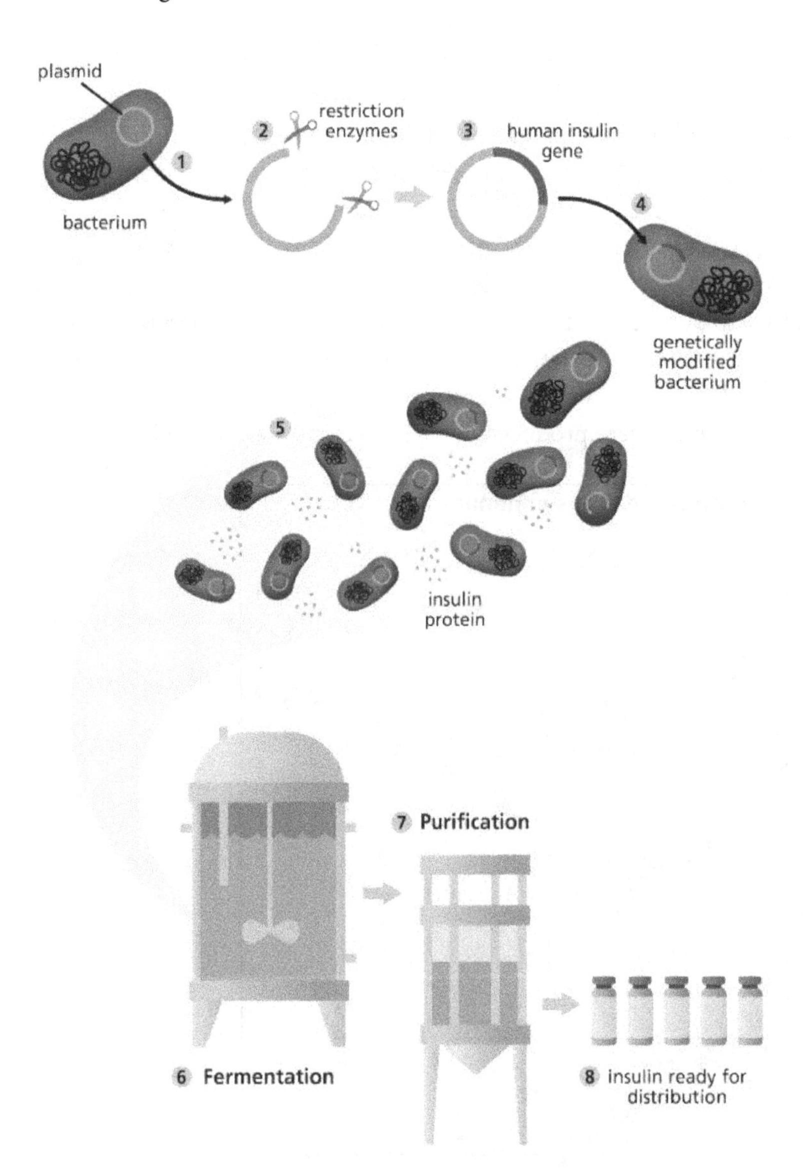

FIGURE 6.24 One of the first genetically engineered proteins was recombinant insulin in 1982.

Medical Biotechnology

Medical biotechnology has become useful in producing pharmaceutical and diagnostic products to help with the treatment and prevention of human diseases. One of the first proteins ever produced using medical biotechnology was recombinant insulin in 1982 (Figure 6.24). Prior to the development of recombinant insulin, diabetics were prescribed insulin extracted from cattle and pigs, which resulted in allergic reactions in a relatively small percentage of recipients. Diabetics receiving recombinant insulin did not experience allergic reactions.

A variety of genetically modified organisms have been used to produce important vaccines, antibodies, and proteins. Plants have been genetically modified in order to produce important vaccines and antibodies. For example, vaccines against disease-causing viruses and bacteria resulting in hepatitis B, rabies, and diarrhea have been produced in genetically modified plants. In addition, plants have been used to produce human antibodies against bacteria-causing diseases, such as tooth decay, and in cancer treatments, as seen in clinical trials for non-Hodgkin's lymphoma. Animals have also been genetically modified to help produce important medicines, antibodies, and essential proteins. For example, sheep have been genetically modified to produce milk with important proteins helpful in the treatment of cystic fibrosis. Other important proteins, such as erythropoietin (responsible for the production of red blood cells) and clotting factors, have been produced in the milk of genetically modified sheep. Transgenic bacteria are also important contributors to the production of important therapeutic proteins. The most commonly bacterial-produced proteins include human growth hormone, insulin, and clotting factors.

Medical biotechnology is also an important diagnostic tool for inherited disorders. With the use of medical biotechnology, some of the most common inherited disorders, such as sickle-cell anemia and cystic fibrosis, can be detected in parents who may be carriers of a particular disorder prior to conception or diagnosed shortly after conception within the embryo. In the case of an individual being born with a particular disorder, **gene therapy** is a form of medical biotechnology that can be used to treat or cure the disorder. Gene therapy involves isolating a defective gene within a cell and replacing it with a functional gene. An important aspect of gene therapy is ensuring that the gene is continually being expressed within the cell, without action from the immune system. Gene therapy can be used to treat a variety of diseases and disorders, from cardiovascular disease to cancer.

CRISPR-*Cas9*

In 2012, American biochemist Jennifer Doudna (Figure 6.25, middle) and French biochemist Emmanuelle Charpentier (Figure 6.25, right) were the first to propose a new genome editing process that could be used by molecular biologists to alter the genetic material of cells or organisms to study gene function or correct genetic mutations that lead to disease. This genome editing process is now

FIGURE 6.25 Francisco Mojica (left) is considered one of the first scientists to discover genome editing associated with CRISPR-*Cas9*, while Jennifer Doudna (middle) and Emmanuelle Charpentier (right) were responsible for implementing its use as a form of biotechnology to study gene function and altering genomes that cause disease.

referred to as CRISPR-*Cas9*, and for their accomplishments, both women received the Nobel Prize in Chemistry in 2020. However, the discovery of genome editing employed by CRISPR-*Cas9* was first reported in microbial gene sequence studies led by Spanish molecular microbiologist Francisco Mojica (Figure 6.25, left) in the early 1990s.

In 1993, a group of scientists led by Mojica were studying *Haloferax mediterranei*, an archaeal microbe isolated from the salt marshes of Santa Pola, a Spanish coastal town located near the Mediterranean Sea. The team discovered the microbe possessed multiple copies of unusual, but nearly perfect palindromic (reads same forwards and backwards) sequences of functional DNA consisting of thirty nitrogenous bases long, which repeated roughly every thirty-six bases. For the next decade, Mojica's investigative studies into other halophilic archaea, such as *Haloferax volcanii*, and different bacterial strains like *Salmonella typhi* and *Clostridium difficile*, revealed similar repetitive sequences. By 2000, Mojica had proposed these numerous repetitive palindromic sequences be called *clustered regularly interspaced palindromic repeats* or **CRISPR**, and within two years, nearly forty microbes with palindromic repeats were discovered and catalogued.

The purpose of microbes possessing CRISPR intrigued Mojica; therefore, he diverted his investigative studies into the thirty-six bases that separated the repetitive sequences, known as *spacers*. In 2003, Mojica discovered a spacer sequence from *Escherichia coli* matched that of a viral DNA sequence from a bacteriophage (virus that infects bacteria) commonly known to infect the bacterium. This finding perpetuated subsequent investigations into the spacers, revealing sequential DNA matches with bacteriophages and plasmids received from another bacteria during conjugation. Thus, Mojica deduced that microbial CRISPR must contain instructions responsible for initiating an immune response in the presence of genetic invaders to protect them from specific infections.

Today, the use of CRISPR-*Cas9* as a genome editing process as proposed by Doudna and Charpentier, involves disabling a specific gene to study how it works (Figure 6.26). This is accomplished by using a bacterial protein called *Cas9*, produced by the bacteria *Streptococcus pyogenes*, which uses the protein to defend itself by eliminating foreign DNA from an invading bacteriophage or parasitic plasmid DNA. *Cas9* is similar to a restriction enzyme because it can cut double-stranded DNA; however, it differs in that it does not recognize a particular sequence but is rather directed to a specific sequence. *Cas9* binds to a sequence of RNA called "guide RNA" (gRNA) produced from the CRISPR region of the bacterial genome. Upon binding, the *Cas9* protein and gRNA form a complex, which is responsible for cutting the desired sequence of DNA that complements the gRNA sequence. *Cas9* cuts the targeted DNA strands, initiating the DNA repair system. But with no template for the repair enzymes to use, the enzymes instead introduce or remove nucleotides, rejoining both ends. This process results in an altered DNA sequence and a gene that no longer works, allowing researchers to study the gene.

The CRISPR-*Cas9* system has been recently modified to treat genetic disorders resulting from a defective gene. Researchers combine a functional gene with the *Cas9* protein, and after the protein cuts out the defective gene, repair enzymes begin repairing and editing the DNA strand, using the functional gene as its template. An example of how this technology has been used to treat genetic disorders occurred in 2018 to correct the defective genes causing sickle-cell anemia. Genes were corrected in the human cells of sickle-cell patients and injected into mice, and after 19 weeks, the gene remained corrected in about 20 to 40 percent of the injected cells. The success

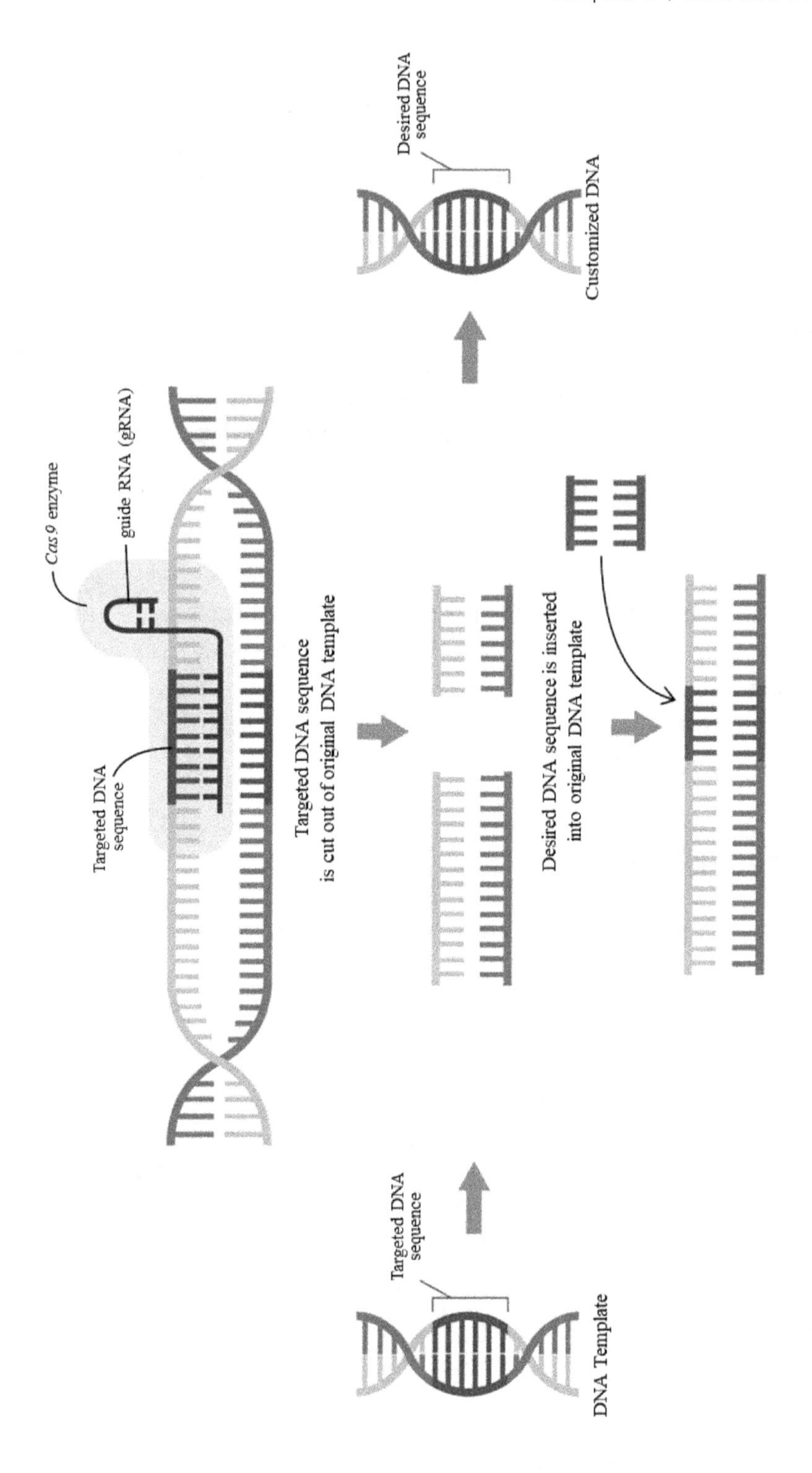

FIGURE 6.26 CRISPR-*Cas9* is a genome editing process that has given molecular biologists the ability to better study the structure and function of genes. The process involves altering or editing the genes of different cells and organisms, including repairing defective genes in humans such as the gene that causes sickle-cell anemia. The use of the technique in treating other genetic disorders and altering the genes of disease-carrying insects seems promising.

of CRISPR-*Cas9* has the scientific community and the public hopeful that it can be used to treat other genetic disorders, such as Parkinson's, Alzheimer's, and cancer. CRISPR-*Cas9* has also been used to alter genes in disease-carrying insects. Despite the current successes, some scientists have concerns about the effects on genes that are not targeted and the misapplication of the technology beyond its primary purpose.

Ethical Issues

The production and consumption of genetically modified organisms or products has raised some important questions and concerns among the general public. One of the most common questions is whether genetically modified organisms or products are hazardous to humans. According to studies performed by the United States Society of Toxicology, transgenic plants show no potential risks to human health. However, other studies have shown that some transgenic organisms can cause allergies, which in some cases can result in death. Other concerns about genetically modified organisms or products lie within the health of the environment. There are some who worry that important ecological organisms, such as the Monarch butterfly (*Danaus plexippus*), will be killed inadvertently. Others suggest that certain insects may develop resistance to the pesticide-resistant crops and will no longer be killed, increasing the loss of genetic diversity due to becoming more vulnerable to environmental changes or pests. Finally, genetically modified crops are not tested or regulated as adequately as some individuals would like, in addition to hidden costs that may reduce the financial advantages of genetically modified crops.

Chapter Summary

Below is a summary of the main ideas from each section of the chapter.

6-1. Discovery of the Structure of DNA

- There were several scientists responsible for the discovery of the structure of DNA and its components. Albert Kossel isolated nucleic acid and identified the five nitrogenous bases: adenine, guanine, cytosine, thymine, and uracil. Phoebus Levene isolated and identified the basic components of DNA, including its nitrogenous bases, adenine, guanine, cytosine, and thymine; the sugar deoxyribose; and the phosphate group, which are all linked together to form nucleotides. Nikolai Koltsov proposed the idea that the inheritance of traits occurred by means of a large molecule that replicates in a semiconservative fashion, with each strand acting as a template. Frederick Griffith demonstrated that an unidentified transforming factor was responsible for the transfer of traits in two forms of *Pneumococcus* bacteria. Jean Brachet discovered that DNA was located within the chromosomes of cells. William Astbury produced the first X-ray diffraction pattern of the DNA molecule, which he suggested could be used to easily deduce its structure. Oswald Avery, Colin MacLeod, and Maclyn McCarty discovered that DNA was the transforming factor resulting in the transfer of traits in the *Pneumococcus* bacteria in Frederick Griffith's experiment. Alfred Hershey and Martha Chase concluded from their bacteriophage experiments that DNA was the genetic material.

6-2. DNA Structure and Function

- Deoxyribonucleic acid (DNA) is composed of a five-carbon sugar, deoxyribose; a phosphate group; and nitrogenous bases.
- DNA is a double helix molecule containing two alternating, covalently bonded sugar and phosphate molecules that form the DNA backbone, which helps support the molecule. The DNA backbone structure was deduced by Watson and Crick in 1953, which was allegedly based on mathematical data collected by Rosalind Franklin.
- The DNA strands exhibit antiparallel arrangement, resulting from one strand ending in a 3' nonbonded sugar and the other in a 5' nonbonded phosphate, running in opposite directions. The numbers reference the carbon to which each molecule is bonded. One strand runs in a 5' to 3' direction, and the other in a 3' to 5' direction. The orientation of the DNA nitrogenous bases is dependent on the antiparallel arrangement and was first suggested by Rosalind Franklin.
- The nitrogenous bases of DNA are located in the center of the DNA backbones and carry the genetic sequence responsible for helping build a living organism. Two types of nitrogenous bases exist, including purines, double-ring structures consisting of adenine and guanine; and pyrimidines, single-ring structures consisting of thymine and cytosine. Hydrogen bonds are responsible for complementary base-pairing one purine with one pyrimidine. Three hydrogen bonds occur between adenine and thymine. Two hydrogen bonds occur between guanine and cytosine. Complementary base pairing is important in helping determine gene function. Erwin Chargaff introduced Chargaff's rule, which states that the number of adenine equals that of thymine, while the number of guanine equals that of cytosine
- A gene is a unique sequence of up to 3,000 DNA bases and outlines the code for synthesizing proteins. Genes are located on chromosomes, first discovered by Thomas Hunt Morgan in 1910.
- Chromosomes compose a genome, which consists of an organism's entire set of DNA. Bacterial genomes consist of a single, circular piece of DNA attached to the plasma membrane called a nucleoid. Eukaryotic cell genomes are found within the nucleus, wrapped around histone proteins. Histone proteins keep DNA from tangling and enable it to be tightly packed with the nucleus. The human genome contains 23 pairs of chromosomes, with over three million base pairs. An insect-dwelling bacteria, *Nasuia deltocephalinicola*, has the smallest genome, at about 112,091 base pairs. Other organismal genomes, such as dogs, have 39 pairs of chromosomes, and fruit flies have four pairs. The length of the DNA molecule is determined by the number of base pairs in an organism's genome.
- Noncoding DNA does not code for proteins and may exist as gene fragments, duplicate genes, or mutated genes. About 25 percent of noncoding DNA is found within genes, and these are called introns. Introns were discovered in 1977, but the term was not used until 1978. There are between 65 and 100,000 DNA bases found in an intron. The remaining 75 percent of noncoding DNA is found between genes. The function of introns is not fully known; however, some scientists suggest they may be involved in the regulation of mRNA formation during translation, crossing over during meiosis, or switching genes on or off, which influences protein synthesis

- DNA that codes for proteins is found within a gene, and these are called exons. Exons contain between 100 and 300 bases.

6-3. DNA Replication

- Upon discovering DNA in 1953, Watson and Crick proposed a replication mechanism based on the specific structural arrangement of the molecule.
- Cell division is an important process that a cell undergoes. Prior to division, a cell must synthesize additional cellular materials and nucleotides and generate energy before DNA replication can occur. Cellular division must proceed when a cell begins replicating its DNA. Cellular division produces two daughter cells, with each receiving the same amount of genetic material
- Three models describe the process of DNA replication. The semiconservative model states that the replicated DNA strand contains one parental strand and one newly formed strand. The semiconservative model proposed by Watson and Crick best describes DNA replication. The conservative model states that the parental DNA strand was preserved, coming back together once replication ends. The dispersive model states that each DNA strands consists of a mixture of both parental and newly formed DNA. In experiments performed by Matthew Meselson and Franklin Stahl in 1958, the semiconservative model was supported, while the conservative and dispersive models were disproved.
- DNA replication requires replication enzymes, enzymes that ensure the accuracy of the process.
- The origin of replication is a short, specific region of DNA and indicates the starting point for DNA replication.
- Helicase recognizes and attaches to the origin of replication and begins unwinding and separating the DNA molecule into two parental strands by breaking the hydrogen bonds holding together the nitrogenous bases. Each parental DNA strands acts as a template where DNA replication can occur, called the replication fork. Single-strand binding proteins attach to each parental strand, helping keep each strand separated.
- RNA primer is a short sequence of nucleotides produced by the enzyme DNA primase. The RNA primer attracts DNA polymerase, an enzyme responsible for complementary base-pairing the correct nitrogenous base to the exposed base on the template strand, using the base-pairing rules A–T and G–C.
- DNA replication occurs in an antiparallel direction along the replication fork, due to its antiparallel arrangement. The new strand created by complementary base pairing in the 5' to 3' direction is called the leading strand. The synthesis of the lagging strand along the other template strand is created discontinuously in a series of fragments called Okazaki fragments. Okazaki fragments contain between 100 and 2,000 nucleotides, depending on the cell type and are joined by DNA ligase.
- During the course of DNA replication, each new strand's complementary base pairs are proofread by different types of DNA polymerase. Upon finding an error, nuclease cuts out the erroneous DNA fragment from the new strand, while DNA polymerase and DNA ligase are used to fix the DNA strand.

- The elongation rate of a new DNA strand in human cells is about 50 nucleotides added per second. The leading and lagging strands are replicated in both directions and at the same rate, until the entire original DNA molecule has been replicated. DNA replication ends when DNA polymerase removes the RNA primer and adds the correct DNA nucleotides to the RNA primer region.
- Telomeres are short sequences of repeated nucleotides that lack genes and cannot code for proteins. Telomeres have a function in protecting genes from losing DNA during each replication and in limiting the number of cellular divisions.

6-4. Gene Expression and Protein Synthesis

- The sequence of nitrogenous bases not only provides the genetic instructions for building an organism but is also responsible for protein synthesis, or gene expression. Protein synthesis occurs in ribosomes, which are located within the cytoplasm of a cell. The sequence of nitrogenous bases only provides the code for protein synthesis but cannot initiate it.
- Ribonucleic acid (RNA) is the molecule that helps decipher the code outlined by the DNA sequence of nitrogenous bases. RNA's role in deciphering the DNA sequence code was first described in the sequence hypothesis, introduced by Francis Crick in 1958. The sequence hypothesis states that a piece of nucleic acid is expressed by the bases, and the sequence codes for the amino acids in a protein. The central dogma of molecular biology explains the relationship between DNA, RNA, and proteins. There are several differences between RNA and DNA. RNA exists as a single strand of nucleotides, contains the sugar ribose, the nitrogenous base thymine is replaced with uracil, and RNA can function as an enzyme. There are three different types of RNA, which are responsible for converting the DNA base sequence into proteins. Messenger RNA (mRNA) deciphers the DNA code used to make a protein and carries the message from the nucleus to the ribosomes. Ribosomal RNA (rRNA) is a major component of a ribosome, making up nearly 60 percent. Transfer RNA (tRNA) translates the mRNA code and carries a particular amino acid to the ribosome to synthesize the protein.
- There are 20 different amino acids that can be used to build a protein, and which amino acid is specified by the DNA base sequences is determined by the genetic code. The genetic code is a tool consisting of 64 three-base sequences called codons that each specify a particular amino acid. In 1961, Francis Crick was the first to introduce the idea that the genetic code consisted of triplets that could code for an amino acid. During an experiment called "poly-U," two researchers, Marshall Nirenberg and J. Heinrich Matthaei, were the first to demonstrate that codons could specify amino acids, when they observed that the codon UUU specified phenylalanine. By 1966, Marshall Nirenberg and his team had revealed all the codons responsible for coding the 20 amino acids. Only 61 of the codons specify for amino acids, while the remaining three codons are called STOP codons, which terminate the process of protein synthesis. Several words are used to describe the genetic code, including degenerate (refers to multiple amino acids being specified by more than one codon), universal (refers to all living organisms using the same mRNA codons to specify amino acids), and unambiguous (refers to each codon specifying only one amino acid).

- Protein synthesis begins with transcription, which is a three-step process by which the DNA base sequences of a gene are transcribed into an mRNA strand. Each step of transcription is associated with a particular part of a gene. The promoter is a nontranscribed region of DNA, containing two short sequences of DNA bases, which initiates the binding of RNA polymerase to begin transcription and indicates the direction and which DNA strand is undergoing transcription. The gene body is the region of a gene where elongation occurs. The end of the gene contains a specific sequence of DNA nucleotides that terminates transcription. There are three steps of transcription. 1. Initiation begins when a complex of transcription factors and RNA polymerase recognizes and binds to the gene promoter, initiating transcription. RNA polymerase unwinds the DNA molecule and begins breaking the hydrogen bonds between the nitrogenous bases, producing two separate DNA strands. The template strand contains the necessary information required to be transcribed and can produce hundreds of mRNA molecules, depending on the amount of protein required. The nontemplate strand is not transcribed but may be a template for another gene. 2. Elongation occurs as RNA polymerase travels down the template strand in the 5' to 3' direction, complementary base-pairing the correct RNA nucleotide with the exposed DNA nucleotide. Upon reaching an adenine nucleotide, RNA polymerase pairs the nitrogenous base uracil. After the mRNA strand reaches about ten nucleotides in length, the strand separates from the DNA template strand, with hydrogen bonds being able to reform between the DNA nucleotides, and the DNA molecule begins rewinding. 3. Termination of transcription occurs when a termination signal on the DNA template strand is reached by RNA polymerase. The newly formed mRNA strand undergoes two types of modifications. Capping involves adding a modified guanine cap on the 5' end of the mRNA strand and a sequence of adenine nucleotides on the 3' end forming a tail. Capping is important because it helps protect and transport the mRNA, while also allowing it to correctly align and bind within the ribosome. Editing is a process that involves removing the noncoding introns and combining the protein-coding exons. Once mRNA is edited, it travels out of the nucleus through the nuclear pores into the cytoplasm. The mRNA strand will remain in the cytoplasm for a few minutes to a few days prior to binding to the ribosome
- Protein synthesis ends with translation, which is also a three-step process wherein the mRNA strand is translated into an amino acid sequence of a protein. Translation requires transfer RNA (tRNA), a molecule responsible for reading the mRNA code and delivering the appropriate amino acid to the ribosome. tRNA is a three-dimensional molecule shaped similar to a clover leaf, containing a sequence of three bases that complements the mRNA codon, called an anticodon. The anticodon deciphers the mRNA codon, initiating tRNA to deliver the correct amino acid, attaching it to the mRNA using ATP. There are about 40 different tRNA molecules, with at least one for each of the 20 amino acids, suggesting that some tRNA molecules can recognize multiple codons. This phenomenon, called the wobble effect and observed by Francis Crick in 1966, is important because it ensures the synthesis of a functional protein. There are three steps of translation. 1. Initiation begins with the start codon AUG, which codes for the amino acid methionine. Once translation is initiated, an initiation complex forms that binds to the start codon, consisting of a small ribosomal subunit with mRNA binding sites and an initiator tRNA molecule. The initiation complex properly aligns the ribosome

and mRNA strand. 2. Elongation involves building the amino acid sequence of the eventual protein. The ribosome travels down the mRNA strand in the 5' to 3' direction, translating each mRNA codon. As tRNA recognizes the codon, it delivers the appropriate amino acid, releasing it and moving to the next codon, then the next. Once an amino acid is released, it bonds to the preceding amino acid by way of peptide bonds, which helps maintain the amino acid chain. 3. Termination occurs when the ribosome reaches a stop codon—UGA, UAG, or UAA—on the mRNA strand, thereby terminating protein synthesis. Releasing factors bind to the ribosome, and the amino acid sequence and mRNA strand are released. The amino acid sequence begins to fold into a functional, three-dimensional protein structure with the help of chaperone proteins after being released. Chaperone proteins influence correct molecular interactions and stabilize the folding of the protein's final structure. The protein may require additional modifications, such as amino acid rearrangement or attachment to other types of molecules.

6-5. Mutations

- A mutation is an alteration in DNA or RNA base pair sequences that can affect the overall structure and function of proteins.
- Mutations can occur spontaneously. Spontaneous mutations are the most common type of mutation. Spontaneous mutations arise due to accidental errors that can occur during DNA replication, resulting in substituted, inserted, or deleted base nucleotides. During and after DNA replication, DNA polymerase proofreads the newly formed DNA strands, and upon its finding an error, DNA repair enzymes cut out and repair the incorrect sequence or mistake. This helps minimize the possibility of spontaneous mutations occurring, making these types of mutations rare.
- Induced mutations occur as a result of toxic environmental factors called mutagens. There are two types of mutagens that can give rise to induced mutations. Chemical mutagens, such as cigarette smoke, interfere with the atoms of the DNA molecule, affecting its ability to replicate. Cigarette smoke contributes to 33 percent of all cancer deaths, including lung cancer, which is the most common and most lethal in the United States. Radiation mutagens include ionizing and nonionizing radiation. Ionizing radiation includes X-rays and gamma rays. The use of X-rays and how it contributes to mutations was first observed in the 1920s by Hermann Muller. Exposure to ionizing radiation can affect the atomic structure of DNA or break apart chromosomes. Nonionizing radiation includes ultraviolet (UV) radiation. Nonionizing radiation affects DNA structure in two ways. When the pyrimidines, thymine and cytosine, absorb nonionizing radiation, it results in bond rearrangement, which can prevent complementary base pairing. Kinks can form within the DNA structure when adjacent thymine bases covalently link, producing thymine dimers, which affects DNA replication. Thymine dimers are removed by DNA repair enzymes.
- There are two types of mutations that can occur as a result of exposure to mutagens. Large-scale mutations are called chromosomal aberrations, which affect the overall organization of a gene. There are four different types of chromosomal aberrations. Inversion occurs when DNA is removed, inverted 180°, and reinserted into the chromosome (hemophilia results from

inversion). Translocation involves the removal and relocation of a DNA segment either into the original chromosome or to another chromosome (down syndrome is a result of translocation). Deletion occurs when a segment of DNA is removed or lost from a chromosome (Williams syndrome is an example of a disease that can result from deletion). Duplication occurs when a DNA segment is duplicated either on the original chromosome or a different chromosome (Charcot-Marie-Tooth (CMT) is an example of a disorder that results from duplication). Small-scale mutations are called point mutations and result when one or more nucleotide base sequences have been changed, affecting the overall amino acid sequence. There are two different types of point mutations. Substitution mutations occur when a single nucleotide base has been substituted for another. Three different types of mutations arise from substitution mutations, including silent mutations (does not affect the amino acid sequence), missense mutations (the amino acid sequence is affected, producing a nonfunctional protein), and a nonsense mutation (substitution results in a stop codon, terminating translation, producing a nonfunctional protein). Frameshift mutations occur when DNA nucleotides are either inserted or deleted, shifting the mRNA codons into a new sequence that produces nonfunctional proteins. Frameshift mutations are more harmful than substitution, because they affect the overall sequence of amino acids.

- Mutations are typically harmful, producing nonfunctional proteins that can lead to serious diseases or death. The nonfunctional proteins produced from mutations are usually enzymes that cannot help catalyze important chemical reactions. The inability of these enzymes to catalyze chemical reactions results in a buildup of molecules that rely on functional enzymes, leading to illness or death. However, some mutations have little to no detrimental effect, while others are beneficial to organisms and are favored by natural selection.

6-6. Biotechnology and Genetic Engineering

- Biotechnology is a field of study that focuses on modifying biological molecules, cells, and/or organisms for a practical benefit. An example of biotechnology is genetic engineering. Genetic engineering involves producing genes of interest by means of manipulating an organism's genetic material through the addition, insertion, or deletion of genes. Organisms produced through genetic engineering are called transgenic or genetically modified organisms. Genetic engineering is an important process that can help researchers understand how genes work, improve agriculture, and develop better treatments for disease.
- Manipulation of DNA involves five main steps.
    - The first step of DNA manipulation is cutting, which requires the use of restriction enzymes, discovered by researchers in the 1960s. Paul Berg is the most notable, discovering that the enzymes were used by bacteria to protect themselves from invading organisms and viruses. Paul Berg observed that the enzymes would cut and destroy the invader's DNA, restricting growth, and could also recognize a specific DNA sequence and cut across the DNA molecule at this location without damaging the gene. Restriction enzymes can recognize a short, specific sequence of DNA called a restriction site. At the restriction site, the

enzymes cut the DNA strand to produce restriction fragments. Restriction fragments have one single-stranded end called a "sticky end." Restriction enzymes produce a specific set of restriction fragments that can recognize each other, which are able to complementary base-pair along the "sticky ends" to form a double-stranded DNA molecule.

- The next step is amplification and uses a process known as the polymerase chain reaction (PCR). PCR was developed in 1985 by Kary Mullis. PCR is a process that can quickly produce an unlimited number of copies of the desirable DNA sequence. The process of PCR begins in a test tube, where small amounts of DNA are mixed with DNA primers and DNA polymerase. The DNA is placed within the PCR machine and exposed to numerous heating and cooling cycles. Heating causes the DNA to separate, while cooling allows DNA polymerase to base-pair free nucleotides to form new strands of the desirable DNA. After 30 minutes, billions of copies are produced from a single DNA molecule. The resulting DNA strands can be used in forensics, identifying genetic disorders, and making transgenic organisms. Although PCR is able to produce specific sequences from small samples, it is also quite sensitive to contamination, which can produce false positives. Once amplified, the DNA must be analyzed. At one time, restriction enzymes were used to produce different-sized DNA fragments that were separated using gel electrophoresis. Gel electrophoresis utilizes an agarose gel, consisting of shallow wells where the DNA fragments are loaded. The gel is then placed within a chamber with a positive and negative electrode, which provides electrical current. The DNA fragments separate as they move toward the positive electrode (DNA is a negatively charged molecule), with the smaller fragments moving faster than the large. The DNA fragments appear as bands within the gel upon completion. More recently, short sequences of no more than five DNA bases repeated ten to 15 times, called short tandem repeats (STRs), have become the preferred method of DNA analysis. STRs have been useful to identify people at crime scenes or after natural disasters, because each person has a different number of repeats. These different numbers of repeats are used to identify a person's genetic profile, or DNA fingerprint.
- After the DNA has been amplified and analyzed, the next step is inserting the DNA into a host cell, known as recombinant DNA technology (rDNA). The introduction of the desired DNA sequence into the host cell requires a vector, which is typically small circular pieces of DNA called plasmids obtained from bacterial cells. Plasmids were first discovered by researchers studying antibiotic resistance in *E. coli* in the 1970s. Plasmids contain a small number of genes that are not part of the bacterial chromosome. The most common plasmids used are those extracted from *E. coli*, because they are easily grown and multiple publications exist about their genetics. The bacterial plasmid is cut using restriction enzymes, and the desired DNA sequence is inserted and sealed with DNA ligase. The new plasmid with recombinant DNA is inserted into a bacterial host cell, although multicellular hosts, including viruses, gametes, or a fertilized ovum, may be used.
- Growing the bacterial cell containing the new plasmid with recombinant DNA is the next step. The bacterial cell undergoes a series of repeated cell divisions to produce clones,

which are cells that are genetically identical. Gene cloning is the process in which the bacterial cell divides and the recombinant DNA replicates to produce millions of copies of the desired DNA sequence. Gene cloning is important because it can produce several copies of a desired gene that can be used in a variety of ways.

- The final step is identifying the desired DNA sequence. One way in which the desired DNA sequence is identified is through the use of a radioactive DNA probe. A DNA probe is a short, single-stranded sequence of radioactive nucleotides that contains the same DNA sequence as the desired DNA strand. Upon identifying the desired DNA sequence in the bacterial host cell, the sequence glows. DNA probes have also been used to identify STRs after they have been separated by gel electrophoresis, although this is not typically necessary, because the STRs were labeled during PCR. DNA sequencing is another way to identify a DNA sequence of interest. This method has become more common, faster, and cheaper. DNA sequencing was introduced in 1977 by Frederick Sanger, who won the 1980 Nobel Prize in Chemistry. Dideoxy sequencing (the Sanger method) uses DNA primer, DNA polymerase, and chain-terminating dideoxynucleotides in order to identify the DNA sequence. Dideoxynucleotides are modified versions of the four basic DNA nucleotides, which are fluorescently labeled and responsible for terminating DNA elongation. Dideoxy sequencing involves adding DNA primer to the region of interest, which begins to extend as DNA polymerase complementary base-pairs the primer until the enzyme reaches a dideoxynucleotide. The resulting DNA fragments are identified by their fluorescent dye, separated by gel electrophoresis, and analyzed by computer software to produce a pool of DNA fragments. Newer generations of DNA sequencing are beginning to be developed, improving efficiency and cost. "Next-generation" sequencing involves a process known as sequencing by synthesis, in which each base is electronically monitored as a constructed DNA strand is complementary base-paired. "Third-generation" sequencing technology identifies each base on the DNA strand as it moves through a membranous pore and passes through an electrical current.

6-7. Modern Uses of Biotechnology

- Biotechnology has been used in a variety of areas, including agriculture, medicine, and genome editing. The modification of plants and animals by means of biotechnology has helped increase agricultural productivity while lowering overall cost.
- Genetically engineered crops were first introduced in the United States in the 1990s, and there are ten different types. Genetically modified corn, cotton, and soybeans are the most prominent and have been modified for increased herbicide tolerance and insect resistance. Herbicide-tolerant crops were modified with genes that can code for an enzyme that can function in the presence of an herbicide. Herbicides are chemicals that are used to eliminate unwanted plants by affecting their ability to synthesize amino acids and functional proteins. The advantage of herbicide-tolerant crops is that it allows farmers to eliminate weeds, which compete with their crops for water, light, and nutrients. Ninety-four percent of domestic soybean acres, 98 percent of cotton acres, and 90 percent of corn acres planted

domestically in 2019 used herbicide-tolerant seeds. Insect-resistant crops have been modified with a gene found in the bacterium *Bacillus thuringiensis* (*Bt*). The *Bt* gene codes for a protein that produces crystal-containing spores that are poisonous to insects but have no effect on humans. The spores affect an insect's digestive system by causing gut wall paralysis, which prevents them from feeding, resulting in death. The most common genetically modified insect-resistant crops are corn and cotton. Some crops are called "stacked" varieties because they contain genetic modifications that allow for both herbicide tolerance and insect resistance. In 2019, 80 percent of corn and 89 percent of cotton acres were planted using "stacked" seeds.

- Biotechnology has also been used to help improve the medical field, with the production of pharmaceutical and diagnostic products. In 1982, the first genetically produced protein was recombinant insulin. Genetically modified plants and animals have been used in order to produce important vaccines, antibodies, medicines, and essential proteins to be used against a variety of diseases and disorders. The use of biotechnology in the medical field has also been used as a diagnostic tool for inherited disorders. Commonly inherited disorders such as sickle-cell anemia or cystic fibrosis can be easily diagnosed with the use of biotechnology. Gene therapy is a process used to help treat or cure different types of genetic disorders. Gene therapy involves isolating a defective gene and replacing it with a functional gene.
- CRISPR-*Cas9* is a new form of genome editing that allows for the alteration or editing of genetic material in order to study gene function or correct genetic mutations. *Cas9* is a protein produced by bacteria in order to protect itself from viruses. *Cas9* is similar to a restriction enzyme in that it can cut a specific sequence of DNA; however, instead of recognizing the sequence, it is directed to the sequence by a type of RNA called "guide RNA" (gRNA). gRNA is produced from the CRISPR region of the bacterial genome. The combination of the *Cas9* protein and gRNA forms a complex responsible for cutting the desired DNA sequence that complements the gRNA sequence. *Cas9* cuts the desired DNA strand, which initiates the DNA repair system. The repair enzymes are useless without a template from which to work, resulting in an altered DNA sequence and a nonfunctional gene that can be studied. CRISPR-*Cas9* has also been modified in order to help fix defective genes that lead to genetic disorders. A functional gene is combined with *Cas9*, and the protein cuts out the defective gene; then, with the use of repair enzymes, the DNA strand is edited, using the functional gene as a template. One of the first uses of CRISPR-*Cas9* to treat a genetic disorder occurred in 2018, when it was used to correct the defective gene causing sickle-cell anemia. It is hopeful that other genetic disorders, such as Parkinson's, Alzheimer's, and cancer, can be treated using CRISPR-*Cas9*. There are some concerns that CRISPR-*Cas9* may affect nontargeted genes or be misapplied beyond its purpose.
- Despite its successes, the use of biotechnology has raised some concerns, such as its safety. Some studies have shown that transgenic organisms pose no risk to humans. Many of the major concerns with the use of transgenic organisms include the increased risk of allergies, its effect on the health of the environment, the development of resistance to herbicide-tolerant or insect-resistant crops, the lack of testing and regulation, and hidden costs.

End-of-Chapter Activities and Questions

Directions: Please apply what you learned in this chapter to complete the following activities.

Define Each Term in Your Own Words

1. *Pneumococcus*
2. Antiparallel
3. Semiconservative
4. Central Dogma of Molecular Biology
5. CRISPR-Cas9

Chapter Review

1. Explain how experimental data collected by Rosalind Franklin and Erwin Chargaff helped Watson and Crick determine the structure of the DNA molecule.
2. Using the following group of terms, summarize in a paragraph the process of DNA replication: semiconservative, origin of replication, helicase, replication fork, RNA primer, primase, DNA polymerase, leading strand, lagging strand, Okazaki fragments, DNA ligase, and nuclease.
3. Transcribe the following DNA sequence: TAC ACG ATG GTT and translate the resulting mRNA into to its corresponding amino acid sequence (see Figure 6.11).
4. A mutation has occurred in the DNA sequence from the question above, shown here: TAC ACG ATG GT**C**. Perform the following three activities on this DNA sequence: transcription, translation, and indicate and explain what type of mutation occurred (hint: silent, missense, or nonsense mutation).
5. When creating a genetically modified organism, researchers must use the following five steps: cut, amplify, insert, grow, and identify. Using these five main steps, create your own genetically modified organism, then indicate whether your genetically modified organism would benefit the field of agriculture or medicine.

Multiple Choice

1. The DNA molecule consists of two strands running in alternating directions, with each strand differing from the other because of the location of a phosphate group and sugar molecule. As a result, the DNA strand is said to exhibit a(n)__________ arrangement.

   a. semiconservative

   b. antiparallel

   c. dispersive

   d. universal

2. Which of the following enzymes is not associated with the process of DNA replication?

   a. restriction enzyme

   b. DNA polymerase

   c. Nuclease

   d. primase

3. During the transcription process, RNA polymerase comes to an exposed adenine (A) on the DNA strand. The resulting complementary nitrogenous base would be ________.

   a. guanine

   b. cytosine

   c. thymine

   d. uracil

4. A DNA sequence contains a mutation that results in a STOP codon after being transcribed and translated. Which of the following diseases may result from this type of mutation?

   a. sickle-cell anemia

   b. Charcot-Marie-Tooth disorder

   c. cystic fibrosis

   d. Williams syndrome

5. Which of the following enzymes is used both during the process of DNA replication and in genetic engineering?

   a. DNA ligase

   b. DNA polymerase

   c. nuclease

   d. primase

References

Soyfer, V. N. (2001). The consequences of political dictatorship for Russian science. *Nature Reviews. Genetics, 2*(9), 723–729.

Crick, F. H. (1958). On protein synthesis. *Symposia of the Society for Experimental Biology, 12,* 138–163.

Watson, J. D., & Crick, F. H. C. (1953) Genetical implications of the structure of deoxyribonucleic acid. *Nature, 171,* 962–967.

Image Credits

Fig. 6.8: Source: https://commons.wikimedia.org/wiki/File:DNA_replication_en.svg.

Fig. 6.9: Copyright © by Daniel Horspool (CC BY-SA 3.0) at https://commons.wikimedia.org/wiki/File:Central_Dogma_of_Molecular_Biochemistry_with_Enzymes.jpg.

Fig. 6.10: Copyright © by Sponk (CC BY-SA 3.0) at https://commons.wikimedia.org/wiki/File:Difference_DNA_RNA-EN.svg.

Fig. 6.11: Source: https://commons.wikimedia.org/wiki/File:06_chart_pu3.png.

Fig. 6.12: Copyright © by Kelvin13 (CC BY 3.0) at https://commons.wikimedia.org/wiki/File:Transcription_Factors.svg.

Fig. 6.13: Copyright © by Kelvinsong (CC BY 3.0) at https://commons.wikimedia.org/wiki/File:MRNA.svg.

Fig. 6.13: Copyright © by OpenStax (CC BY 4.0) at https://commons.wikimedia.org/wiki/File:OSC_Microbio_11_04_tRNA.jpg.

Fig. 6.15: Copyright © by Kelvinsong (CC BY 3.0) at https://commons.wikimedia.org/wiki/File:Protein_synthesis.svg.

Fig. 6.17: Source: https://commons.wikimedia.org/wiki/File:-Types-of-mutation.png.

Fig. 6.18: Copyright © by OpenStax (CC BY 4.0) at https://commons.wikimedia.org/wiki/File:Figure_14_06_05.jpg.

Fig. 6.19: Copyright © by Simon Caulton (CC BY-SA 4.0) at https://commons.wikimedia.org/wiki/File:BamHI2.png.

Fig. 6.20: Copyright © by Enzoklop (CC BY-SA 3.0) at https://commons.wikimedia.org/wiki/File:Polymerase_chain_reaction.svg.

Fig. 6.21: Copyright © by Mckenzielower (CC BY-SA 4.0) at https://commons.wikimedia.org/wiki/File:Gel_Electrophoresis.svg.

Fig. 6.22: Copyright © by OpenStax (CC BY 4.0) at https://commons.wikimedia.org/wiki/File:OSC_Microbio_12_01_MolCloning.jpg.

Fig. 6.23: Copyright © by OpenStax (CC BY 4.0) at https://commons.wikimedia.org/wiki/File:Figure_14_02_04ab.jpg.

Fig. 6.24: Source: https://commons.wikimedia.org/wiki/File:-Adoption_of_Genetically_Engineered_Crops_in_the_US.png.

Fig. 6.25: Copyright © by Genome Research Limited (CC BY 3.0) at https://medium.com/@squarecog/data-science-at-zymergen-1f1fdc1feaf8.

Fig. 6.25a: Copyright © by Manuel Castells (CC BY-SA 2.0) at https://commons.wikimedia.org/wiki/File:FCMM_2019.jpg.

Fig. 6.25b: Copyright © by (CC BY-SA 4.0) at https://commons.wikimedia.org/wiki/File:Dr._Jennifer_Doudna.jpg.

Fig. 6.25c: Copyright © by (CC BY-SA 4.0) at https://commons.wikimedia.org/wiki/File:Dr_Emmanuelle_Charpentier_at_York_University,_Toronto.jpg.

Fig. 6.26: Copyright © 2020 by istockphoto/Trinset.

CHAPTER

7

Chromosomes and Cellular Division

PROFILES IN SCIENCE

FIGURE 7.1 Wilhelm von Waldeyer-Hartz.

Wilhelm von Waldeyer-Hartz, considered a great forefather in the fields of anatomy, pathology, and embryology, was born on October 6, 1836, in Hehlen, Germany, and died in Berlin, Germany, in 1921. After quickly completing grammar school in 1856, he began attending the University of Göttingen upon obtaining an eligibility certificate for university studies. While at the University of Göttingen, his intent was to study mathematics and natural sciences; however, after meeting anatomist Jakob Henle (who discovered the loop of Henle in the kidneys), Waldeyer-Hartz decided to instead study medicine. As a result of a law stating that individuals of Prussian birth were required to complete their studies at a Prussian university, he was forced to transfer to the University of Greifswald in 1859. He remained at the University of Greifswald for two years before transferring to the University of Berlin, where he received his doctorate in 1861, followed by passing the state medical exam in 1862. After receiving his medical degree, he taught anatomy and physiology in a number of distinguished academic institutions, including the universities of Königsberg (1862–1864), Breslau (1864–1872), and Strasbourg (1872–1883). In 1883, he accepted a full professorship position teaching anatomy and was named the director of the Anatomical Institute at the University of Berlin, where he remained for over 33 years. Waldeyer-Hartz was also an accomplished scientific writer, publishing over 250 articles between 1862 and 1920 encompassing a variety of subjects, including cancer, hernias, and the human pelvis. Over the course of these publications, he coined several terms, some of which are still used in anatomy textbooks today, including Waldeyer's tonsillar ring (the lymphoid tissue ring of the naso- and oropharynx) and Waldeyer's glands (of the eyelids). In addition to these anatomical terms, Waldeyer-Hartz is also credited with coining the term *chromosome*, in his 1888 publication "Karyokinesis and Its Relation to the Process of Fertilization," and the term *neuron* in 1891, when proposing the neuron theory in "On Some Recent Research in the Field of Anatomy of the Central Nervous System." In 1916, after his resignation from the University of Berlin, the Prussian House of Lords raised Waldeyer-Hartz to nobility.

Chapter Introduction

Reproduction is one of the defining characteristics of life. The ability of a living organism to produce another organism much like itself ensures the stability of the population. Regardless of the mode of reproduction an organism takes, whether asexual, a method performed by prokaryotes and some unicellular eukaryotes, or sexual, which involves the fusion of genetic material from two distinct parents, all new life originates at the cellular level. Cells perform an integral process that includes a cell cycle, consisting of carefully regulated events necessary for reproduction, growth, development, or body maintenance and repair, called cell division. During cell division, genetic material from a progenitor, housed within a unique structure called a chromosome, is replicated, rearranged, and evenly distributed into each new cell produced. These new cells will divide, giving rise to more cells, which helps sustain the continuity of life. In this chapter, the detailed structure of chromosomes and what occurs when these chromosomes are replicated and begin moving through the cell cycle will be introduced. Detailed descriptions of the series of events associated with the cell cycles of binary fission, mitosis, and meiosis will also be discussed. Finally, the chapter will conclude with cancer, how it develops, its unique characteristics, and the treatment options available to patients to help combat the disease.

Chapter Objectives

In this chapter, students will learn the following:

7-1. Chromosomes are unique structures composed of a complex of DNA and proteins that are responsible for the inheritance of traits.

7-2. Prokaryotic cells divide by binary fission, a process in which one individual cell divides to produce two genetically identical daughter cells.

7-3. Eukaryotic cell division involves a cell cycle that includes cellular growth and cellular division by mitosis, wherein a parental cell divides to produce two genetically identical daughter cells.

7-4. Sexual reproduction involves the fertilization of gametes, cells produced in the gonads of sexually reproducing organisms, during meiosis.

7-5. Cancer is a lethal disease where unregulated cell division produces a mass of unique cells that can spread, invade, and cause harm to other parts of the body.

7-1. Chromosome Structure

In 1879, German anatomist and founder of the field of cytogenetics (study of chromosomes) Walther Flemming (Figure 7.2, left) was the first to observe and describe chromosomes, a term not used until almost ten years later. With the innovation of a new staining technique, Flemming discovered that when using a group of dyes known as aniline dyes, structures within a cell's nucleus strongly absorbed one in particular, a basophilic dye. Basophilic dyes are acidic pH stains, which carry a positive charge, and upon application, Flemming observed that the structures within the nucleus were dyed red. (Note: recall that the sugar-phosphate groups of DNA are negatively charged molecules, which would react with the basophilic dye due to its positive charge.) Flemming named the mass of long, threadlike fibers within the nucleus he observed **chromatin** (Greek for color), the DNA and protein complex that condenses and coils to form chromosomes. Flemming observed the coalescing and shortening of the chromatin into chromosomes and outlined the sequence of changes, movement, and distribution chromosomes made when the nucleus divided. He first described this process in 1878 in his paper "Zur Kenntniss der Zelle und ihrer Theilungs-Erscheinungen" ("To the Knowledge of the Cell and Its Phenomena of Division") and again later in 1882 in his historic book, *Zell-substanz, Kern und Zelltheilung* (*Cell Substance, Nucleus, and Cell Division*), which contained several hundred drawings of this process he would coin **mitosis** (Greek for "thread"), which will be discussed in section 7-3.

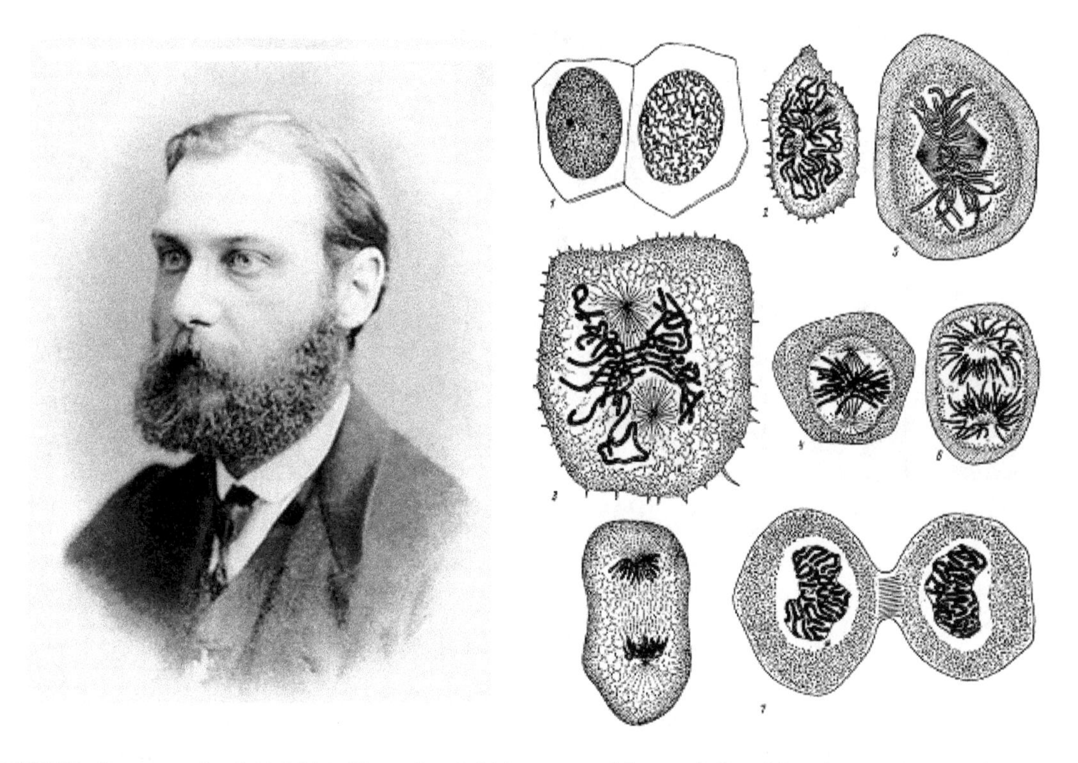

FIGURE 7.2 German scientist Walther Flemming (left) began studying and describing the structure and movement of chromosomes in 1879 after observing a mass of stained long, threadlike fibers within the nucleus he called chromatin. He produced several drawings (right) based upon his observations on how chromosomes changed, moved, and were distributed during a process he would call mitosis.

The term **chromosome**, which refers to the observable structure that results when chromatin coalesces in the nucleus, was not introduced until 1888 by another German anatomist and colleague of Walther Flemming, Wilhelm von Waldeyer-Hartz. Waldeyer-Hartz, along with Flemming, was also studying the basophilic-stained fibers within the nucleus, and used the term in his publication "Über Karyokinese und ihre Beziehungen zu den Befruchtungsvorgängen" ("Karyokinesis and Its Relation to the Process of Fertilization").

Chromosomes consist of long, continuous pieces of DNA and a variety of associated proteins. These unique structures compose an organism's genome and are responsible for transmitting inherited traits, as independently discovered and outlined by Walter Sutton (Figure 7.3, left) and Theodor Boveri (Figure 7.3, middle) in 1902 in the **chromosomal theory of inheritance**. Each chromosome in an organism varies in length and contains distinct banding patterns when stained, and the number of chromosomes varies among different species. For example, a mosquito has as few as six chromosomes, while humans have 46 chromosomes. As detailed in Thomas Hunt Morgan's (Figure 7.3, right) 1910 paper, "Sex Limited Inheritance in Drosophila," chromosomes contain an organism's genes, confirming the chromosomal theory of inheritance. As each organism has a distinct number of chromosomes, each chromosome has a distinct number of genes. In humans, chromosome 1, the largest of all, consists of 3,000 different genes, while the smallest, chromosome 22, contains only 600 total genes.

FIGURE 7.3 The chromosomal theory of inheritance, which states that chromosomes are responsible for the transmission of genetic traits, was proposed by Walter Sutton (left) and Theodor Boveri (middle) in 1902. In 1910, Thomas Hunt Morgan (right) confirmed the hypothesis by stating that chromosomes contain the genes of an organism.

Due to their unique shape and staining patterns, chromosomes can be visually displayed using a **karyotype**. Karyotypes (Figure 7.4) are generally used as a diagnostic tool for chromosomal disorders, as in the case of trisomy 21 or Down syndrome. When prepared from a single, somatic cell, a karyotype can display an organism's entire set of chromosomes, including the sex chromosomes, revealing the sex of the fetus.

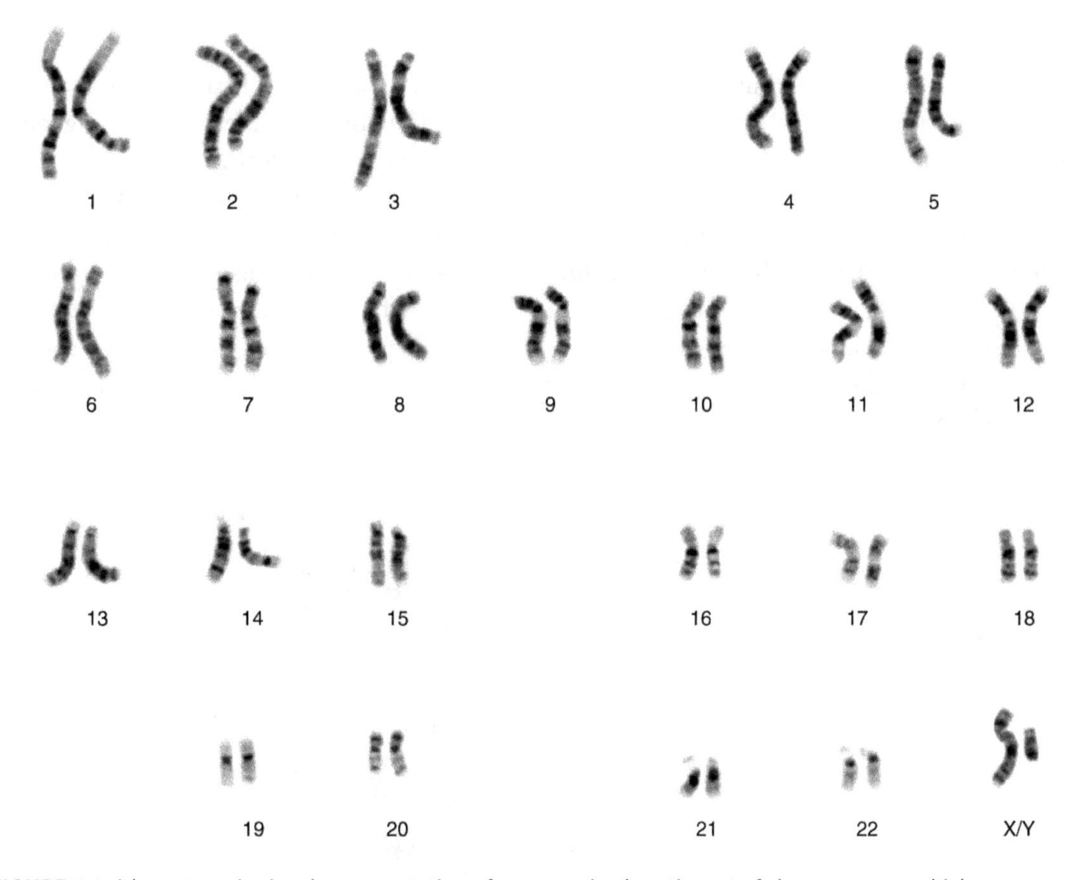

FIGURE 7.4 A karyotype is visual representation of an organism's entire set of chromosomes, which are arranged by shape and staining patterns.

Karyotypes are prepared using five main steps:

1. **Cell collection.** Cells are collected in one of two ways. In adults and children, cell collection is accomplished by taking a blood sample. In a fetus, cells are either collected through amniocentesis or chorionic villus sampling. Amniocentesis involves the extraction of amniotic fluid from the amniotic sac, while chorionic villus sampling requires the removal of tissue from the placenta, each of which can be performed within the first trimester of a pregnancy.
2. **Cell division.** The collected cells are placed within a nutrient-filled culture tube and treated with a chemical, which encourages the cells to undergo mitosis for several days.
3. **Arrested cell division.** As the cells are undergoing mitosis, the cells are treated with a chemical that stops cell division at the metaphase stage, the stage in which the chromosomes are highly condensed and more visible.
4. **Chromosomal staining**. Once chromosomes are visible, they are stained, demonstrating their unique staining pattern.
5. **Chromosomal arrangement.** Using a microscope and computerized digital software, the chromosomes are identified and arranged into pairs using chromosomal characteristics such as size and shape, staining patterns, and centromere position.

Prokaryotic Chromosome Structure

A prokaryotic chromosome consists primarily of DNA and a variety of proteins, known as nucleoid-associated proteins, or NAPs. The DNA portion of the chromosome carries the genetic information essential to the cell, while the NAPs perform two main functions, gene regulation and chromosome organization. The structure of the chromosome, in most prokaryotic cells, exists as a single, circular piece of DNA, first discovered by Australian biochemist John Cairns in 1963 when working with *Escherichia coli*. The prokaryotic chromosome is located within the **nucleoid** (Figure 7.5), an irregularly shaped region devoid of a membrane attached to the cell membrane. The prokaryotic chromosome is quite large, in comparison to the overall dimensions of the cell; therefore, specific proteins within the nucleoid densely coil the prokaryotic chromosome, resulting in a genome with genes closely packed together.

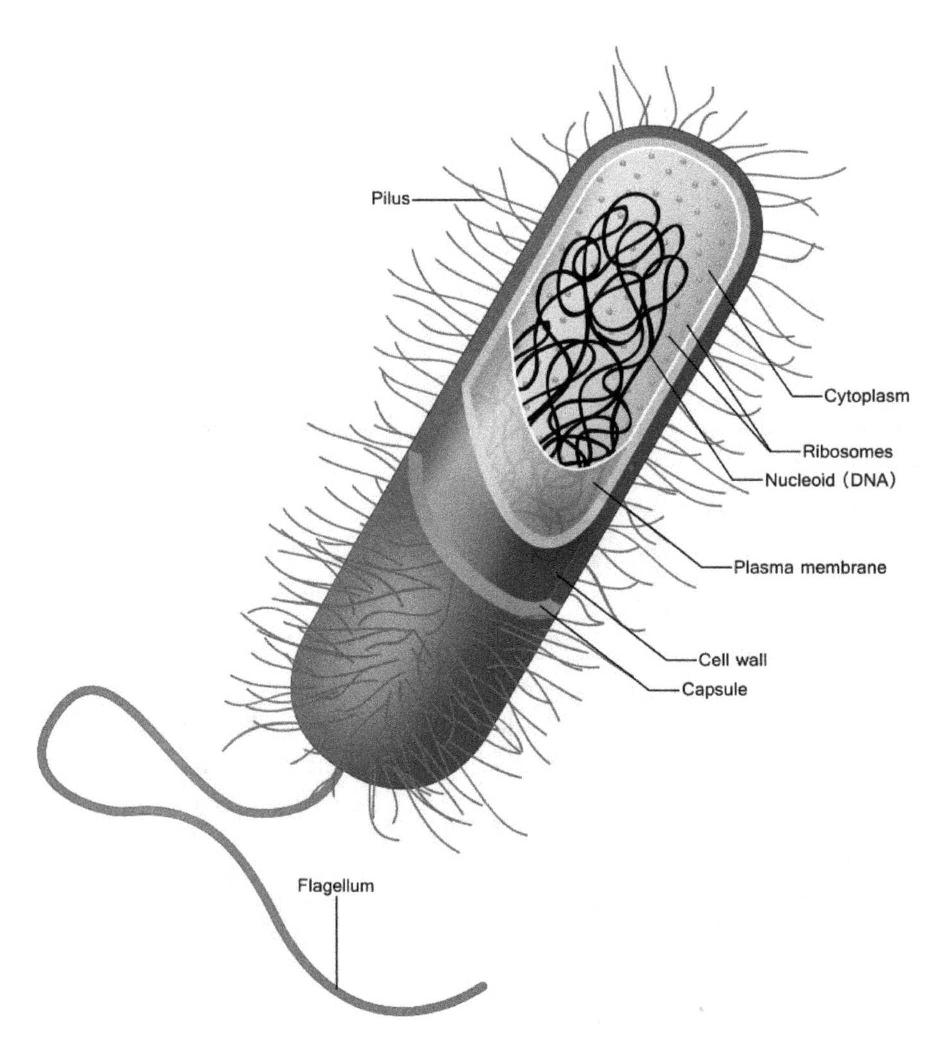

FIGURE 7.5 A prokaryotic cell consists of a chromosome that exists as a single, circular structure consisting of densely packed genes located within the cell's nucleoid.

Eukaryotic Chromosome Structure

Eukaryotic chromosomes are considerably more complex and elaborate than prokaryotic chromosomes. Recall that DNA of a eukaryotic cell is located within a double-membrane organelle called the nucleus. Inside the nucleus, DNA combines with a large number of proteins to form a mass of long, threadlike fibers called chromatin, first observed by Walther Flemming in 1879.

The eukaryotic chromosome (Figure 7.6) exists as chromatin within the nucleus as free-floating, linear molecules composed of a single DNA molecule and proteins. Proteins make up more than 50 percent of the structural complex and are involved in a variety of roles, such as DNA and RNA synthesis, but a majority play a structural role. These proteins are called **histones**. DNA is wrapped around histone proteins, which play several important roles, including keeping the DNA from becoming tangled within the nucleus and organizing the chromatin into chromosomes. However, the most important role histone proteins play is tightly packing DNA into chromosomes within the nucleus. There are five different histone proteins in chromatin responsible for tightly packing the DNA into the nucleus. This is important, because if DNA from a single cell were laid end-to-end, it would stretch about six feet! Continual packing of DNA involves the formation of a **nucleosome**, a "bead-like" structure consisting of DNA wrapped twice around a group of eight histone proteins. Nucleosomes are then combined as interactions between adjacent nucleosomes occur and DNA begins to coil and fold. As additional coiling and folding of DNA continues with the help of additional proteins, the chromatin is compacted, forming the chromosome.

Once a eukaryotic chromosome has formed, it consists of two regions: a single centromere and two telomeres. These regions are important in maintaining the structure, function, and stability of the chromosome. Despite its meaning, "middle body," the centromere can be located anywhere on the chromosome. When DNA replicates, it forms a duplicated chromosome, and the centromere is the attachment point for both double helices, forming what is known as **sister chromatids** (Figure 7.6). It is at the centromere where the sister chromatids are pulled apart during the process of cell division.

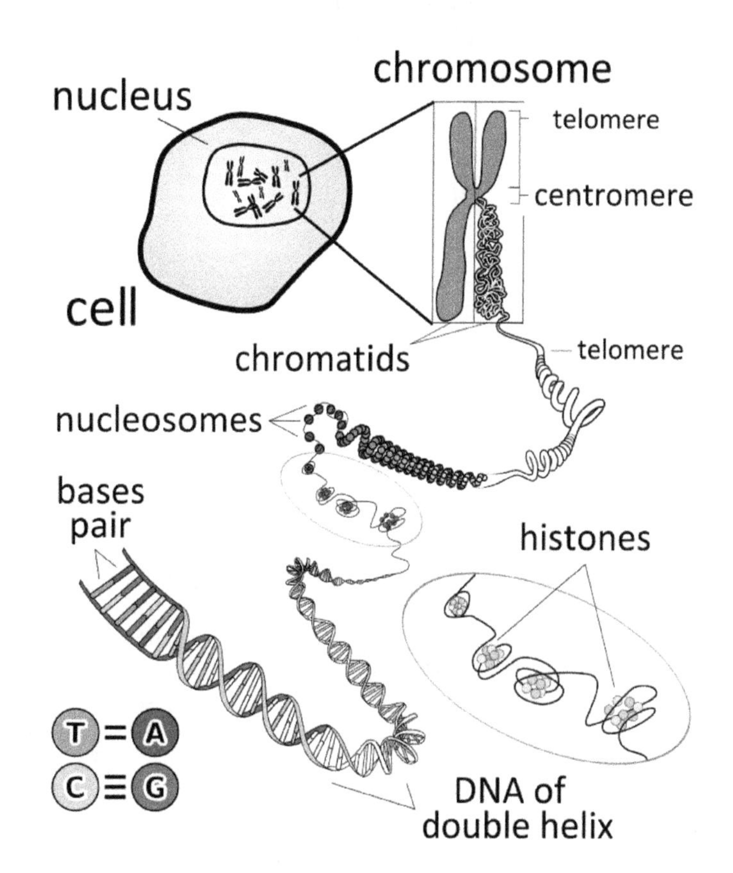

FIGURE 7.6 A eukaryotic chromosome, found in a cell's nucleus, exists as DNA wrapped around histone proteins that becomes tightly packed into "bead-like" structures called nucleosomes. A condensed chromosome consists of a centromere, which attachs the sister chromatids after DNA replication, and two telomeres that protect the end of the chromosome and help regulate cellular division.

A **telomere** consists of a repeating six-nucleotide sequence, such as TTAGGG, and is located at both ends of a eukaryotic chromosome. Although telomeres may contain hundreds to thousands of repeating DNA nucleotide sequences, these sequences do not contain genes; therefore, they are unable to code for proteins. However, telomeres serve two main functions. First, telomeres help protect the ends of chromosomes. Without the presence of telomeres on the ends of chromosomes, DNA repair enzymes might remove coding portions of a chromosome, or the ends of two or more chromosomes might combine, forming long, connected structures unable to be distributed during cell division. This helps maintain the overall stability of the chromosome by preventing the loss of DNA during each replication. Second, in 1961, Leonard Hayflick, an American anatomist, discovered that cultured animal cells contained an "internal clock," specifying a limited number of divisions a cell can undergo before it stops. This idea became known as the Hayflick limit, which refuted popular belief at the time that normal cells would divide continuously. It was not until the 1990s, however, that scientists discovered the location of this "internal clock"—the telomeres of chromosomes.

According to the Hayflick limit (Figure 7.7), each time a cell divides, a small portion (50 to 200) of DNA nucleotides from the telomeres is removed, resulting in shorter and shorter chromosomes. Upon a cell division in which the removal of telomere nucleotides results in a loss of functional DNA, the cell may remain active; however, the cell will not proceed with any additional divisions

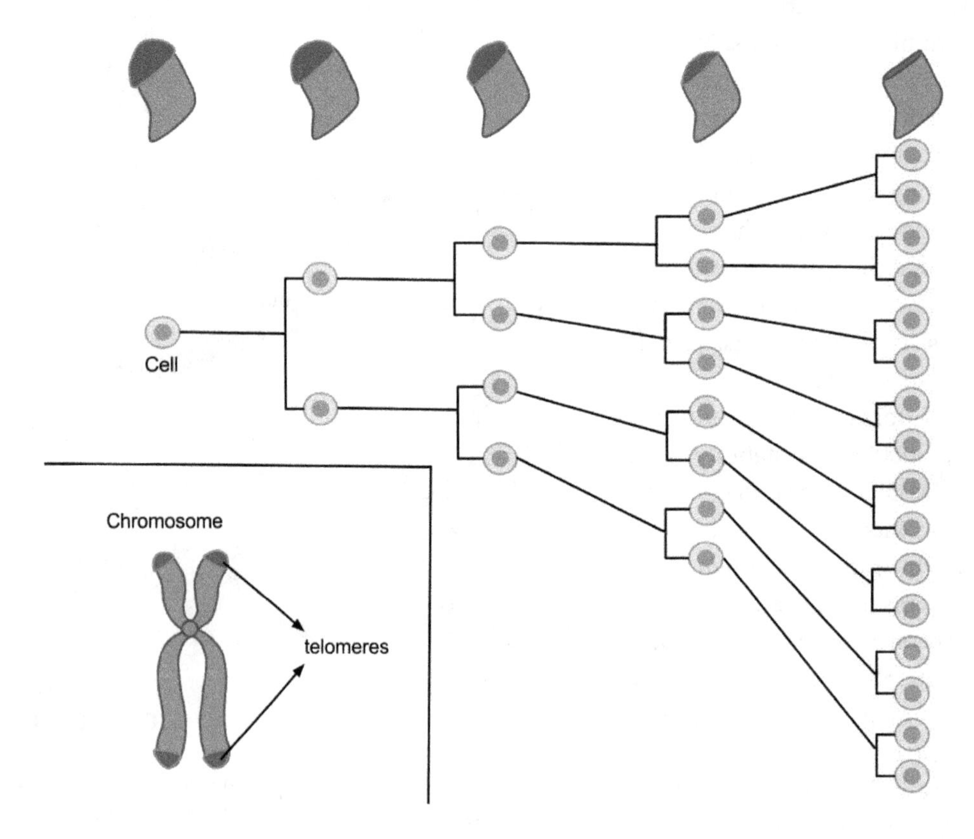

FIGURE 7.7 The number of divisions a cell can undergo is regulated by the telomeres located at the end of the chromosome. After each cell division, a portion of the telomere is removed and upon dividing and a portion of functional DNA is lost, the cell undergoes apoptosis. A cell containing a specified number of cell divisions or "internal clock" is known as the Hayflick limit.

thereafter. In addition, the cell may also undergo a process known as **apoptosis**, or programmed cell death. In human cells, Hayflick demonstrated that the number of cell divisions it may undergo prior to ceasing division was about 50. This was an important discovery by Hayflick, who was also the first to report that cancer cells were immortal cells that would continue to divide. The ability of most cancer cells to continually divide has been linked to an enzyme known as **telomerase**. The enzyme is responsible for rebuilding and extending telomeres, which prevents shortening, thereby providing their ability for indefinite cellular division. Cancer will be discussed further in section 7-5.

Prokaryotic cells divide by means of **binary fission** (division in half), a form of **asexual reproduction**, in which a single individual produces two genetically identical offspring. In order for binary fission to occur, certain conditions must be met, including an adequate supply of nutrients and optimal temperatures. When conditions are met, most prokaryotic cells can complete cell division in one to three hours. However, there are some prokaryotic cells, such as *Escherichia coli*, which can complete the process in about 20 minutes.

Prokaryotic cell division involves a cell cycle consisting of two main stages: 1. cell growth and chromosomal replication and 2. binary fission.

Cell Growth and Chromosomal Replication

Prior to binary fission, the prokaryotic cell undergoes a long period of growth in which the cell begins to increase in size and DNA replication occurs. The replication of DNA initiates the beginning of binary fission. Recall that the prokaryotic chromosome exists as a single, circular piece of DNA attached to the cell membrane, and during DNA replication, the chromosome will be duplicated. During DNA replication, the double-stranded DNA molecule unwinds, or unzips, exposing the nitrogenous bases on each strand of the molecule. Free-floating nucleotides then complementarily pair with the exposed bases, producing two exact duplicates of the double-stranded, circular DNA molecule. The newly duplicated chromosomes are attached to different, specialized locations on the cell membrane (Figure 7.8).

Binary Fission

Once the DNA replicates, the cell begins the process of binary fission, which consists of three main steps: cell elongation, cell membrane and cell wall formation, and daughter cell formation (Figure 7.8). During cell elongation, the cell grows, doubling in length. This growth increases the distance between the attachment points of the chromosomes, and the distance continues to increase as new portions of cell membrane and cell wall are added. As elongation and growth of the cell membrane and cell wall continue between the attachment points, the duplicated chromosomes separate and are pulled apart from each other. As the new cell membrane and cell wall continue to grow in length, proteins pinch the cell inward across its midline, dividing the parental cell. Binary fission is complete when the inward pinching of the cell membrane and cell wall of the parental cell fuse, producing two new daughter cells, each with one of the replicated chromosomes. The resulting daughter cells are not only identical in size but are also genetically identical to each other and to the original parent cell.

Binary fission is a relatively fast and easy process, leading to the growth and success of prokaryotic populations. In addition, this process is efficient, because the prokaryotic daughter cells that are produced contain exact copies of the genes of the parent cell. This is an important advantage to prokaryotes, because it not only allows them to establish new populations but also enables them to reproduce in relatively isolated habitats. Despite binary fission enabling prokaryotic cells to produce genetically identical offspring, this advantage can also be considered a disadvantage. Because the offspring are genetically identical, the genetic diversity of the population is greatly reduced, meaning that the population may be less adaptable and more susceptible to environmental changes.

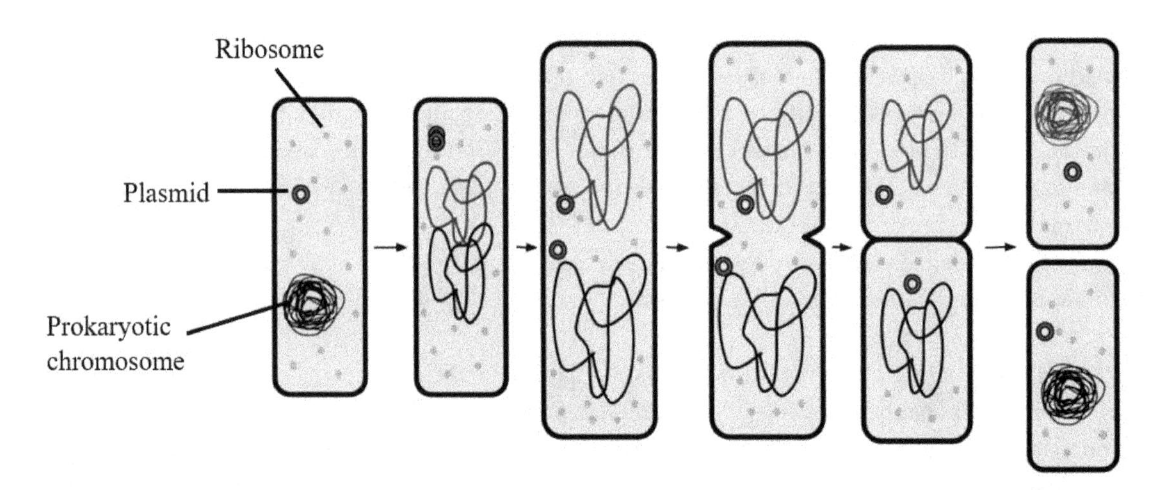

FIGURE 7.8 Prokaryotic cell division involves a three step process: cellular growth, chromosomal replication, and binary fission. After chromosomal replication, the cell begins to elongate, pulling the replicated chromosomes apart. A new cell membrane and cell wall starts to form separating the cell, and continues until the cell divides into two genetically identical daughter cells.

7-3. Cell Cycle and Cellular Reproduction

Cell division in eukaryotic cells involves a **cell cycle**, a series of events beginning with an individual cell performing normal cellular processes, ending with its division into two daughter cells. The cell cycle of eukaryotic cells consists of three main stages: interphase, mitosis (M phase), and cytokinesis.

Interphase

In the 1950s and 1960s, **interphase** was once considered the "resting phase" of a cell because, when observed, the cells did not appear active. However, interphase is a very active and busy period for the cell. While in interphase, a cell is performing normal cellular processes, including basic metabolic activities, synthesizing new organelles, duplicating chromosomes, producing proteins, and growing. The amount of time a cell spends in interphase varies; however, some cells may remain in interphase for nearly 90 percent of their cell cycle.

Interphase is divided into three main phases: cell growth, DNA replication, and synthesis of additional organelles and proteins responsible for cell division. The times outlined in each phase are based on a human cell undergoing a single division within 24 hours.

G_1 Phase

The **G_1 phase** of interphase (Figure 7.9) typically lasts between five and six hours and begins with the cell recovering from its preceding division. Once recovered, the cell begins to perform normal daily functions, including cell growth, protein synthesis, and cellular respiration. Although most cells continue their progression through the subsequent phases of interphase, some cells like muscle and nerve cells, enter a distinct dormant stage upon reaching maturity, where they remain physiologically active, but no longer proceed with the necessary preparations for further divisions. Scientists recognize this period in the cell cycle as the G_0 phase (Figure 7.9).

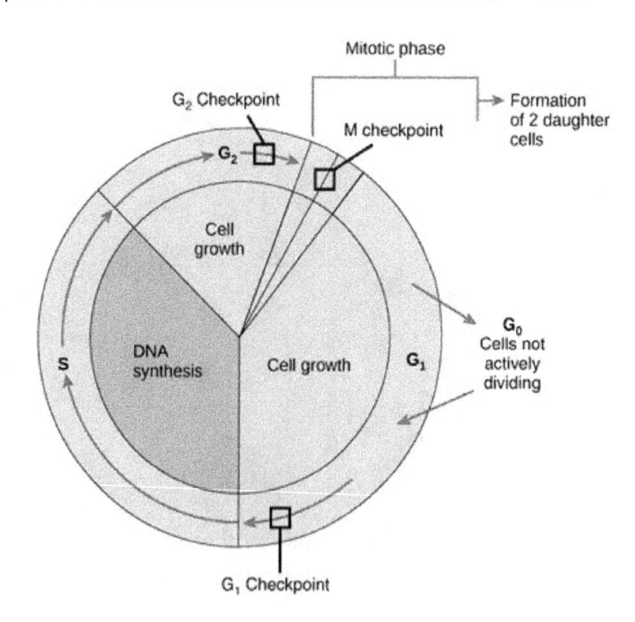

FIGURE 7.9 Generalized representation of the cell cycle, which consists of an order of events in which a cell proceeds through prior and during cellular division. The cell cycle consists of three stages: interphase (G_0, G_1, S, and G_2), mitotic phase, and cytokinesis.

In preparation for cell division, the cell uses the G_1 phase to begin synthesizing additional organelles, as well as the necessary materials required for building new cell membranes. In addition, the cell also begins to acquire and/or synthesize the materials required for DNA replication. Prior to DNA replication, a cell must decide if it should proceed forward with division, at a part of the cell cycle known as the **G_1 checkpoint** (Figure 7.9). This decision is made by receiving internal and external cellular signals, and if positive, the cell initiates DNA replication, or the **S phase**.

S Phase (DNA Replication)

Upon receiving the positive signal to initiate cell division, the cell enters the **S phase** (Figure 7.9). This particular phase can take between eight and 12 hours and involves DNA replication, the process by which chromosomes are duplicated within the cell's nucleus. Recall that each chromosome consists of one DNA double helix, but after the S phase, each DNA double helix within the chromosome is doubled and identical to the original. This process results in two identical structures known as sister chromatids, joined together at the centromere.

In addition to the replication of DNA, an organelle known as the centrosome also replicates during the S phase. Centrosomes were first discovered simultaneously and independently in the 1880s by Edouard van Beneden and Theodor Boveri while studying the mitotic division of fertilized eggs in nematodes. Centrosomes are important in organizing microtubules, which are thick, hollow tubes composed of the globular protein tubulin. Microtubules are an important component of the mitotic spindle, a structure that will be discussed shortly. Centrosomes are only found in animal cells and contain a pair of centrioles, cylindrical organelles composed of the protein tubulin, which are important in organizing the mitotic spindle and finalizing cytokinesis, or cytoplasmic division.

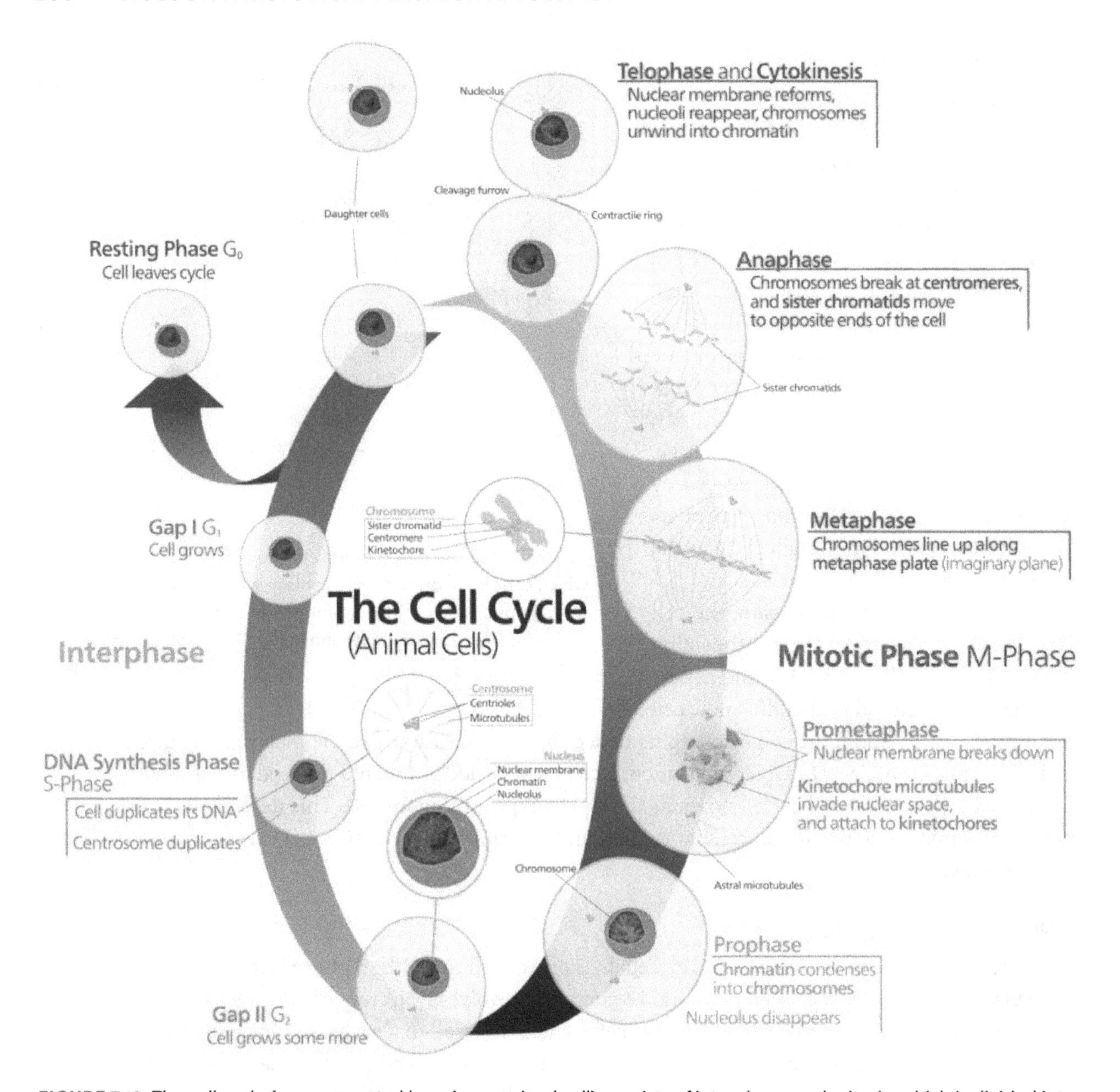

FIGURE 7.10 The cell cycle (as represented here in an animal cell) consists of interphase and mitosis, which is divided into four main stages: prophase, metaphase, anaphase, and telophase. Mitosis is shortly followed by cytokinesis resulting in two genetically identical daughter cells.

G_2 Phase

The final phase of interphase is the **G_2 phase** (Figure 7.9), a phase that takes between four and six hours. During the G_2 phase, the cell is synthesizing the necessary proteins required for cell division to occur, including the proteins associated with the microtubules and the formation of the mitotic spindle. Since cell division will be occurring soon, the cell also begins assembling and storing additional cell membrane materials in vesicles beneath the cell membrane. Before the cell enters the mitosis phase, it must pass through another checkpoint, the **G_2 checkpoint** (Figure 7.9), to assess whether all chromosomes have been replicated and ensure the chromosomal DNA was not damaged during the S phase.

Mitosis (M Phase)

Upon deciding it is ready to proceed with division, the cell enters **mitosis** (Greek for "thread"), or the **M phase** (Figure 7.9), the phase of the cell cycle in which the nucleus and genetic material of a parent cell divide, producing two genetically identical cells called daughter cells. Mitosis was first observed and described by Walther Flemming in 1879, with a more extensive description and the name "mitosis" being given to the process in his 1882 book, *Zellsubstanz, Kern und Zelltheilung* (*Cell Substance, Nucleus, and Cell Division*). His work built upon observations made by other scientists at the same time, including Anton Schneider, who was the first to observe and describe the mitotic phases of metaphase and anaphase in 1873. In Flemming's 1882 publication, he described the rearrangement of chromosomes and their distribution into daughter cell nuclei occurring in two phases. The progressive phase began once the chromosomes could be observed in the nucleus, until they aligned in the middle of the cell, while during the regressive phase, the chromosomes separated, leading to the formation of two daughter nuclei.

Today, the process of mitosis is outlined in four main stages (Figure 7.10), occupying only a small portion of the cell cycle, lasting about an hour.

Prophase

Prophase is characterized by several different events (Figure 7.10). The first main event of prophase is the condensing of the duplicated chromatin fibers into distinct, visible chromosomes. The duplicated chromosomes appear as sister chromatids attached at their centromeres.

After the chromosomes have condensed, the mitotic spindle, an important structure in helping with the movement of chromosomes, begins to form and assemble itself within the cytoplasm. The mitotic spindle is primarily composed of microtubule fibers, which in animal cells are organized and begin to assemble within the centrosomes, organelles that replicated during the S phase of interphase. The formation of the mitotic spindle begins as microtubule fibers within the centrosomes extend and lengthen, causing the centrosomes to move farther away from each other, toward opposite ends of the cell. In addition, a group of small, radiating microtubule fibers also extend from the centrosomes, called *asters*. It was once thought that the asters may play a supportive role for the centrioles; however, this may not be the case. Recent experiments have demonstrated that centrioles are not necessary for mitotic spindle formation, because when they are removed, the structure still forms. Asters play a critical role in correctly orienting the mitotic spindle and inducing the cleavage furrow in animal cells, which will be discussed shortly.

As the mitotic spindle continues to form, the cell enters a *prometaphase* stage, where the nuclear envelope, the double-membrane structure surrounding the nucleus, begins to fragment and break down, releasing the chromosomes. After they are released into the cytoplasm, a group of highly positively charged Ki-67 proteins, found in the nucleoli between cell divisions, bind to the chromosomes. One end of the protein attaches to the chromosome, while the other end extends out into the cytoplasm, similar to bristles on a brush. This arrangement provides a physical and electrostatic barrier that prevents the chromosomes from coalescing during mitotic division. The chromosomes, which are even more condensed, begin to attach to the mitotic spindle by way of a specialized protein structure called a *kinetochore,* which forms near the centromere of the sister chromatids. As the kinetochore of the sister chromatids attaches to the mitotic spindle, the chromosomes are pulled back and forth until they are aligned.

Metaphase

Once the centrosomes are located on opposite ends of the cell, the chromosomes are pulled back and forth by the mitotic spindle, aligning them in a straight line along an imaginary plane in the center of the cell, known as the **metaphase plate** (Figure 7.10). While in metaphase, the cell reaches an important checkpoint in the cell cycle known as the **M checkpoint** (Figure 7.9). It is at this checkpoint that the cell determines whether the chromosomes are attached and aligned correctly prior to moving to anaphase.

Anaphase

When the cell clears the M checkpoint, it enters anaphase, the shortest of the four phases of mitosis. During anaphase, sister chromatids are pulled apart at their centromeres and begin moving toward opposite ends of the cell as the microtubules attached to the kinetochores shorten and disassemble (Figure 7.10). Upon the completion of anaphase, each end of the cell contains a complete collection of independent chromosomes.

Telophase

Once the independent chromosomes have reached opposite ends of the cell, telophase begins; this is the final phase of mitosis. Telophase is characterized by events opposite to those outlined in prophase (Figure 7.10). The chromosomes start to uncoil, returning to chromatin fibers. The mitotic spindle and its associated microtubule fibers disintegrate and are no longer visible. Finally, remnants of the original nuclear envelope and components of the endomembrane system are used to form new nuclear envelopes around each group of independent chromosomes, giving rise to two new daughter nuclei.

Cytokinesis

Cytokinesis (a term coined by American biologist Charles Otis Whitman in 1887) is the third and final stage of the eukaryotic cell cycle. This stage is defined by the division of the cytoplasm and additional cellular contents into the resulting daughter cells. In some cells, cytokinesis can begin as early as anaphase, but generally, it begins at the end of telophase and continues until the resulting cells enter the G_1 *phase* of interphase. Although cytokinesis involves the division of the cytoplasm, the process differs in animal and plant cells.

Animal Cells

Cytokinesis in animal cells is known as cleavage and begins during anaphase. As the sister chromatids begin to separate, an interaction occurs between two proteins—actin and myosin—forming a structure known as the **contractile ring**, located along the old metaphase plate. As the contractile ring constricts, it forms a shallow indentation known as the **cleavage furrow**. As the ring continues to constrict, the ring tightens around the parent cell, much like a drawstring. The continual tightening begins to pinch the parent cell into two daughter cells, with each new cell consisting of its nucleus, cytoplasm, and organelles.

Plant Cells

Plant cell cytokinesis differs from that of animal cells because it does not involve the formation of a cleavage furrow. Cytokinesis in plant cells begins along the old metaphase plate with the formation of a **cell plate** during telophase. The cell plate forms when vesicles containing important carbohydrates and other components required for building new cell walls and cell membranes are released from the Golgi apparatus and begin congregating and fusing together. As more vesicles congregate and fuse, the cell plate begins to grow, until its edges fuse with the original cell membrane. As important cell membrane molecules and carbohydrates collect within the cell plate, new cell walls form between the two daughter cells. The new cell walls continue to grow and become stronger with the accumulation of cellulose.

Mitosis is an important process in both unicellular and multicellular eukaryotic organisms. In unicellular organisms, such as yeast, *Amoeba*, and *Paramecium*, mitosis is used as a means of asexual reproduction. In multicellular organisms, mitosis is used in the reproduction of stem cells, for growth and maintenance of tissues, and body repair. In biotechnology, mitosis generates nuclei for cloning and may produce stem cells, specialized cells that differentiate into various types of cells, including muscle and nerve cells.

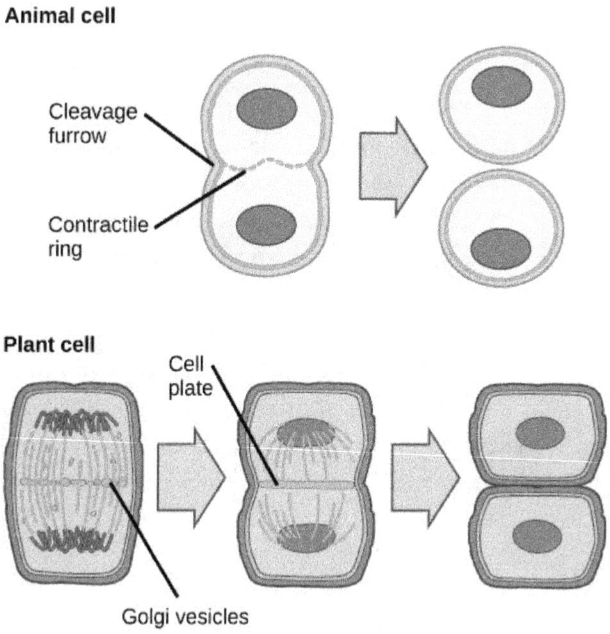

FIGURE 7.11 Cytokinesis is the division of cytoplasm and additional cellular materials into the newly formed daughter cells following mitosis. In animal cells, a contractile ring forms and constricts along the old metaphase plate, creating a shallow identation that continues to constrict until the parent cell is pinched into two daughter cells. In plant cells, a cell plate forms along the old metaphase plate as vesicles from the Golgi apparatus fuse together depositing their contents to form new cell walls between the daughter cells.

7-4. Sexual Reproduction and Meiosis

Sexual Reproduction

Sexual reproduction is a form of reproduction that involves the fusion of sex cells, or **gametes**, from two distinct parents during a process known as **fertilization**. Fertilization was first understood through studies performed on sea urchins in 1876 by German scientist Oskar Hertwig, wherein he revealed that fertilization occurred when the nuclei of the gametes fused together. In 1883, Edouard van Beneden showed that the gametes involved in fertilization were unique cells containing only half the number of chromosomes as the rest of the cells in an organism. In other words, gametes consist of only one pair of chromosomes, or what is called a **haploid** number of chromosomes, designated by the letter **n**. Body cells, or **somatic cells**, have two copies of each chromosome and are known as **diploid** or **2n**, with one copy inherited from the mother and the other inherited from the father. For example, fruit flies have a diploid number of eight and haploid number of four, while dogs have a diploid number of 78 and haploid number of 39.

Human somatic cells contain a diploid number of 46 chromosomes, 23 chromosomes inherited from each parent, while human gametes consist of half the number of chromosomes, a haploid number equaling 23. Of the 23 chromosomes within a human gamete, 22 of these are known as autosomes, and the one remaining chromosome is a **sex chromosome**, either an X or a Y, as seen in Figure 7.4. Sex chromosomes differ in size and genetic makeup, and the combination of these chromosomes determines the sex of the offspring. Fertilization between haploid gametes results in a diploid **zygote**, another observation made by Edouard van Beneden in 1883. The resulting zygote contains a mixture of genetic material from both parents, then proceeds through a series of developmental phases to become a genetically unique adult organism.

In order to produce gametes, cells within the ovaries and testes of sexually reproducing organisms undergo a type of cell division known as meiosis. Each of the gametes produced contains a haploid number of chromosomes, including one pair of autosomal chromosomes and one sex chromosome. The process of meiosis follows interphase, a period of cellular activity similar to that prior to mitosis.

Interphase

Meiotic interphase is similar to mitotic interphase in that it consists of the same three phases outlined earlier in this chapter: G_1 phase, S phase, and G_2 phase. As a review, the G_1 phase is associated with daily cellular functions, including growth and protein synthesis, as well as organelle duplication and the acquisition and synthesis of materials required for the S phase, or DNA replication phase. During the S phase, each pair of chromosomes replicates, producing identical sister chromatids joined together at a corresponding centromere. In addition, the centrosomes, organelles that organize the spindle apparatus microtubules, also replicate. During the G_2 phase, proteins are synthesized that are responsible for the formation of the spindle apparatus, along with the assembly and storage of additional cell membrane materials.

The chromosomes replicated during the S phase of meiotic interphase are called **homologous chromosomes**. Homologous chromosomes are chromosomes inherited from each individual parent. For example, in humans, each cell consists of a diploid number of 46 chromosomes, 23 maternal and 23 paternal, with 22 autosomes and one sex chromosome. Both homologous chromosomes in a pair are similar in the following ways: length, shape, staining pattern, centromere position, genetic composition, and gene arrangement. Upon the completion of interphase, meiosis begins.

FIGURE 7.12 In 1883, Edouard van Beneden discovered the process of meiosis.

Meiosis

The process of meiosis was discovered in 1883 by Edouard van Beneden (Figure 7.12), using the horse intestinal roundworm *Ascaris megalocephala*. This group of roundworms only have a total of four chromosomes in their body cells, a low number that proved beneficial for van Beneden to study. Van Beneden discovered that the gametes produced by the roundworm only contained two chromosomes, but the original chromosome number of four was restored in the zygote after fertilization, and the chromosome number remained unchanged in successive embryonic cells. His research led him to ascertain that the body cells of organisms contain a constant, fixed number of chromosomes, while the gametes contain only half this number. Consequently, after fertilization occurred between the gametes, the fixed number of chromosomes was restored. These observations and subsequent studies by van Beneden led to the discovery of the process of meiosis.

Meiosis is a form of nuclear division occurring in the gonads (testes and ovaries) of sexually reproducing organisms and is responsible for gamete formation. Meiosis contains two nuclear divisions and reduces the diploid number of chromosomes in a parent cell into four haploid, genetically distinct, daughter cells. Each nuclear division involves four phases, with each phase containing structures and events similar to those discussed in mitosis.

The first nuclear division of meiosis is called **meiosis I** and involves the separation of the homologous chromosomes.

Prophase I

Prophase I is a unique, complex, and active phase of meiosis I due to a variety of specialized events that can last for several days. Prophase I follows meiotic interphase and begins when the replicated chromosomes condense, become visible, and start aligning themselves with their homologous counterparts (Figure 7.13). The maternal and paternal homologues are attached by proteins, which are responsible for aligning the entire length of the homologues side by side along similar genes, forming **tetrads**. Tetrads are so named because the homologous chromosomes consist of two chromosomes (replicated maternal and paternal chromosomes), with four chromatids each.

Once the homologous chromosomes are aligned as tetrads along an X-shaped region called the **chiasma** (plural *chiasmata*), a unique DNA recombination process occurs, known as crossing over. **Crossing over** is a random exchange of genetic material that occurs between the maternal and paternal homologous chromosomes (Figure 7.13). This exchange results in genetic recombination, a term used to describe chromosomes consisting of a unique mixture of maternal and paternal genes. Depending on the size and length of the chromosome and the position of its centromere, crossing over

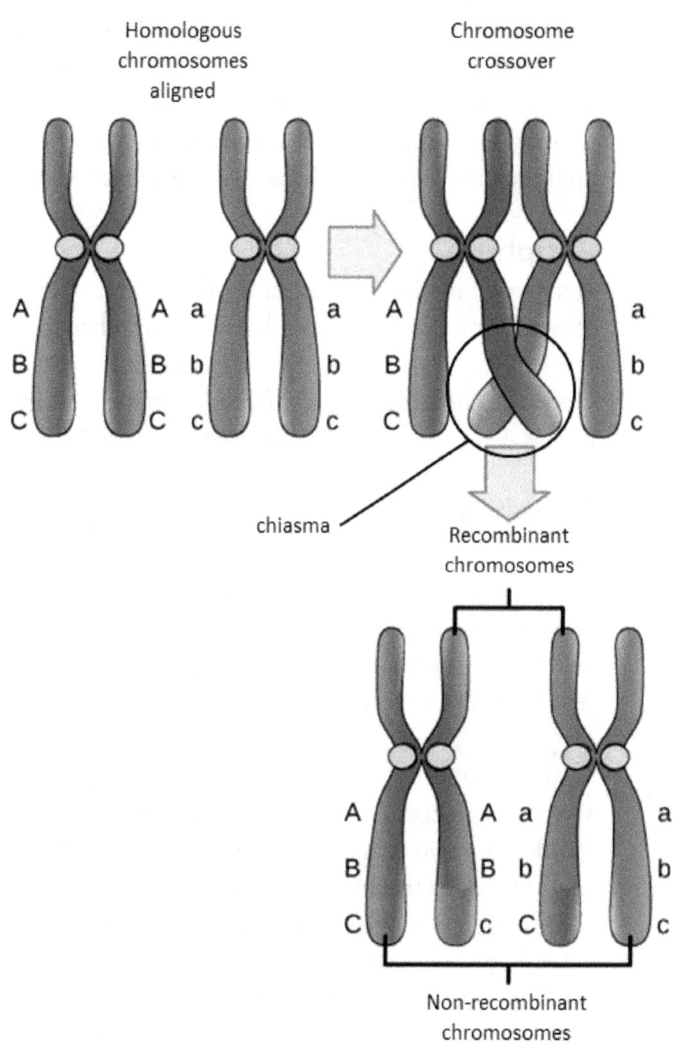

FIGURE 7.13 During prophase I of meiosis, homologous chromosomes align and crossing over occurs between the parental chromosomes along the chiasma. Crossing over produces chromosomes that consist of a mixture of unique genes, known as genetic recombination.

can occur on average two to three times along a variety of locations on the chromosome. Crossing over is responsible for assisting in the separation of chromosomes during meiosis but also, more importantly, generating genetic diversity. The recombination of genetic material during crossing over is the primary reason as to why siblings are not genetically identical to each other.

The condensed homologous chromosomes are released shortly after the completion of crossing over, when the nuclear envelope breaks down (Figure 7.14). The kinetochores of the homologous chromosomes attach to the spindle apparatus, which began to form after the chromosomes condensed and the centrosomes started moving toward the cellular poles. Once attached, the microtubules of the spindle apparatus move the homologous chromosomes toward the center of the cell, initiating metaphase I.

Metaphase I

The homologous chromosomes attach to the spindle fibers by way of the kinetochores, and the spindle fibers move the homologous chromosomes toward the metaphase plate. Once at the metaphase plate, the homologous chromosomes are randomly aligned along the center of the cell, independently of each other (Figure 7.14). This random and independent alignment of homologous chromosomes is an important aspect of meiosis because it generates genetic diversity.

Anaphase I

Anaphase I marks the beginning of the first division of nuclear material, as the homologous chromosomes are randomly separated from one another as the spindle fibers pull the pairs to opposite cellular poles (Figure 7.14). Unlike mitotic anaphase, the sister chromatids remain attached at their centromeres. Upon the completion of anaphase I, a mixture of maternal and paternal homologous chromosomes are located at each cellular pole, with each group consisting of the haploid number of chromosomes.

Telophase I

The end of meiosis I begins with the breakdown of the spindle apparatus, after the homologous chromosomes, now containing two sister chromatids, arrive at opposite cellular poles (Figure 7.14). In some species, the chromosomes decondense as nuclear envelopes reform around each haploid set of chromosomes; however, there are some cases in which the nuclear envelope does not reform and the cells proceed with meiosis II.

Cytokinesis

Cytokinesis occurs simultaneously with telophase I (Figure 7.14), and the cytoplasm divides as the cell divides into two daughter cells. In animal cells, the cleavage furrow forms, and in plant cells, the cell plate forms. At the end of cytokinesis, the cell has divided, with each resulting daughter cell nucleus having its own haploid set of genetic material. Once cytokinesis finishes, the cells may return to a brief interphase, but the chromosomes do not undergo additional replication, and the cells begin meiosis II.

The second nuclear division of meiosis is called **meiosis II** and is similar to the process of mitosis, in which the sister chromatids are separated. In some organisms, once the cells complete telophase I and cytokinesis, the cells proceed with a brief interphase. During this time, the chromosomes temporarily uncoil, but DNA replication does not occur. In addition, some proteins are manufactured, which are responsible for leading the cells through meiosis II. In other organisms, meiosis II occurs immediately, without an intervening interphase.

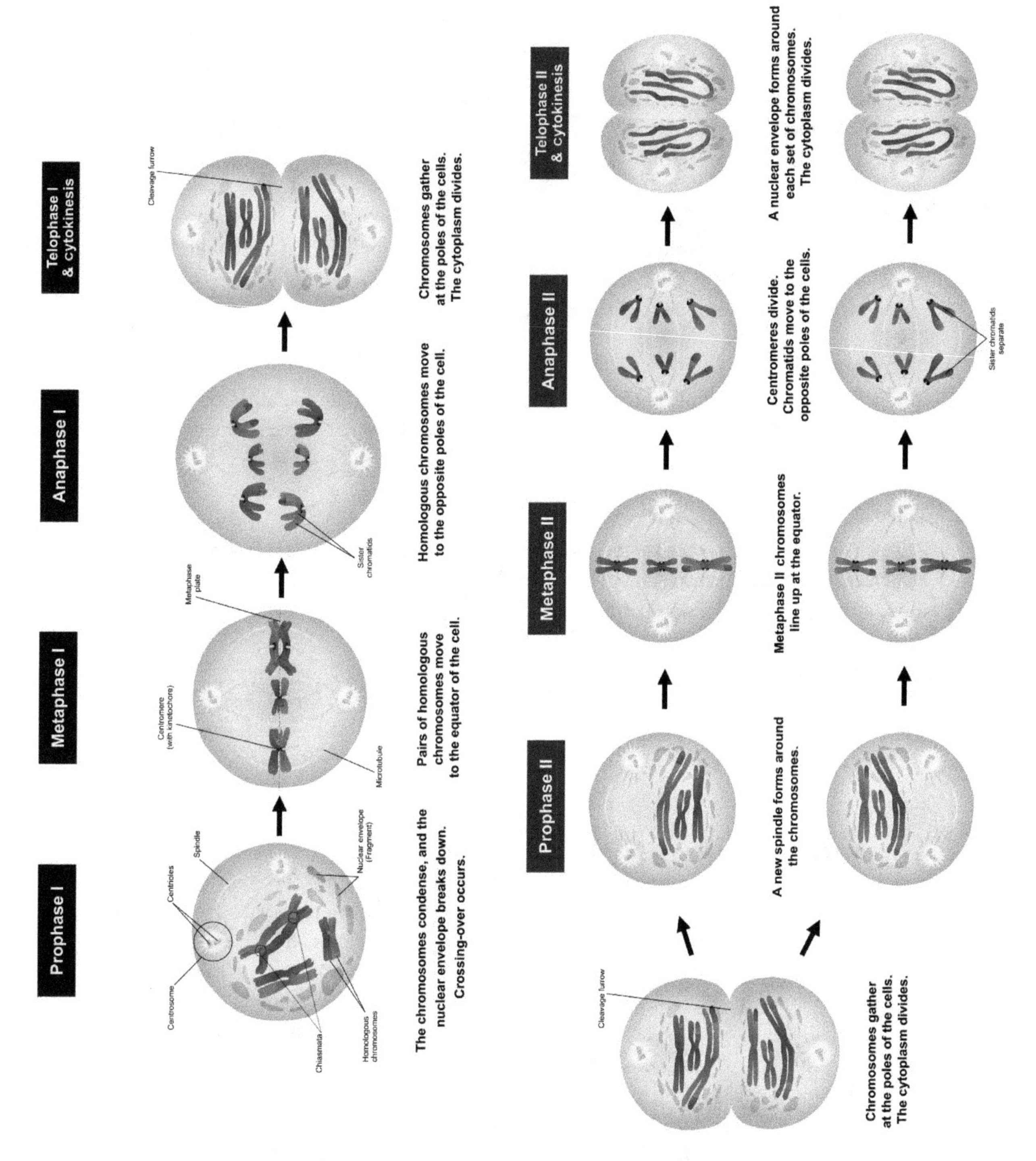

FIGURE 7.14 The process of meiosis consists of two nuclear divisions that is responsible for producing four, genetically distinct, haploid daughter cells called gametes. Each nuclear division contains prophase, metaphase, anaphase, and telophase, shortly followed by cytokinesis.

Prophase II

Prophase II marks the beginning of meiosis II as the nuclear envelopes, if formed following telophase I and cytokinesis, break down. In addition, the chromosomes begin to condense and become visible as a new spindle apparatus forms. Kinetochores attach the condensed sister chromatids to the spindle fibers, and the spindle fibers start to move the chromosomes toward the metaphase plate (Figure 7.14).

Metaphase II

During metaphase II, the spindle fibers move the sister chromatids and align them in the center of the cell (Figure 7.14). As a result of crossing over during prophase I, the sister chromatids along the center of the cell are not genetically identical.

Anaphase II

In anaphase II, as in mitotic anaphase, the spindle fibers separate the attached sister chromatids into individual chromosomes and pull the individual chromosomes toward opposite cellular poles (Figure 7.14).

Telophase II

Once all of the chromosomes are located at the cellular poles, telophase II begins. During telophase II, nuclear envelopes form around the separated chromosomes, the chromosomes decondense, and the spindle apparatus breaks down (Figure 7.14).

Cytokinesis

The events of telophase II are shortly followed by cytokinesis and the division of the cytoplasm. Upon the completion of cytokinesis and meiosis II, four genetically unique, haploid daughter cells are produced from a diploid parental cell (Figure 7.14).

Gametes

Meiosis produces two types of gametes: sperm and eggs (Figure 7.15).

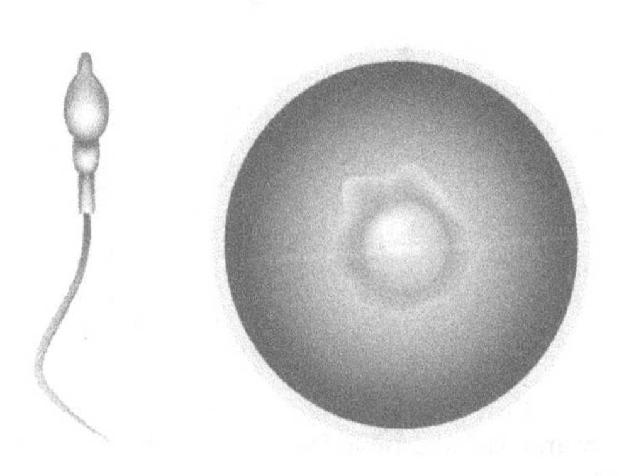

FIGURE 7.15 Two different types of gametes are produced upon the completion of meiosis, sperm cells (left) and an egg cell (right).

Sperm are small, motile cells with little cytoplasm. Four sperm cells of equal size are produced in the testes during meiosis, with each cell containing a haploid number of chromosomes. When sperm cells were first observed in 1678 by Antonie van Leeuwenhoek, he believed they were parasites. However, in 1685, Leeuwenhoek outlined a different approach, suggesting that sperm contained a premade human, an idea known as **preformation** (Figure 7.16).

In order for the premade human to develop into a new life, it required the nurturing nature of a female. In 1779, an experiment performed by Italian biologist Lazzaro Spallanzani with amphibian semen demonstrated that seminal fluid alone was not responsible for fertilization. Although his experiment indirectly established that sperm was important for fertilization, he still believed, much like Leeuwenhoek, that they were microbial contaminations. It was not until the 1800s that scientists demonstrated that sperm were not contaminants but rather an integral part of the fertilization process.

Eggs are produced in the ovaries of sexually reproducing females and are considerably larger than sperm, with more cytoplasm. During meiosis in the ovaries, four cells are produced; however, only one of these cells will develop into the egg, while the others are called **polar bodies** (Figure 7.17).

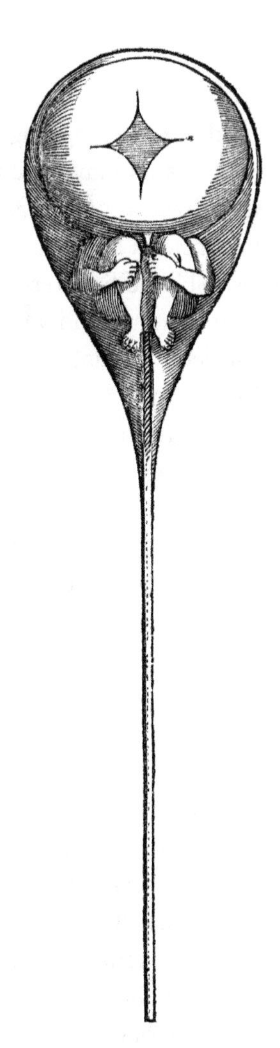

FIGURE 7.16 Preformation, or the idea that sperm contained a tiny, pre-made human, was first suggested by Antonie van Leeuwenhoek in 1685. This illustration of preformation was drawn by Nicolaas Hartsoeker in 1695.

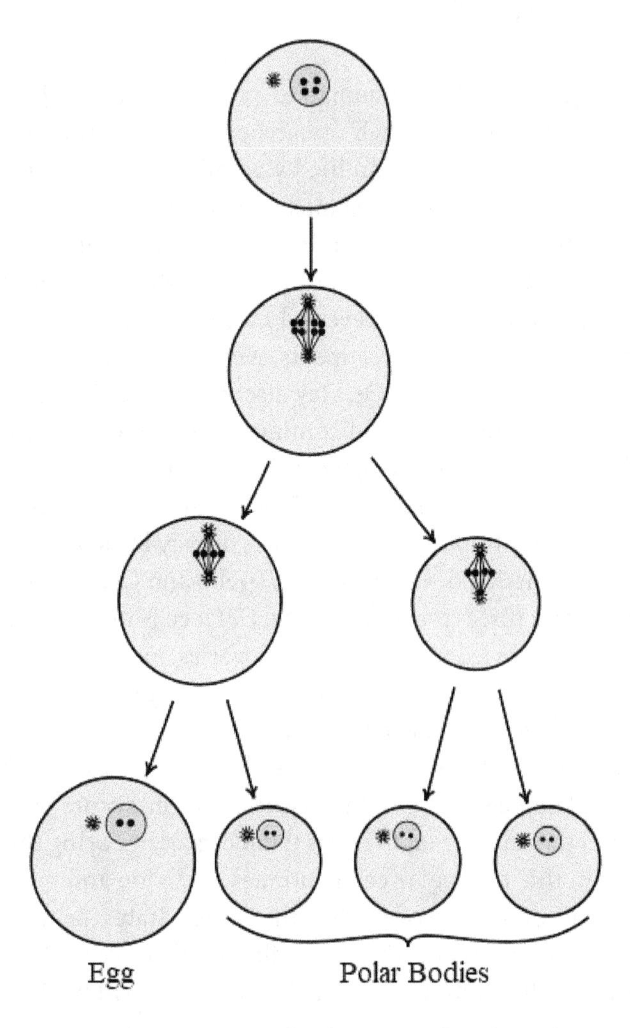

FIGURE 7.17 During gamete development in females, only one of the cells produced will develop into the egg. The remaining cells are called polar bodies, which serve no developmental purpose. These cells are quickly disintegrated and reabsorbed by the female body upon formation.

Although each cell receives the same amount of genetic material, the cell that develops into an "egg" receives a majority of the cytoplasm, organelles, nutrients, and a variety of biochemicals, including RNA and proteins. The biochemicals are important for the development of the zygote and cell specialization in the embryo. The remaining smaller polar bodies are not associated with zygote or embryo development and therefore disintegrate immediately upon formation and are absorbed by the body of the female.

7-5. Cancer

The physical ability of a human to live beyond their biological means would require stopping the body from aging. Although researchers believe future technological advancements may have the potential to extend human life by an additional sixty to eighty years, the physical body has an expiration date, and thus remains a mortal entity. However, despite death, Henrietta Lacks, a black woman born in Roanoke, Virginia, has remained an immortal contributor to biomedical research since 1951. Upon a visit to John Hopkins Medical Center for a "knot" in her womb and vaginal bleeding, numerous tests revealed Lacks had cervical cancer. The diagnosis led to radiation treatment and during these treatments, American cell biologist, Dr. George Gey, acquired a sample of cells from Lack's cervix. Dr. Gey discovered under specific, laboratory conditions, these cells were immortal, therefore, would continuously divide and nearly double in number every day. These cells, known as *HeLa cells*, were the first group of immortal cells to be cultured and grown in a laboratory setting, and in over 70 years since their acquisition, the cells are continuing to thrive and multiply. Today, Henrietta Lacks' legacy lives on as her immortal cells have been widely used to study genetics, develop vaccines (polio and COVID-19), and determining treatments for AIDS, Parkinson's disease, and infertility. *HeLa* cells are especially useful and important in the study of cancer and has led to numerous discoveries, including the link between the human papilloma virus (HPV) and cervical cancer by German virologist Harald zur Hausen, who received the Nobel Prize in Physiology or Medicine in 2008.

Cancer is a disease characterized by uncontrollable cellular growth and division that can spread into and damage surrounding tissues. Cancer occurs when one cell in the body loses its ability to control the number of divisions it undergoes, ignoring the cellular signal prompting apoptosis. As a result, this particular cell continues to divide and grows into a cancerous mass. Cancer is the second leading cause of death in the United States, behind heart disease, and represents more than 20 percent of total deaths. According to data from 2013–2015 compiled by the National Cancer Institute, almost 40 percent of all Americans will be diagnosed with some form of cancer during their lifetime.

Cancer develops when DNA is disrupted within a normal cell, affecting its ability to regulate the cell cycle. The disruption of cellular DNA occurs as a result of a buildup of mutations propagated through either chemical exposure, radiation, bacteria, or viruses. A specific mutation in the DNA leads to an enzyme commonly found in cancer cells called **telomerase**, which is responsible for rebuilding telomeres. Recall that telomeres regulate the number of cell divisions a normal cell can undergo; however, if telomeres are being rebuilt and maintained at a constant length in cancer cells, they are able to divide indefinitely.

Characteristics of Cancer Cells

Cancer cells have several unique characteristics that differentiate them from normal, healthy cells.

1. **Nonspecialized cells.** Cancer cells are nonspecialized, abnormally shaped cells that are able to divide indefinitely when supplied with adequate space and nutrition. Cancer cells disregard normal cellular signals regulating cell division, which is then passed on each time the cancer cell divides. Although some cancer cells are not considered fast-dividing cells, they do divide faster and more times than the normal cell in which they originated. Some of the fastest cancer cells can proceed through a cell cycle in as little as 18 hours. If a cancer cell does stop dividing, it does so at random checkpoints within the cell cycle, not at the normal mitotic checkpoints outlined in section 7-3.
2. **Large, abnormal nuclei.** The nuclei of cancer cells are larger than those found in normal cells, consisting of several chromosomal abnormalities. For example, some nuclei of cancer cells may contain extra copies of chromosomes, giving the nuclei an abnormal number, while some possess portions that are either duplicated or deleted.
3. **Immortal.** Normal cells can only divide a specified number of times until a loss of functional DNA triggers apoptosis. Although abnormal cells are generally recognized by the immune system, triggering apoptosis, cancer cells bypass this cellular signal, which is why they are often referred to as being immortal.
4. **Lack of contact inhibition.** Normal cells exhibit contact inhibition, which refers to the point at which cellular division ceases when the cells come into contact with a neighboring cell. Cancer cells, however, lack contact inhibition and continue dividing, producing multiple layers of overlapping cells that develop into a tumor. Tumors will be discussed in more detail shortly.
5. **Metastasis.** Adhesion molecules, proteins found within the cell membranes of normal, healthy cells, are responsible for holding cells together. However, cancer cells lack adhesion molecules and can separate from the original mass of cells. Once removed from the mass, the cancer cells travel through the circulatory and lymphatic systems to other regions of the body, in a process known as metastasis.
6. **Angiogenesis.** Angiogenesis is the formation of new blood vessels. When metastasized cancer cells anchor themselves onto healthy tissues or organs, additional mutations within the cancer cell initiate angiogenesis, which provides the cancer cells with the required nutrients and oxygen for sustained growth in this new area.
7. **Transmissible.** In a 2016 study, it was discovered that not only can cancer cells be transferred between individuals of the same species, but transference can also occur between individuals of different species. The transmissibility of cancer is rare, especially in humans, expect in the cases of organ transplants or surgical accidents. However, its occurrence has been well-documented in Tasmanian devils and soft-shell clams. The 2016 study revealed the presence of different forms of disseminated neoplasia, a transmissible cancer, found in various species of bivalves, including mussels, clams, and cockles. Upon analyzing the DNA of cancer cells in a host, the researchers discovered that the DNA was genetically different from the host DNA but similar to cancer cell DNA found in other individuals of the same species.

This led researchers to suggest that cancer cells must have been transferred between hosts. In addition, the researchers also examined the DNA of cancer cells in one specific species of clam, revealing that although the DNA was genetically distinct from the host, it was similar to the cancer cell DNA found in a completely different clam species. Consequently, the researchers suggested that cancer cells have the ability to be transmitted between completely different species.

Tumors

A tumor is mass of abnormal cells found within normal tissue that are no longer regulated by the cell cycle. Tumors can be one of two types: benign or malignant.

Benign tumors are masses of abnormal cells that remain confined to their original location and do not spread (Figure 7.18, left). A majority of benign tumors are not associated with serious problems and can be safely removed without issue by surgery. Common examples of benign tumors include uterine fibroids and lipomas in the skin. Although benign tumors are noncancerous, most cancers begin as benign tumors, when mutations generate both genetic and cellular changes, resulting in cells that ignore the signals regulating the cell cycle. When this occurs, the benign tumor becomes a malignant tumor, which can metastasize.

A malignant tumor consists of a mass of unrestrained growing and dividing cancer cells, which can metastasize and spread to other parts of the body (Figure 7.18, right). Metastasis occurs when genetic and cellular changes cause the cell surface of cancer cells to lose their ability to remain attached to the mass. As a result, the cancer cells separate, enter the circulatory system or lymphatic system, and travel to another part of the body. Once there, the cancer cells invade healthy tissue and begin proliferating, producing a new tumor while also initiating angiogenesis. This invasion begins to impair the physiological function of the organ or organs.

Malignant tumors cause death by growing larger, using energy and nutrient supplies from the body and invading the spaces normally occupied by normal tissues and organs. Tumor growth is

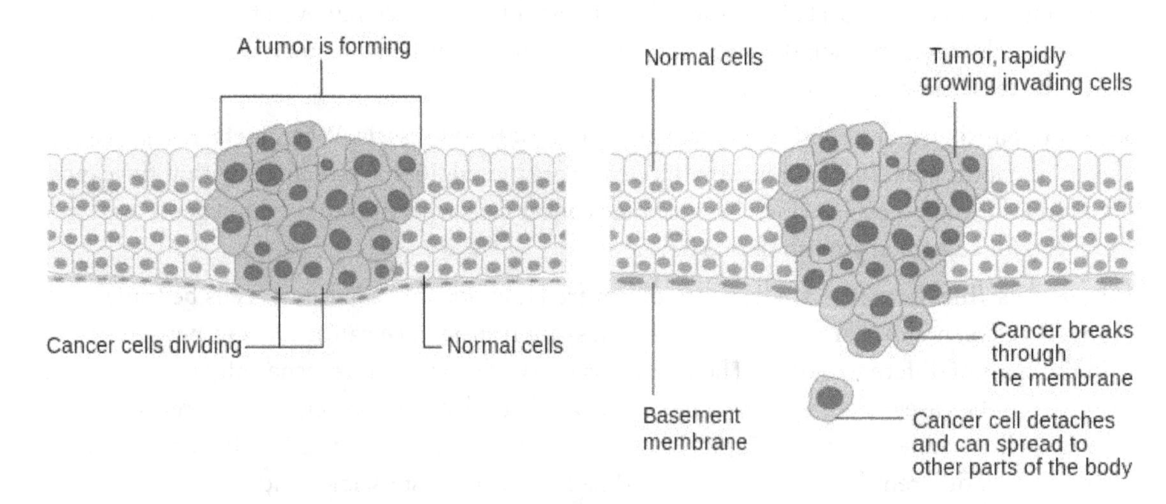

FIGURE 7.18 There are two different types of tumors. Benign tumors (left), such as fibroids or lipomas, are a mass of non-cancerous cells that do not spread. A malignant tumor (right) is a mass of cancerous cells that can metastasize or spread to other parts of the body.

dependent on the cell affected and the overall size of the tumor. A smaller tumor grows relatively slowly due to fewer cells dividing; however, as the tumor gets larger, growth rate increases, due to the larger number of cells dividing. When tumors invade a new region of the body, they affect the ability of these tissues and organs to perform necessary physiological processes, including those critical to life, such as heart function and breathing, leading to the death not only of the tissue or organ but of the individual as well.

Treatment

The various different types of cancer treatments primarily focus on surgically removing, slowing down, or killing rapidly dividing cells. The first step in removing a cancerous tumor is through surgery, and this is for the most part successful. However, there are instances in which surgery may prove difficult because it may not result in the removal of all cancerous tissue, especially cancerous tissue that has become metastatic. Therefore, two other treatments options may be used in order to slow or kill the fast-growing, dividing cells: radiation and chemotherapy.

High-energy radiation is used in order to treat localized cancerous tumors. This method involves directing high-energy radiation to a particular part of the body where the tumor is located. The use of high-energy radiation is effective because it damages the DNA in the cancerous cells, disrupting their cell division and growth. Unfortunately, the use of high-energy radiation can also be harmful to normal cells; however, normal cells have the ability to repair their damaged DNA, while cancer cells lack this ability, which makes high-energy radiation an effective treatment. Although beneficial in treating localized cancer, the use of high-energy radiation is not effective against metastatic cancer.

Chemotherapy is the most frequently used cancer treatment and is primarily used to treat metastatic cancers (Figure 7.19). When administered, powerful chemotherapy drugs are intravenously introduced into the circulatory system. Chemotherapy drugs are called analogs because they are structurally designed to resemble one of the four types of DNA nucleotides. Once the drug is introduced, the cancer cells mistakenly use the DNA nucleotide analogs to synthesize new DNA molecules. This disrupts DNA replication and affects the ability of the cancer cells to divide, ultimately slowing tumor growth. Unfortunately, the use of chemotherapy can also affect healthy cells, including immune system cells, which makes a person more vulnerable to infection. In addition, cancer patients undergoing chemotherapy also report bruising, nausea, and hair loss.

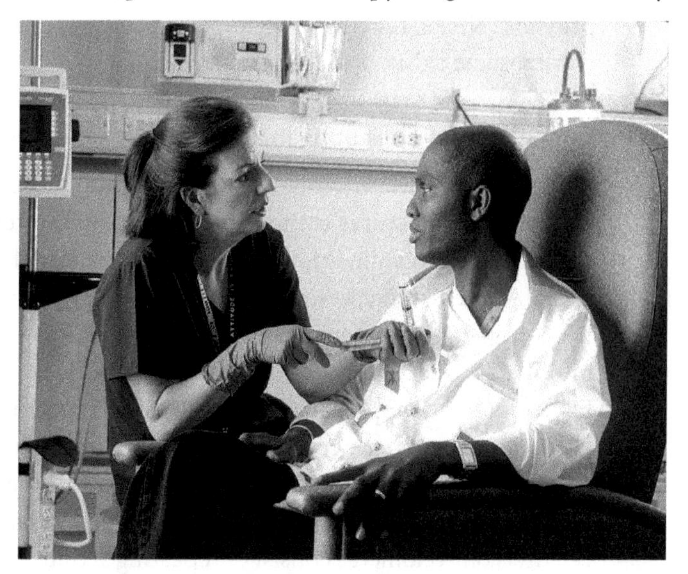

FIGURE 7.19 One of the most widely used treatments for cancer is chemotherapy. Chemotherapy uses DNA nucleotide analogs to disrupt the ability of cancer cells to replicate their DNA, which helps slow their division and growth.

Chapter Summary

7-1. Chromosome Structure

- Walther Flemming was the first to observe and describe chromatin, while using a new staining technique in 1879. Chromatin is a mass of long, threadlike fibers composed of DNA and proteins found in the nucleus, which shortens and coalesces into chromosomes. The changes, movements, and distribution of chromosomes when the nucleus divides were observed and described by Flemming in a process he called mitosis.
- Chromosomes are composed of long, continuous pieces of DNA and various types of proteins that make up an organism's genome. The term *chromosome* was first introduced in 1888 by Wilhelm von Waldeyer-Hartz. Chromosomes transmit inherited traits, as outlined in the chromosomal theory of inheritance, introduced by Theodor Boveri and Walter Sutton in 1902, later confirmed by Thomas Hunt Morgan in 1910. Chromosomes vary in shape, length, banding patterns, the numbers of them in different species, and the number of genes contained. A karyotype is a visual representation of an organism's chromosomes, arranged by shape and staining patterns. Karyotypes can be used as a diagnostic tool in identifying chromosomal disorders. A karyotype is prepared using five main steps: cells are collected by means of a blood sample, amniocentesis, or chorionic villus sampling; cells are encouraged to divide in a nutrient-filled culture tube; cell division is stopped when the chromosomes are highly condensed and visible during metaphase; the chromosomes are stained, which illustrates their distinct banding patterns; he chromosomes are arranged by size, shape, staining pattern, and location of centromeres.
- The prokaryotic chromosome has two main components, DNA and nucleoid-associated proteins (NAPs). Prokaryotic DNA carries the genetic information. Nucleoid-associated proteins (NAPs) function in gene regulation and chromosome organization. The prokaryotic chromosome exists as a large, single, circular piece of DNA, located within the nucleoid region of the cell. Due to its size in comparison to the overall dimensions of the cell, the chromosome is tightly packed, resulting in a genome with densely packed genes.
- The eukaryotic chromosome is located within the nucleus of a cell. Eukaryotic chromosomes are complex structures composed of DNA and proteins in a mass of threadlike fibers called chromatin. Chromatin is found free-floating within the nucleus, consisting of a single molecule of DNA and histone proteins. Histone proteins primarily play a structural role in preventing DNA from tangling, helping organize chromatin into chromosomes and tightly packing DNA into the nucleus. DNA packing results in nucleosomes, a "bead-like" structure containing DNA double-wrapped around eight histone proteins. Additional folding and coiling of the chromatin results in a chromosome. The formed chromosome contains a single centromere and two telomeres. The centromere helps attach sister chromatids (replicated DNA) to one another after DNA replication and is the location at which the chromatids separate during cell division. Telomeres consist of repeating six-nucleotide sequences located at the ends of the chromosome. Telomeres lack genes and cannot code for proteins. Telomeres help maintain chromosomal stability by preventing the removal of coding portions of the chromosome. The telomeres act as an "internal clock," specifying the maximum number of divisions the cell can undergo before apoptosis (cellular death). This idea became known as the Hayflick

limit, named after Leonard Hayflick, who discovered this phenomenon in 1961. According to the Hayflick limit, a human cell can divide about 50 times before the loss of functional DNA results in apoptosis. Cancer cells continually divide because they can rebuild and extend telomeres using the enzyme telomerase.

7-2. Prokaryotic Cell Division

- Prokaryotic cells divide by binary fission, a form of asexual reproduction. Asexual reproduction is a mode of reproduction in which one cell divides into two genetically identical cells. Prokaryotic cell division can only occur with an adequate supply of nutrients and optimal temperatures. A majority of prokaryotic cells can complete cell division in one to three hours, while others can perform the process in as little as 20 minutes.
- Prokaryotic cell division involves two main stages. During the cell growth and chromosomal replication stage, the prokaryotic cell grows, increasing in size, and chromosomal replication occurs. The duplicated chromosomes are attached to different, specialized locations on the cell membrane. Binary fission occurs after DNA replication and consists of three main steps. Cell elongation increases the length of the cell, and as this occurs, the attachment points of the chromosomes begin to increase, pulling the chromosomes apart. As the cell continues to elongate, a new cell membrane and cell wall forms, while proteins begin pinching the cell along the midline. The formation of daughter cells completes binary fission as the parental cell membrane and cell wall continue to pinch and fuse together. Each new daughter cell produced contains a replicated chromosome, making them genetically identical to the parental cell. Binary fission provides prokaryotic cells with advantages: fast and easy to perform and each daughter cell is genetically identical to the parent cell. This allows prokaryotic cells to establish new populations and enables them to reproduce in isolated habitats. Binary fission produces genetically identical offspring, which reduces their genetic diversity, a disadvantage to prokaryotic cells. Less genetically diverse prokaryotic cells make them less adaptable to potential environmental changes.

7-3. Cell Cycle and Cellular Reproduction

- The cell cycle is a series of three main events, in which a cell performs normal cellular processes that end in the cell dividing into two daughter cells.
    - Interphase is the first step in the cell cycle. Once thought to be a "resting phase," interphase is an active period for a cell, in which it performs normal cellular functions. Cellular functions include metabolism, synthesis of organelles and proteins, duplication of chromosomes, and growth. A cell may spend about 90 percent of its time in interphase (about 20 hours). Interphase has three main phases. The G_1 phase is the period in which a cell grows, synthesizes proteins, and carries out cellular respiration, in addition to synthesizing and gathering the materials required for DNA replication and cell division. All cells must pass through the G_1 checkpoint prior to entering the S phase of interphase. The S phase is associated with DNA and centrosome replication. DNA replication involves duplicating

the cell's chromosomes, resulting in sister chromatids joined together at the centromere. Centrosomes are organelles that help organize the microtubules, structures made of tubulin that are responsible for constructing the mitotic spindle. Centrosomes were independently discovered in the 1880s by Edouard van Beneden and Theodor Boveri. Centrosomes are only found in animal cells and contain centrioles. Centrioles are organelles that contain tubulin and function in organizing the mitotic spindle and cytokinesis. The G_2 phase is the final phase, when the cell is actively synthesizing proteins and cell membrane materials in preparation for cell division. All cells must pass through the G_2 checkpoint prior to entering mitosis to ensure that all chromosomes were replicated correctly.

- Mitosis (M phase) is the next step of the cell cycle. The entire process of mitosis was first observed and formally described by Walther Flemming in 1879. Anton Schneider first observed and described metaphase and anaphase in 1873. In 1882, Walther Flemming gave a more extensive description of mitosis, which he described as occurring in two phases. The progressive phase is the phase where chromosomes are visible and aligned in the center of the cell. The regressive phase is the phase where chromosomes separate and the cells divide. Mitosis can now be described in four main stages, which last no more than an hour. Prophase—chromatin condenses into chromosomes, the mitotic spindle forms, the nuclear envelope fragments, and the chromosomes attach to the mitotic spindle. The fragmentation of the nuclear envelope releases the chromosomes into the cytoplasm, which are quickly bound with Ki-67 proteins. Ki-67 proteins prevent the chromosomes from coalescing once released from the nuclear envelope. Metaphase—condensed chromosomes align in the center of the cell along a region known as the metaphase plate. All cells must pass through the M checkpoint prior to entering anaphase to ensure the chromosomes are attached to the mitotic spindle and properly aligned. Anaphase—the sister chromatids are pulled apart at their centromeres to opposite ends of the cell as the mitotic spindle shortens. Telophase—the chromosomes uncoil, the mitotic spindle breaks down, and new nuclear envelopes assemble around the independent chromosomes, forming two new daughter nuclei.
- Cytokinesis is the last step in the cell cycle. Cytokinesis is the division of the cytoplasm and cellular contents into the newly formed daughter cells. Cytokinesis will generally continue until the G_1 phase of the next interphase. The process of cytokinesis differs in animal and plant cells. In animal cells, cytokinesis begins in anaphase and is known as cleavage. The proteins actin and myosin form a contractile ring that constricts to create an indentation known as a cleavage furrow along the old metaphase plate. As the contractile ring continues to constrict, the parent cell is pinched into two daughter cells, each with their own nucleus and cellular components. In plant cells, cytokinesis begins with the formation of the cell plate, along the old metaphase plate. Vesicles containing carbohydrates and components for a new cell walls and cell membranes are released from the Golgi apparatus. The vesicles congregate and fuse together, forming the cell plate. As more vesicles congregate along the cell plate, new cell walls begin to form, dividing the parent cell into two daughter cells.

- Mitosis is an important reproductive method for some unicellular organisms such as yeast, *Amoeba*, and *Paramecium*. In multicellular organisms, mitosis is important for the growth, maintenance, and repair of body tissues and for the reproduction of stem cells. Mitosis is also used in biotechnology for purposes of cloning and generating stem cells.

7-4. Sexual Reproduction and Meiosis

- Sexual reproduction is a mode of reproduction that involves the fusion of gametes from two parents during fertilization. The fusion of gametes during fertilization was first described by Oskar Hertwig in 1876. Gametes contain only half the number of chromosomes as somatic cells, a discovery made by Edouard van Beneden in 1883. Gametes are haploid (n) and somatic cells are diploid (2n). Human somatic cells have 46 chromosomes (diploid), while the gametes have 23 chromosomes (haploid). Of the 23 chromosomes in a human gamete, 22 are called autosomes, and the remaining chromosome is a sex chromosome, either X or Y. Sex chromosomes differ in size and genetic makeup and determine the sex of the offspring. When haploid gametes fuse during fertilization, a diploid zygote is formed, a phenomenon observed by Edouard van Beneden in 1883. A zygote contains genetic material from both parents, then develops into a genetically unique adult. Gametes are produced in the ovaries and testes of sexually reproducing organisms by meiosis.
- Meiosis is preceded by interphase, a three-phase process similar to that of mitotic interphase. The G_1 phase includes cellular growth, protein synthesis, organelle replication, and synthesizing materials required for DNA replication. The S phase is a replication phase. Each pair of chromosomes replicates to produce sister chromatids attached at the centromere. These particular group of chromosomes are called homologous chromosomes. Homologous chromosomes are inherited from each parent. In humans, the offspring inherit 23 chromosomes from the mother and 23 from the father. Homologous chromosomes are similar in length, shape, staining pattern, centromere position, genetic composition, and gene arrangement. Centrosomes, the organelles responsible for organizing the spindle apparatus microtubules, also replicate. The G_2 phase is the final phase, and in it, the cell is preparing for division by synthesizing proteins associated with the spindle apparatus and producing materials required for building new cell membranes.
- Meiosis was observed and described by Edouard van Beneden in 1883 while studying the horse intestinal roundworm *Ascaris megalocephala*. *Ascaris megalocephala* have a total of four chromosomes, but van Beneden discovered that the gametes contained only two chromosomes. Once the gametes fused during fertilization, the original chromosome number was restored.
- Meiosis occurs in the ovaries and testes of sexually reproducing organisms and produces gametes. Meiosis consists of two nuclear divisions, each containing four phases, and is responsible for reducing a diploid cell into four genetically distinct haploid cells.
    - Meiosis I is the first nuclear division, consisting of four phases, which separates the homologous chromosomes. Prophase I consists of several specialized events. Homologous chromosomes Tetrads contain replicated maternal and paternal chromosomes, equaling four chromatids condense, become visible, and align and attach with each other along similar genes to form tetrads. Crossing over, a DNA recombination process wherein genetic material is randomly exchanged between the homologous chromosomes, occurs along a region known as the chiasma. Crossing over assists with the separation of chromosomes and helps generate genetic diversity. Nuclear envelope fragments, spindle apparatus forms, and the homologous chromosomes attach to the spindle by the kinetochores. Metaphase I—condensed homologous chromosomes randomly and independently align

in the center of the cell along a region known as the metaphase plate. The random and independent alignment of homologous chromosomes generates genetic diversity. Anaphase I—homologous chromosomes randomly separate as the spindle shortens, pulling a mixture of maternal and paternal chromosomes to opposite poles of the cell. Unlike in mitotic anaphase, the sister chromatids remain attached. Telophase I—homologous chromosomes uncoil, the spindle apparatus breaks down, and the nuclear envelope may or may not reform around each haploid set of chromosomes. Cytokinesis—cytoplasm divides, cleavage furrow forms in animal cells, and cell plate forms in plant cells to form daughter cell nuclei with a haploid number of chromosomes. Cells may return to a short interphase after cytokinesis.

- Meiosis II is the second nuclear division, consisting of four phases, which separates the sister chromatids in a process similar to mitosis. During the short interphase, the chromosomes temporarily uncoil but do not replicate, and proteins are synthesized that help the cells proceed through meiosis II. Prophase II—sister chromatids condense, a new spindle apparatus forms, the nuclear envelope fragments, and sister chromatids attach to the spindle by the kinetochores. Metaphase II—the sister chromatids, which are genetically distinct, are pulled by the spindle apparatus until they align along the metaphase plate. Anaphase II—the sister chromatids separate into individual chromosomes and are pulled to opposite poles of the cell. Telophase II—chromosomes uncoil, spindle apparatus breaks down, and the nuclear envelope reforms around the chromosomes. Cytokinesis—cytoplasm is divided, and four genetically distinct, haploid daughter cells are formed.

- Two types of gametes are produced after meiosis. Sperm are small, motile, equal-sized cells containing little cytoplasm. Four cells are produced in the testes during meiosis, with each containing a haploid number of chromosomes. Sperm cells were first observed in 1678 by Antonie van Leeuwenhoek, who believed they were parasites. Preformation was an idea developed by Leeuwenhoek in 1685 that suggested that sperm contained a premade human that relied on a female for nourishment. Experiments performed by Lazzaro Spallanzani in 1779 suggested sperm was responsible for fertilization. Sperm was determined not to be a microbial contaminant, but an important part of fertilization, in the 1800s. Eggs are larger gametes in comparison to sperm, containing more cytoplasm. Four cells are produced in the ovaries during meiosis, but only one of these cells will receive the majority of the cytoplasm, organelles, nutrients, and biochemicals during meiosis to become the egg. The remaining cells are called polar bodies and will disintegrate and be reabsorbed by the female body after forming.

7-5. Cancer

- Cancer is defined as uncontrollable cellular growth and division, resulting in cells that can spread into surrounding tissues, damaging them. Cancer occurs when a cell loses its ability to control its cell cycle as a result of a mutation, which leads the cell to ignore apoptosis signals. Telomerase is an enzyme produced by cancerous cells that rebuilds telomeres. Telomeres are responsible for regulating the cell divisions a cell can undergo. Cancer is the second leading cause of death in the United States. The National Cancer Institute predicts that about 40 percent of all Americans will be diagnosed in their lifetime.

- Cancer has several different characteristics. Nonspecialized cells with an abnormal shape that can divide indefinitely and faster than normal cells. Cancer cells contain large nuclei, with different types of chromosomal abnormalities. They are immortal, which means they have lost the ability to trigger cellular apoptosis. The cells lack contact inhibition and are unable to stop cellular division upon coming into contact with another cell. Cancer cells can metastasize from the original mass, because they lack the adhesion molecules that hold normal cells together. When cancer cells metastasize and relocate to healthy tissues or organs, they can form new blood vessels, known as angiogenesis. Cancer cells are transmissible between individuals of the same species and individuals of different species.
- An abnormal mass of cells no longer regulated by the cell cycle is called a tumor. There are two different types of tumors: benign tumors are noncancerous and nonmetastasizing and malignant tumors are masses of cancerous cells that can metastasize and invade healthy tissues or organs. Once cancerous cells have invaded healthy tissues or organs, they initiate angiogenesis and begin to grow into new tumors, affecting the normal physiology of these tissues and organs.
- Various different options are available for treating cancer. Surgery is the first step in removing a cancerous tumor and is typically the most successful. Surgery is not effective against metastatic cancer. High-energy radiation treats localized cancerous tumors. High-energy radiation is effective in damaging the DNA in cancerous cells, affecting their ability to divide and grow. The use of high-energy radiation also damages the DNA of normal, healthy cells, but these cells can repair their damaged DNA, a characteristic that cancer cells lack. High-energy radiation is not effective against metastatic cancer. Chemotherapy is the most frequently used cancer treatment. Chemotherapy involves the use of drugs called analogs, which are structural derivatives of DNA nucleotides. Cancer cells use the analogs for DNA replication, but this disrupts the process leading to the inability of the cells to divide, thereby slowing tumor growth. The use of chemotherapy can affect healthy immune system cells, making the person receiving the treatment more vulnerable to infection. Patients exposed to chemotherapy have reported bruising, nausea, and hair loss. Chemotherapy is effective against metastatic cancer.

End-of-Chapter Activities and Questions

Directions: Please refer back to what you learned in this chapter to complete the following activities.

Define Each Term in Your Own Words

1. Hayflick Limit
2. Binary Fission
3. Ki-67
4. Crossing Over
5. Metastasis

Chapter Review

1. Describe the relationship between the Hayflick limit, telomeres, telomerase, and cancer.
2. Explain how binary fission in prokaryotes and mitosis in unicellular eukaryotes are similar.

3. Identify and describe the cell cycle checkpoints. Why is it important for eukaryotic cells to be regulated by these checkpoints?
4. How are the events of meiosis II similar to those of mitosis?
5. What are the differences between benign and malignant tumors? How does a benign tumor become malignant?

Multiple Choice

1. What is the major difference between a prokaryotic and eukaryotic chromosome?
   - **a.** prokaryotic chromosome contains histone proteins; eukaryotic chromosome has nucleoid-associated proteins
   - **b.** prokaryotic chromosome is circular; eukaryotic chromosome is linear
   - **c.** prokaryotic chromosome is found in the nucleus; eukaryotic chromosome is found within the nucleoid
   - **d.** prokaryotic chromosome is free-floating; eukaryotic chromosome is attached to the cell membrane
2. Why is binary fission considered a disadvantage to prokaryotic cells?
   - **a.** reduces genetic diversity of the population
   - **b.** enables them to establish new populations
   - **c.** prevents reproduction in isolated habitats
   - **d.** relatively fast and easy process
3. What structure forms along the old metaphase plate during cytokinesis in plant cells?
   - **a.** contractile ring
   - **b.** cell plate
   - **c.** cleavage furrow
   - **d.** spindle apparatus
4. Upon the completion of meiosis, what immediate structure(s) is (are) produced?
   - **a.** zygote
   - **b.** sex chromosomes
   - **c.** gametes
   - **d.** all of the choices are correct
5. What enzyme enables cancer cells to become immortal and ignore the cellular signals triggering apoptosis?
   - **a.** primase
   - **b.** polymerase
   - **c.** telomerase
   - **d.** helicase

Image Credits

Fig. 7.1: Source: https://commons.wikimedia.org/wiki/File:Wilhelm_von_Waldeyer-Hartz.jpg.

Fig. 7.2a: Source: https://commons.wikimedia.org/wiki/File:Walther_flemming_2.jpg.

Fig. 7.2b: Source: https://commons.wikimedia.org/wiki/File:-Cell_division_according_to_W._Flemming_(1882).png.

Fig. 7.3a: Source: https://commons.wikimedia.org/wiki/File:Walter_sutton.jpg.

Fig. 7.3b: Source: https://commons.wikimedia.org/wiki/File:-Theodor_Boveri_high_res-2.jpg.

Fig. 7.3c: Source: https://commons.wikimedia.org/wiki/File:Thomas_Hunt_Morgan.jpg.

Fig. 7.4: Copyright © by OpenStax (CC BY 3.0) at https://commons.wikimedia.org/wiki/File:2923_Male_Chromosomes.jpg.

Fig. 7.5: Copyright © by Ali Zifan (CC BY-SA 4.0) at https://commons.wikimedia.org/wiki/File:Prokaryote_cell.svg.

Fig. 7.6: Copyright © by KES47 (CC BY 3.0) at https://commons.wikimedia.org/wiki/File:Chromosome_en.svg.

Fig. 7.7: Copyright © by Azmistowski17 (CC BY-SA 4.0) at https://commons.wikimedia.org/wiki/File:Hayflick_Limit_(1).svg.

Fig. 7.8: Copyright © by Ecoddington14 (CC BY-SA 3.0) at https://commons.wikimedia.org/wiki/File:Binary_Fission.png.

Fig. 7.9: Copyright © by OpenStax (CC BY 4.0) at https://commons.wikimedia.org/wiki/File:Figure_10_02_01.jpg.

Fig. 7.10: Source: https://commons.wikimedia.org/wiki/File:-Animal_cell_cycle-en.svg.

Fig. 7.11: Copyright © by OpenStax (CC BY 4.0) at https://commons.wikimedia.org/wiki/File:Figure_10_02_04.jpg.

Fig. 7.12: Source: https://commons.wikimedia.org/wiki/File:E-douard_van_Beneden.jpg.

Fig. 7.13: Copyright © by OpenStax (CC BY 4.0) at https://commons.wikimedia.org/wiki/File:Figure_12_03_04.jpg.

Fig. 7.14: Copyright © by Ali Zifan (CC BY-SA 4.0) at https://commons.wikimedia.org/wiki/File:Meiosis_Stages.svg.

Fig. 7.15a: Copyright © by Database Center for Life Science (DBCLS) (CC BY 3.0) at https://commons.wikimedia.org/wiki/File:Sperm_togopic.png.

Fig. 7.15b: Copyright © by Database Center for Life Science (DBCLS) (CC BY 3.0) at https://commons.wikimedia.org/wiki/File:Ovum.png.

Fig. 7.16: Copyright © by Nicolas Hartsoeker / Wellcome Collection (CC BY 4.0) at https://commons.wikimedia.org/wiki/File:N._Hartsoeker,_Essay_de_dioptrique_Wellcome_M0016638.jpg.

Fig. 7.17: Source: https://commons.wikimedia.org/wiki/File:Gray5.svg.

Fig. 7.18a: Copyright © by Cancer Research UK (CC BY-SA 4.0) at https://commons.wikimedia.org/wiki/File:Diagram_showing_how_cancer_cells_keep_on_reproducing_to_form_a_tumour_CRUK_127.svg.

Fig. 7.18b: Copyright © by Cancer Research UK (CC BY-SA 4.0) at https://commons.wikimedia.org/wiki/File:Diagram_showing_a_malignant_tumour_CRUK_069.svg.

Fig. 7.19: Source: https://commons.wikimedia.org/wiki/File:Nurse_administers_chemotherapy.jpg.

CHAPTER

8

Genetics and Inheritance

PROFILES IN SCIENCE

FIGURE 8.1 Gregor Mendel.

Johann (Gregor) Mendel, the father of modern genetics, was born July 20, 1822, in what is now Hynčice, Czech Republic, and died in 1884 in Brno, Czech Republic (formerly Brünn, Austria-Hungary). Due to financial limitations, his family was unable to send Mendel to school at an early age; therefore, he spent time working on the family farm as a gardener and studying beekeeping. Mendel began grammar school in Opava (Czech Republic) at age 11 and graduated with honors in 1840. Mendel enrolled at the Philosophical Institute of the University of Olmütz (Czech Republic) shortly after graduation in 1840 to study philosophy, physics, and mathematics and offered tutoring to students as a means of financial support. After graduating in 1843, Mendel entered the Augustinian St. Thomas Monastery one year later to begin formal training to become a priest, receiving the name Gregor shortly after being admitted to the order. After being ordained in 1847 and having his own parish one year later, Mendel began substitute teaching in 1849 because he found certain aspects of his work as a parish priest stressful. In 1851, Mendel was sent to the University of Vienna (Austria), where he studied physics under the likes of Christian Doppler (Doppler effect), mathematics, botany, and microscopy. Mendel returned to the monastery in 1853 and began substitute teaching at a secondary school, retaining his position until 1868 despite failing two teacher certification exams (1850 and 1856). In 1856, Mendel began his famed pea experiments in the monastery's experimental garden, primarily focusing his research on the principles of inheritance and publishing his results in 1866 ("Experiments on Plant Hybridization"), one year after presenting his work at the Natural Science Society in Brünn. Mendel's work was largely disregarded by the scientific community until 1900, when four independently working botanists rediscovered his work and replicated his results. Upon becoming abbot in 1868, Mendel's devotion to science was replaced with administrative responsibilities, including disputes with the civil government for imposing taxes on religious institutions. After Mendel's death, many of his personal papers were burned by the succeeding abbot to end these disputes; however, some of his letters and documents survived in the monastery archives.

Chapter Introduction

Art historians are not certain when painters began blending blue and yellow to make green; however, the discovery of art manuals containing recipes, along with pigment analyses, suggests that it may have started as early as the 18th century. Starting in the 19th century, most European painters realized the ease and convenience of the process and began blending their blues and yellows to make green. The idea of blending to form an intermediate was also prominent in the world of science during the same time period. Many scientists in the 1800s accepted the **blending concept of inheritance theory**, which stated that the resulting offspring of parents with contrasting traits would exhibit intermediate traits between those of both parents. For example, the blending concept idea was used to explain the result of a cross between a plant with red flowers and one with white flowers producing plants with pink flowers. One of the main issues with the blending concept of inheritance theory was its failure to explain why the contrasting traits would reappear in subsequent generations. This chapter will introduce an alternative theory proposed by Gregor Mendel, whose experiments helped formulate new ideas about the inheritance of traits, which are governed by the laws of probability. The chapter will conclude with the significance of how these laws apply to various patterns of inheritance, including those that lead to various types of genetic disorders.

Chapter Objectives

In this chapter, students will learn the following:

8-1. The term genetics was first used in the early 1800s, the inheritance of traits was explained by Gregor Mendel in the mid-1800s, and four botanists working independently of one another in the 1900s are responsible for rediscovering Mendel's work.

8-2. Gregor Mendel's experiments on peas helped him formulate the law of segregation and modern genetics has developed useful terms to help better understand the ideas of this law.

8-3. Gregor Mendel's pea experiments involving two different character traits helped him formulate the law of independent assortment; however, the discovery of linked genes on a chromosome was found to violate this law.

8-4. The laws of probability play a central role in genetics and certain tools, such as a Punnett square and a test cross, are useful in helping predict or determine an organism's genotype and phenotype.

8-5. The general laws of inheritance apply to even the most complex patterns of inheritance, such as incomplete dominance, codominance, multiple allelism, polygenic traits, pleiotropy, and sex-linked traits.

8-6. Genetic disorders are easily traced using a pedigree to determine a disorder's pattern of inheritance and whether the condition is dominant or recessive.

8-1. Genetics

Pre-Mendelian Genetics

For thousands of years, farmers exploited the knowledge that living things inherited traits from their parents in order improve their crops and livestock, although the mechanism of action was poorly understood. Between 1816 and 1819, members of the Sheep Breeders' Society in Brno (now in the Czech Republic), which included the most prominent sheep breeders in the region, began debating a variety of topics, such as inbreeding and how to effectively combine different wool traits (e.g., elasticity, color, density, and fineness). Of the topics discussed, the effects of inbreeding on the quality of wool, an important source of wealth for many sheep breeders at the time, became the most controversial. One prominent member of the Society, J. M. Ehrenfels, an Austrian sheep breeder, argued that traits were inherited by means of "physiological laws of nature," maintaining that blood was the main contributor. Ehrenfels also argued that inherited traits were environmentally influenced and was steadfast in the belief that inbreeding decreased the quality of the wool.

FIGURE 8.2 Imre Festetics introduced the term *genetics* in his 1819 publications to describe the inheritance of traits occurring because of the "genetic laws of nature."

Another member, Hungarian sheep breeder Imre Festetics (Figure 8.2), who began inbreeding his sheep in 1803, argued that intrinsic factors were responsible for heredity, and these factors would only become more concentrated and predictable through inbreeding. Prior to the start of the debates in 1816, Festetics had been inbreeding his sheep for over a decade, producing some of the highest-quality wool available and a stock of sheep better than he could buy. As a means of supporting his argument, Festetics was asked to publish the results of his multiple years of inbreeding. In 1819, Festetics published a series of papers introducing a new set of laws to describe the inheritance of traits, known as the "genetic laws of nature," one of which provided the first ideas about the segregation of characters and the reappearance of traits in later generations. Festetics's publication also introduced the term *genetic*, a term used to differentiate his laws from those of Ehrenfels's "physiological laws." Today, scientists use the term *genetics* to refer to the inheritance of traits (*heredity*) and the variation of inherited characteristics observed in organisms.

Mendelian Genetics

Mendelian genetics began in 1854, when Gregor Mendel was commissioned by Abbot C. F. Napp of the Augustinian St. Thomas's Abbey to begin studying plant variation in the experimental garden originally planted on the grounds by Napp in 1830. Prior to beginning his experiments in 1856, Mendel spent two years evaluating roughly 34 distinct varieties of peas he obtained from different breeders. For Mendel to focus his attention on the principles of **inheritance**, the transmission of traits from one generation to the next, he selected 22 varieties, with a majority being *Pisum sativum* (common garden pea), for his experiments (Figure 8.3). There are several different reasons why Mendel may have chosen *Pisum sativum* for his studies. Besides its ease of cultivation and its very short generation time, which allowed it to develop quickly, Mendel could fertilize the *Pisum* plants manually, using an old breeder's trick, a trick for which he was well-known. To fertilize the plants, Mendel pulled back the flower petals, removed the immature stamens (male part of a flower that produces pollen), then, using a brush, transferred pollen from the stamen of one flower to the carpel (female part of a flower) of the altered flower. This allowed Mendel to control fertilization, obtain hundreds of offspring, and determine the outcome of his breeding experiments. Breeders had begun studying *Pisum sativum* several years prior to Mendel's experiments, developing the pea to display differentiating yet observable **characteristics** that Mendel specifically selected for his experiments. These characters included seed form, seed color, seed coat color, pod form, pod color, flower position, and stem length (Figure 8.3). Each of these characters produces two distinguishable **traits** that would "stand out clearly and definitely in the plants," allowing Mendel to focus his attention on specific traits, such as round or wrinkled seeds.

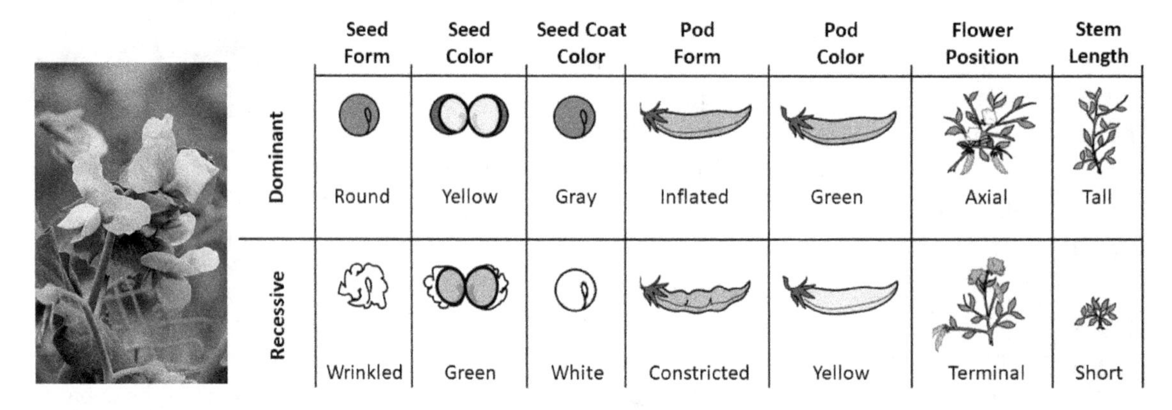

FIGURE 8.3 Mendel's choice of experimental plant was *Pisum sativum* (the common garden pea). There were seven traits Mendel focused on while performing his experiments, including seed form, seed color, seed coat color, pod form, pod color, flower position, and stem length, with each displaying two distinct traits.

Prior to beginning his experiments (explained in more detail in sections 8-2 and 8-3), Mendel developed distinct populations of pea plants consisting of **true-breeding** individuals. Mendel meticulously bred, over many generations, similar plants to obtain parental plants that when crossed produced offspring with the same trait. For example, true-breeding round-seed plants, when self-fertilized over many generations, would produce offspring with round seeds.

Over the course of eight years (1856–1863), Mendel would study nearly 28,000 *Pisum* plants, crossing them and keeping a detailed catalogue of the number of traits he observed in each generation. Mendel was one of the first scientists to apply a quantitative and mathematical approach to analyzing his data, enabling him to propose a theory of the mechanism of inheritance—the

particulate theory—that dispelled the blending concept of inheritance theory of the 1800s. Mendel suggested that the inheritance of traits from parents to offspring occurs because of the exchange and reshuffling of minute particles of hereditary units (which he called *elementen*) from generation to generation. Mendel outlined how *elementen* behaves in two fundamental laws of heredity, known as the law of segregation and the law of independent assortment, in his 1866 publication, "Experiments on Plant Hybridization," which appeared in the *Proceedings of the Science Research Society of Brünn*. Mendel received 40 prints of his publication and sent them to friends, family, distinguished botanists, and those who may not have been regular readers of the Scientific Society's main publication. Unfortunately, his offprint was largely discredited by many in the scientific community to whom he sent copies; however, the *Proceedings* seemed to spread much farther, making its way to many different esteemed libraries spanning central Europe to the United States. Despite this, Mendel's work was only referenced four times between 1869 and 1880, mentioned by name only twice between 1880 and 1892, and thereafter, his work and name were obscure until 1899.

Mendelian Genetics Rediscovered

There are four botanists who have been credited with rediscovering Mendel's laws of inheritance and bringing them to the attention of the scientific community. Although each was working independently of each other in four different countries, it has been speculated that none of the men were aware of Mendel's previous work. While working with a variety of plant hybrids, each botanist replicated Mendel's results, helping confirm, expand, and redefine Mendel's laws. These four botanists are Carl Correns, Erich von Tschermak, Hugo de Vries, and William Jasper Spillman; however, the credibility of one of these botanists aroused suspicion beginning in the 1950s, and as recently as 2016, historians and researchers proclaimed that one of these botanists should be excluded due to reasons outlined shortly.

Carl Correns (Figure 8.4a) was a German botany professor who spent ten years at the University of Tübingen studying the inheritance of traits in corn and peas. He began his experiments on trait inheritance in corn and peas in 1892 and continued examining corn and pea hybrids produced by

FIGURE 8.4 Four botanists have been credited with rediscovering Gregor Mendel's work on the inheritance of traits at the beginning of the 20th century, including a) Carl Correns, b) Erich von Tschermak, c) Hugo de Vries, and d) William Jasper Spillman.

means of cross breeding until 1899. However, when analyzing the results of his experiments, he found that the results of his corn experiments were much more complicated than that of his pea results, which produced simple-to-read ratios. In order to better understand these results, he began reviewing the literature and discovered Mendel's "Experiments on Plant Hybridization" publication, which only verified his results. Despite having known of Mendel's previous experiments on the hawkweed plant, Correns was completely unaware of Mendel's research on pea plants and the laws that he developed about inheritance. In 1900, Correns published the results of his experiments in his paper "G. Mendel's Law Concerning the Behavior of the Progeny of Varietal Hybrids," referencing both Mendel's results and his laws of inheritance.

Erich von Tschermak (Figure 8.4b) was an Austrian botanist who, as of the 1960s, has been removed from the list of individuals credited with rediscovering the work of Mendel, but his experimental work on the garden pea and his results deserve mention. Von Tschermak began experimenting with the garden pea in 1898, and upon his analyzing results, after acquiring a copy of Mendel's paper from the University of Vienna, he discovered that his results closely resembled those of Mendel. In the same year as Correns, but only a few months later, von Tschermak published "On Deliberate Cross-Fertilization in the Garden Pea," in which he outlined his results and placed a footnote to Mendel. After publishing, von Tschermak would utilize Mendelian genetics to develop new varieties of barley, wheat-rye, and oat hybrids. Despite his accomplishments, concerns began to arise in the 1960s about the legitimacy of von Tschermak's role as a rediscoverer of Mendel's work. There are critics who have suggested that von Tschermak was unable to analyze or draw generalizations about his results, while others have mentioned that he misunderstood the underlying rules outlined in Mendel's laws. These accusations were only compounded in 2016 when unpublished portions of von Tschermak's memoires and preserved family letters written between him and his brother Armin von Tschermak from 1898 and 1901 were discovered. Within portions of the unpublished memoires, von Tschermak admitted that his brother was instrumental in guiding him during the mathematical analysis of his results and was responsible for helping him complete most of what he had written. In addition, the rediscovered letters revealed that Armin played a major role as Erich's mentor in every facet of his life, including his career choices, his research topics, and how best to present his work. With these revelations, it would be hard to refute Armin's influence on and involvement in Erich's "rediscovery" and views of Mendel's work.

In 1889, Dutch botanist Hugo de Vries (Figure 8.4c) wrote a book influenced by Charles Darwin's theory of pangenesis (1868) called *Intracellular Pangenesis*. Within its pages, de Vries outlined a particulate theory of inheritance, which stated that the inheritance of traits occurred because of independent particles of hereditary material found in the nuclei of all cells called **pangenes**. De Vries suggested that each nucleus contained an entire set of pangenes that were responsible for influencing an organism's traits, even speculating that they would separate and recombine over the course of successive generations of breeding. Beginning in 1896, de Vries began crossbreeding a variety of different plant species, including corn, peas, and poppies. During his experiments, he began to demonstrate and confirm the existence of segregation laws among all the plant species he studied. Despite the desire of holding on to these results to be published in a large book, de Vries instead published "The Law of Segregation of Hybrids" in 1900, after obtaining Mendel's paper one year earlier, which detailed the patterns of inheritance he observed. Thereafter, de Vries published a more thorough description of his experiments in *The Mutation Theory* (1900–1903), a two-volume publication that included a variety of experimental topics, including details on the segregation laws he observed during his 1896 experiments.

American botanist William Jasper Spillman (Figure 8.4d), typically not included in the list of rediscoverers of Mendel's laws of inheritance for reasons explained shortly, rediscovered Mendel's laws while developing different wheat varieties in Washington State during the late 1890s. As a proponent of the blending concept of inheritance, Spillman demonstrated by experimental results that traits did not blend but rather separated and recombined, leading him to conclude that hybrids "tended to produce certain definite types, and possibly in definite proportions" (Carlson 2005, 20). The primary purpose of Spillman's experiments was to produce a wheat variety that could easily adapt to the environmental conditions of eastern Washington. What Spillman did not know was that his experiments would reveal the inheritance of traits, resulting in him becoming the first American to independently rediscover Mendel's laws. Spillman presented his findings in his 1901 paper, "Quantitative Studies on the Transmission of Parental Characteristics to Hybrid Offspring," a year and a half after the publications of Correns, Von Tschermak, and de Vries. During his presentation of his paper at the Eleventh Annual Convention of the Association of American Agricultural Colleges and Experiment Stations meeting in Washington, DC, it was assumed that in addition to Spillman, other well-established botanists in attendance knew nothing of these prior publications or of Mendel's work, as Mendel's name was never mentioned. Spillman's paper was officially published in 1902, but the obscurity of the journal in which it was published prevented his recognition as a rediscoverer of Mendel's law. However, when his paper was reprinted in 1903 in the *Journal of the Royal Horticultural Society*, the same journal in which an English translation of Mendel's "Experiments on Plant Hybridization" was first published, many British scientists suggested that he, too, be added to the list of rediscoverers.

8-2. Law of Segregation

Mendel performed his first group of experiments after obtaining his true-breeding individuals, the group of parental plants that would produce offspring with the same trait over several generations (true-breeding round-seed plants produce round-seed offspring). In the first group of experiments Mendel performed, he only focused on single characters (such as seed form) with two distinct traits (round vs. wrinkled).

Mendel began his experiment by crossing two different true-breeding varieties, such as pea plants that produced round seeds with those that produced wrinkled seeds (Figure 8.5). The parental plants made up what he called the **parental generation** or **P generation**. If the blending concept of inheritance theory was correct, Mendel would expect the resulting offspring of a cross between both parental plants, the **filial generation** or **F_1 generation**, to exhibit a trait of intermediate form. However, after planting and growing the resulting seeds of the P generation, Mendel observed results that were contrary to popular belief. Instead of producing offspring with an intermediate trait, all the F_1 offspring (known as monohybrids) produced round seeds, a trait resembling only one of the parents, while the wrinkled-seed trait had "disappeared." Mendel then performed a cross of the F_1 offspring, known as a **monohybrid cross**, which refers to following a single character of interest by crossing parental hybrids produced from a previous cross. Upon performing the monohybrid

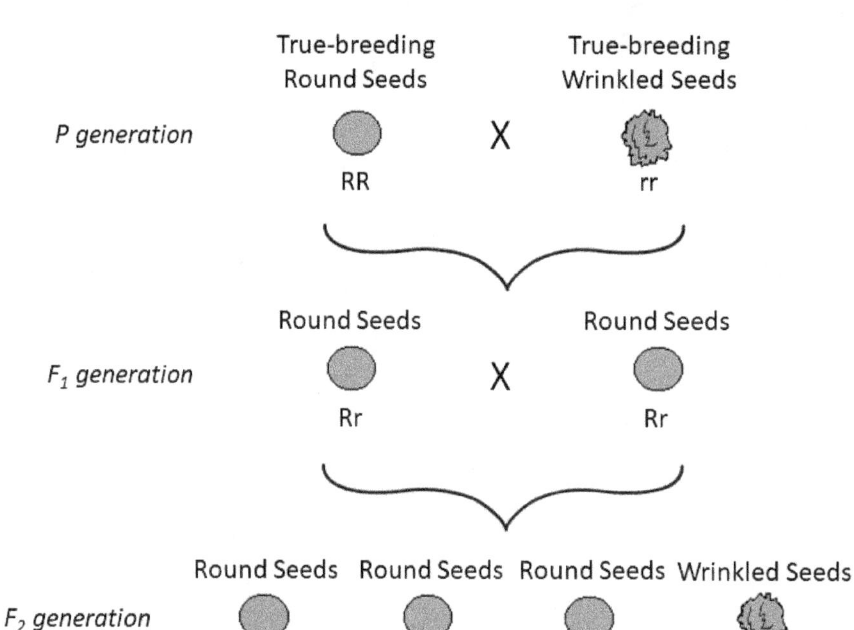

FIGURE 8.5 In his first group of experiments, Mendel focused on single character traits. Mendel crossed two different true-breeding pea plants, one for round seeds with one for wrinkled seeds, in his *P generation*. The resulting offspring in his *F_1 generation* all had round seeds, suggesting that the wrinkled seed trait must have disappeared. Mendel then performed a monohybrid cross with the offspring from the *F_1 generation* and observed that the resulting *F_2 generation* consisted round seeds, as well as the disappeared trait of wrinkled seeds.

cross of the F_1 offspring, the second generation, known as the **F_2 generation** produced a 3:1 ratio of offspring—about 75 percent had round seeds, and about 25 percent had wrinkled seeds, the trait that had "disappeared" in the F_1 generation. Mendel concluded that the "disappeared" trait did not in fact "disappear" but was hidden or masked by the other trait. Mendel described these traits as being either **dominant** or **recessive.** Dominant traits are those that mask the expression of another trait, while the trait being masked is called recessive. Mendel performed similar crosses with the remaining characters, observing the same pattern of inheritance in each. In addition, Mendel would continue carrying out subsequent crosses through the F_2 generation, producing as many as six generations for some traits.

Based on his observations and quantitative analysis, Mendel developed three main conclusions:

1. Each pea plant contains a distinct pair of hereditary units, or *elementen*, for every trait.
2. During gamete formation, *elementen* segregates and each gamete receives one hereditary unit or the other.
3. Upon the fertilization of gametes, the hereditary units recombine to form a new combination of *elementen*.

Mendel's conclusion that *elementen* segregates upon gamete formation laid the foundation for the first part of his particulate theory of inheritance, the **law of segregation** (Figure 8.6).

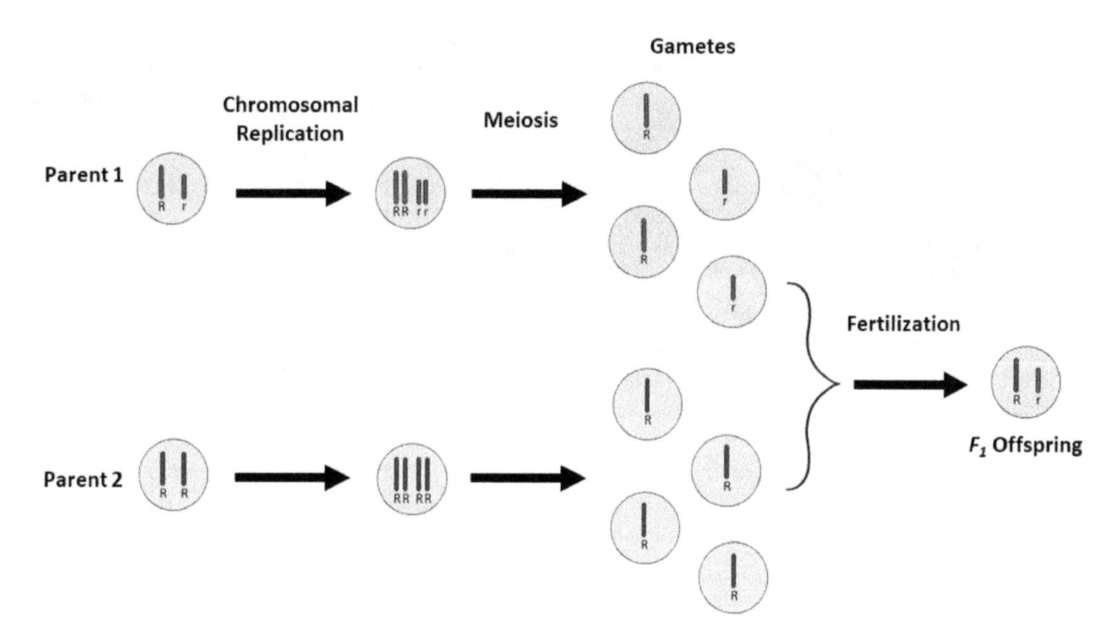

FIGURE 8.6 As a result of his monohybrid cross, Mendel concluded that each plant has two traits for every character, these traits segregate during the formation of gametes, and a new combination of traits form when the gametes are fertilized.

Mendel's Law of Segregation and Modern Genetics

Recall that in 1879, Walther Flemming was the first to observe and describe chromosomes, a term, however, which was not coined until 1888 by Wilhelm von Waldeyer-Hartz. As researchers began to study and describe chromosomes in more detail at the turn of the 20th century, many began to see similarities to Mendel's *elementen*, especially in how chromosomes separate during cell division. Shortly after Theodor Boveri and Walter Sutton introduced their **chromosomal theory of inheritance** in 1902, which stated that chromosomes were responsible for the inheritance of traits, Danish geneticist Wilhelm Johannsen (Figure 8.7a) would introduce the term **gene** in his 1909 publication, *The Elements of an Exact Theory of Heredity*. Johannsen gave a name to Mendel's *elementen*, but gene was a shortened version of pangene, the term used by Hugo de Vries in *Intracellular Pangenesis* (1868) to describe the hereditary particles responsible for the inheritance of traits.

One year later, in 1910, Thomas Hunt Morgan (Figure 8.7b) performed his *Drosophila* experiment that demonstrated genes were located on chromosomes. However, a gene is not randomly dispersed along a chromosome but rather located at a specific region on a chromosome known as a *locus* (plural *loci*). Recall that Mendel described his pea plants as having two hereditary units for every trait. These hereditary units are called **alleles** (a shortened version of the original term, **allelomorph**, introduced by William Bateson (Figure 8.7c) and Edith Rebecca Saunders (Figure 8.7d) in their 1902 publication, "The Facts of Heredity in the Light of Mendel's Discovery"), and each gene has two alleles for every trait.

FIGURE 8.7 a) Wilhelm Johannsen coined the terms *genotype* and *phenotype* in 1903 to describe an organism's genetic composition and the expression of that composition, along with naming the hereditary particles responsible for inheritance, *genes*, in 1909. b) Thomas Hunt Morgan demonstrated that genes were found on chromosomes in 1910. The specific location in which a gene is found on a chromosome is known as a *locus*. c) William Bateson and d) Edith Rebecca Saunders used the term *allelomorph* (now shortened to allele) to describe the two hereditary units for every trait in 1902.

One of these alleles is located at the locus of one homologous chromosome, while the other allele is located on the other homologous chromosome. During his monohybrid crosses, where one trait was hidden by another trait, Mendel described this phenomenon using the terms dominant or recessive. In modern genetics, alleles that are dominant are expressed using a capital letter, while a recessive allele is designated by a lowercase letter. As an example, the gene for seed form in pea plants has two alleles—one for round seeds and one for wrinkled seeds. Since round seeds are dominant to wrinkled seeds, round seeds would receive a capital letter *R* and wrinkled seeds would have a lowercase letter *r*.

If an organism has two identical alleles for a particular gene, the organism is said to be **homozygous** for that particular trait. This was especially evident in Mendel's true-breeding P generation, when he crossed two true-breeding individuals, one with round seeds and the other with wrinkled seeds. Since round seeds are dominant to wrinkled seeds, a true-breeding round-seed pea plant would have two dominant alleles (RR) and would be **homozygous dominant** for this trait. As for wrinkled seeds, the true-breeding individuals would have two recessive alleles (rr) and would be **homozygous recessive** for this trait.

When an organism has two different alleles for a particular gene, the organism is said to be **heterozygous** for that particular trait. A heterozygous trait is represented using both a capital letter and a lowercase letter, with the capital letter typically written first, such as *Rr*. The resulting F_1 offspring from Mendel's cross of true-breeding parents, RR (round) and rr (wrinkled), produced heterozygous individuals (Rr), each expressing the dominant allele. As illustrated, the expressed allele is the dominant allele, while the recessive allele is hidden, as explained by Mendel.

Letters of the alphabet are used to demonstrate whether an organism's alleles are dominant or recessive. However, when an individual expresses the dominant allele, as is the case with round seeds, it can be either homozygous dominant (RR) or heterozygous (Rr). When outlining what specific alleles of a gene an organism has and how these alleles are expressed, geneticists will use the words *genotype* and *phenotype*, which were first used by Wilhelm Johannsen in his 1903 publication, "On Heredity in Society and in Pure Lines." Johannsen first developed these concepts while studying heritable variation in plants such as barley (*Hordeum vulgare*) and the common bean (*Phaseolus vulgaris*). Johannsen would expand on these concepts in later publications, such as *The Elements of Heredity* (1905), *The Elements of an Exact Theory of Heredity* (1909), and *The Genotype Conception of Heredity* (1911).

The term **genotype** refers to the specific combination of alleles an organism inherits from its parents during fertilization. An organism's genotype is distinguished using letters and phrases, such as homozygous dominant (RR), heterozygous (Rr), and homozygous recessive (rr). Although an organism's genotype is a nonobservable characteristic, the expression of an organism's genotype gives rise to that organism's **phenotype**, which is an observable characteristic. A phenotypic characteristic not only includes an organism's outward physical appearance, such as the genotypes RR and Rr being expressed as the dominant trait of round, while the genotype rr is expressed as the recessive trait of wrinkled but can also include an organism's physiological and psychological traits as well. In addition to being influenced by genotype, an organism's phenotype can also be influenced by the environment, such as temperature, or by a combination of the two.

8-3. Law of Independent Assortment

During his first group of experiments, Mendel performed crosses focusing on single character traits such as seed form. The next group of experiments Mendel performed involved the inheritance of two different character traits, such as seed form and seed color, because he hypothesized that all traits were inherited independently of one another and not influenced by the presence of one another.

Mendel used true-breeding parents—pea plants with round, yellow seeds (RRYY) and pea plants with wrinkled, green seeds (rryy)—to produce his P generation. Upon crossing these two parental varieties, all the resulting F_1 offspring were heterozygotes expressing the dominant traits, round and yellow, traits that Mendel had determined to be dominant during his single-character-trait experiments. Mendel then performed a **dihybrid cross**, a cross following two characters of interest, using the offspring produced from the F_1 cross (Figure 8.8). Upon calculating the resulting F_2 offspring, Mendel discovered that four different seed varieties were produced equaling a 9:3:3:1 ratio—nine were dominant for both traits (round and yellow—RRYY, RrYy); three were dominant for one trait, recessive for the other (round and green—RRyy, Rryy); three were recessive for one trait and dominant for the other (wrinkled and yellow—rrYY, rrYy); and one was recessive for both traits (wrinkled and green—rryy). Mendel observed the same 9:3:3:1 ratio upon conducting subsequent experiments using a variety of combinations of F_1 heterozygotes differing in two traits.

Mendel concluded that the 9:3:3:1 ratio could be explained by how the alleles independently segregate during gamete formation in the F_1 generation, resulting in four different types of gametes—ry, RY, rY, and Ry. When all the gametes have an equal chance of fertilization during a dihybrid cross, the presence of four different phenotypes exhibiting a 9:3:3:1 ratio is expected in the F_2 generation. This conclusion was the basis for the second part of Mendel's particulate theory of inheritance, the law of independent assortment. The **law of independent assortment** states that all alleles independently assort during gamete formation and the inheritance of one trait is not dependent on the inheritance of another (Figure 8.9).

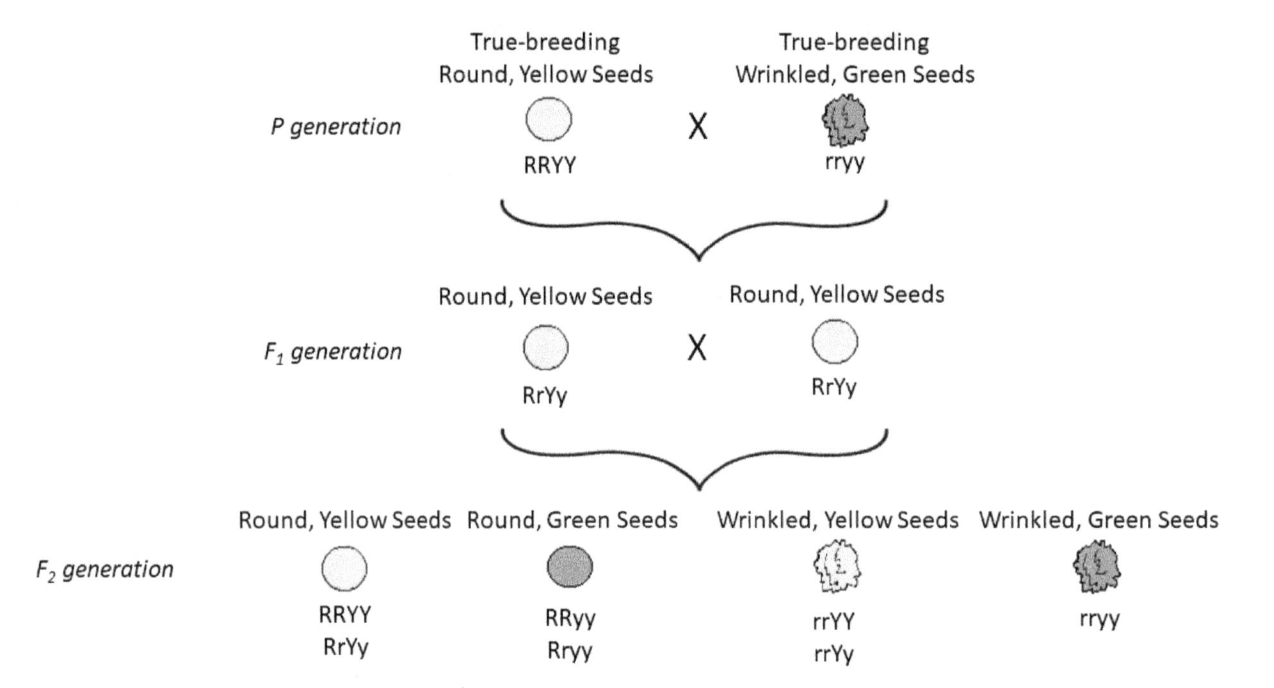

FIGURE 8.8 Mendel's second group of experiments focused on two different character traits. In the *P generation*, Mendel crossed two different true-breeding pea plants, one for round, yellow seeds with one for wrinkled, green seeds. The resulting offspring in his *F_1 generation* were all heterozygotes for round, yellow seeds. Mendel performed a dihybrid cross that produced a *F_2 generation* of offspring exhibiting a 9:3:3:1 ratio of traits.

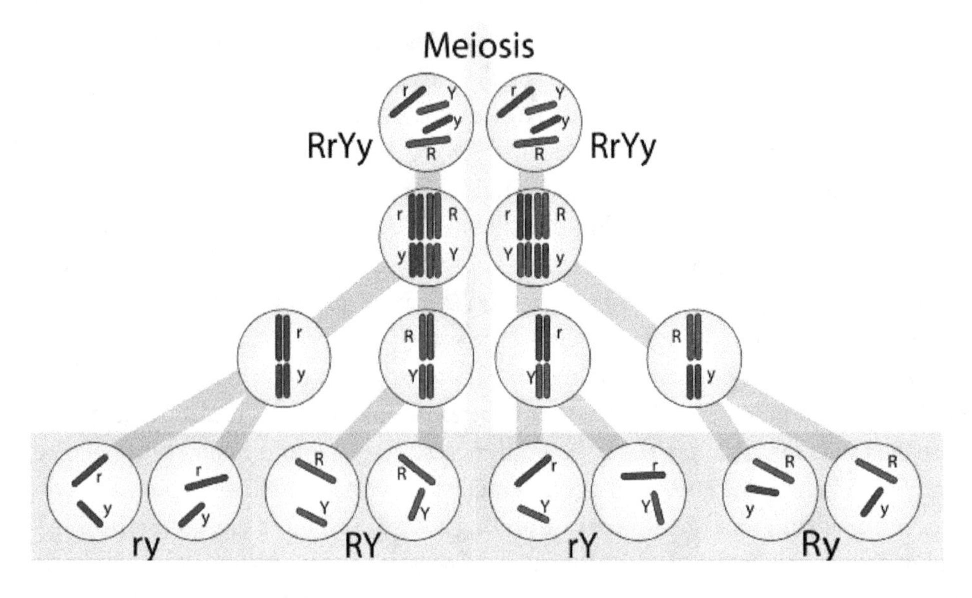

FIGURE 8.9 Mendel's law of independent assortment states that traits for every character assort independently of one another, resulting in a random combination of traits in the gametes. An individual with the genotype RrYy produces four different types of gametes, ry, RY, rY, and Ry.

Linked Genes

All seven traits Mendel studied over the course of his experiments were either located on genes of the same chromosome—yet far apart—or on completely different chromosomes. Therefore, Mendel's law of independent assortment only applies to genes located on nonhomologous chromosomes and does not consider genes located close together on a chromosome. Sometimes genes that are located close together on the same chromosome do not independently assort and are inherited together. These types of genes are called **linked genes**.

The pattern of inheritance associated with linked genes was first observed and described in 1905 by British geneticists William Bateson, Edith Rebecca Saunders, and Reginald Punnett. According to Mendel's law of independent assortment, one would expect an offspring ratio of 9:3:3:1 when performing a dihybrid cross. However, while Bateson, Saunders, and Punnett were examining flower color and pollen shape in sweet pea plants (*Lathyrus odoratus*), their results did not match Mendel's expected ratios. The researchers knew the following going into their experiment about the sweet pea plants they were examining—purple flowers (P) were dominant over red flowers (p) and long pollen grains (L) were dominant over short pollen grains (l). Bateson, Saunders, and Punnett began their experiment by crossing two true-breeding plants, purple flowers and long pollen grains (PPLL) with red flowers and round pollen grains (ppll). The resulting F_1 offspring were all heterozygotes, PpLl, as expected. When the researchers performed the dihybrid cross using individuals from the F_1 generation, they expected the cross to produce a 9:3:3:1 ratio. However, their results deviated from what was expected, with more of the offspring resembling those of the parental generation. This led the researchers to formulate three conclusions about the alleles for these particular traits: they must be linked together on the same chromosome, they do not independently assort, and as a result, these traits are inherited together, all of which violate Mendel's law of independent assortment. Although the researchers discovered that the alleles were linked, they were unable to answer the question as to why.

During many experiments that took place in the "fly room" at Columbia University, Thomas Hunt Morgan's studies with *Drosophila* revealed many different effects that genes had on patterns of inheritance. These studies were about to provide answers to the question laid out by Bateson, Saunders, and Punnett in 1905 as to why the alleles were linked. As early as 1910, Morgan discovered a pattern of inheritance in which certain traits were expressed by genes located on sex chromosomes (discussed further in section 8-5). However, Morgan's experiments in 1912 demonstrated a different pattern of inheritance that occurred when genes linked on autosomal chromosomes are inherited. Working alongside American biologist Clara J. Lynch, Morgan focused on two different *Drosophila* traits—body color and wing size—traits whose alleles are located on autosomal chromosomes. Morgan and Lynch performed their experiment with two different types of flies, flies with gray bodies and normal-sized wings, called wild-type flies, and a group of mutant flies, flies having black bodies with smaller-than-normal wings, called vestigial wings, created in the lab through breeding. Each of the alleles that exist for both body color and wing size for the mutant flies are recessive to wild-type alleles.

The researchers began their experiments by crossing a true-breeding wild-type fly with a true-breeding mutant fly (P generation), which resulted in an F_1 generation of dihybrid heterozygotes, expressing the gray bodies and normal-sized wings (Figure 8.10). To determine whether the genes were linked, Morgan and Lynch crossed a dihybrid heterozygote with a homozygous

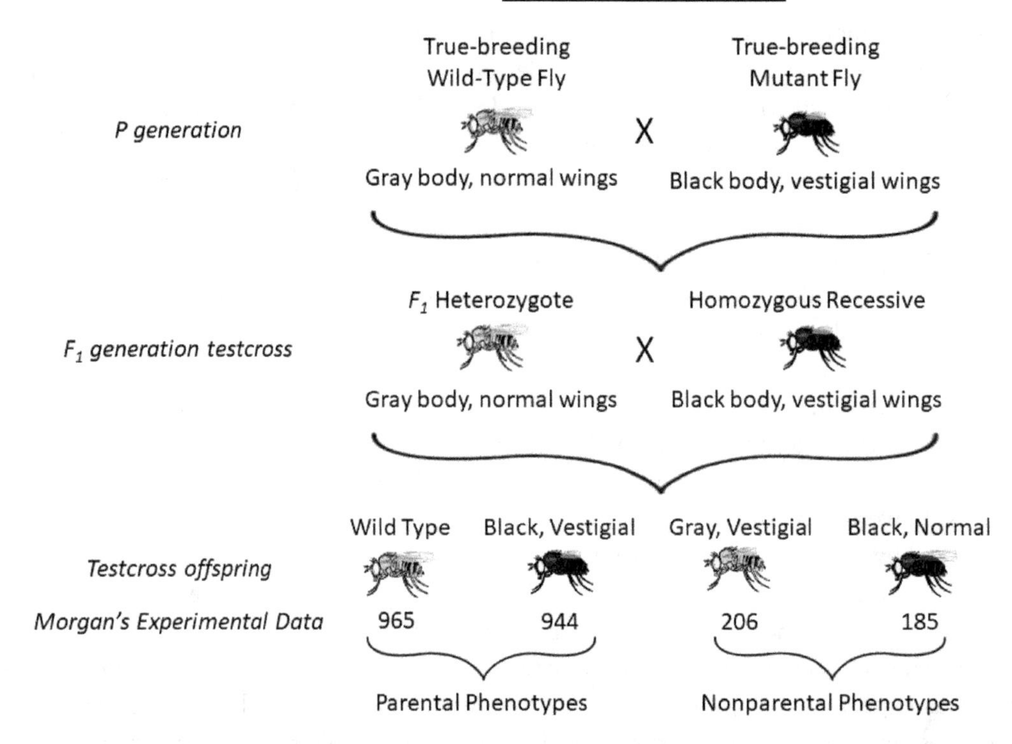

FIGURE 8.10 In 1912, Thomas Hunt Morgan and Clara J. Lynch demonstrated that alleles for body color and wing size in *Drosophila* did not independently assort and must be linked on the same chromosome because they were inherited together. Their conclusion was based upon most offspring expressing parental phenotypes than nonparental phenotypes. The observation of nonparental phenotypes suggested to the researchers that the linked genes must separate during gamete formation.

recessive male (black body with vestigial wings), which is known as a test cross (discussed further in section 8-4). The test cross resulted in most of the offspring exhibiting traits expressed by individuals of the parental generation, known as **parental phenotypes**. Since most offspring expressed parental phenotypes, Morgan and Lynch concluded that the alleles for body color and wing size do not independently assort; therefore, they must be linked on the same chromosome, because they were inherited together. However, some of the offspring expressed a combination of traits that were different from those of the parental phenotypes. These **nonparental phenotypes**, the researchers concluded, must occur when the connection between the linked genes occasionally breaks, producing a combination of alleles not observed in the parental generation. Morgan and Lynch suggested that some process during gamete formation was responsible for the breaking and recombination of alleles. Although not discerned at the time, with subsequent experiments thereafter, Morgan proposed that chromosomes must "exchange" genetic material with one another. This process became known as crossing over.

Understanding specific events of meiosis, especially crossing over, can help one better explain the occurrence of parental and nonparental phenotypes observed in the linked gene experiments. Recall from chapter 7, when homologous chromosomes align during prophase I, they randomly exchange genetic material during crossing over, resulting in genetic recombination (Figure 8.11). Although crossing over can happen along any region of a chromosome, the probability of it occurring increases the farther apart the genes are on the chromosome. This idea was first proposed and described by Morgan's undergraduate assistant, American Alfred H. Sturtevant, in 1911, which led to further studies on gene linkage and mapping specific locations of genes on chromosomes. When the gametes form after meiosis is complete, each cell contains a new combination of alleles, different from those of the parents. These events are the reason why Mendel observed the independent assortment of genes located on the same chromosome. However, when genes are linked, they are less likely to separate during crossing over because of how close they are on the chromosome and are therefore inherited together. Bateson, Saunders, Punnett, Morgan, and Lynch all observed a small number of offspring during their experiments expressing nonparental phenotypes, suggesting that the linked genes must have separated. The most reasonable explanation is that the linked genes separated during crossing over, producing a new combination of alleles, which both groups of researchers observed in a relatively small number of offspring.

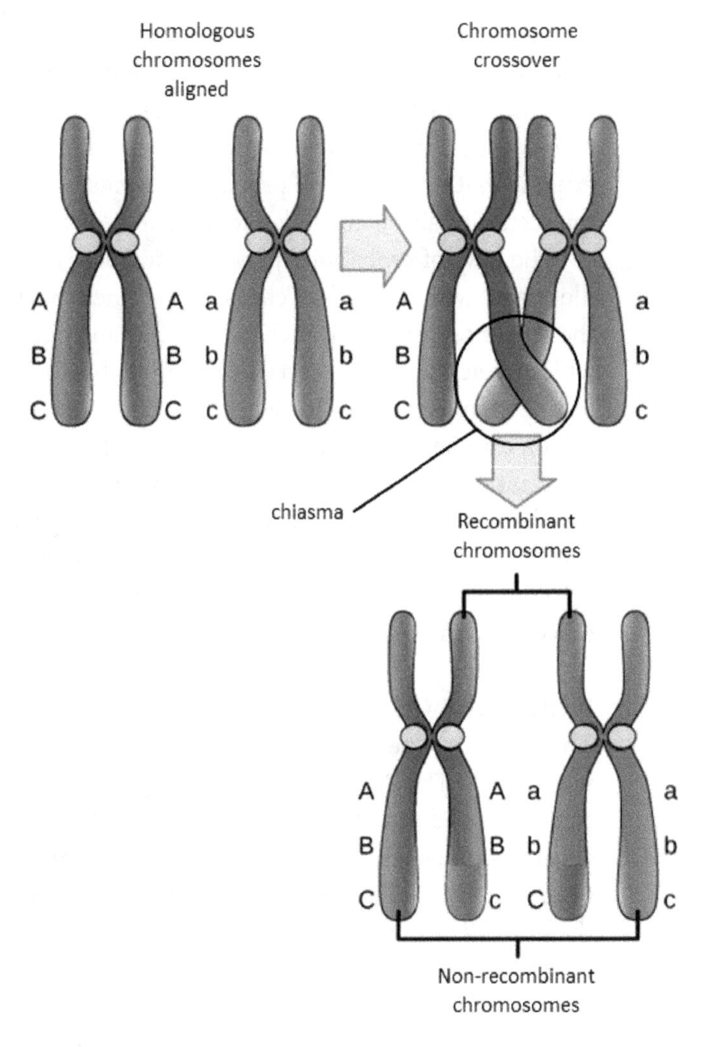

FIGURE 8.11 Crossing over during meiosis results in chromosomes consisting of a new mixture of unique genes, known as genetic recombination. This occurrence provides an explanation as why nonparental phenotypes were observed during Morgan and Lynch's experiment with linked genes in *Drosophila*.

8-4. Laws of Probability

Probability is a group of mathematical laws responsible for predicting the chance that a particular event will occur. The likelihood of an event occurring by chance is measured using a scale that ranges from 0 to 1, with 0 representing the probability of an event not occurring and 1 representing the probability of an event occurring. For example, the laws of probability can be applied to a deck of cards, which contains a total of 52 cards. Here are a couple different scenarios in which the laws of probability can be applied to a deck of cards:

What is the probability of drawing a face card of any suit? = 12/52

What is the probability of drawing a non–face card of any suit? = 40/52

An important aspect of the laws of probability is that one event is not dictated by the occurrence of another event, referred to as independent events. For example, the chance that a face card of any suit will be drawn from the deck of cards is a completely independent event and unaffected by what card was previously drawn. Another important aspect of applying the laws of probability to any event is that the sum of all probabilities must equal 1, as evident when the two probabilities from the scenarios above are added together.

The laws of probability play a central role in understanding Mendel's laws of inheritance, as illustrated with the deck of cards. As outlined in Mendel's law of segregation, when gametes form, the alleles segregate and each gamete has an equal chance of receiving one allele or the other, but which allele the gamete receives is impossible to know. The laws of probability also influence the fertilization of gametes. Recall that each gamete produced by an individual is a set of distinct cells with a unique set of alleles; therefore, each gamete has an equal of chance of being involved in fertilization. Finally, the laws of probability can also be applied to Mendel's law of independent assortment. When alleles segregate, the specific group of alleles a gamete receives is an independent event and is not influenced by another group of alleles.

There are two commonly used tools that allow geneticists to calculate and determine the probability of an offspring having a particular genotype or phenotype—a Punnett square and a test cross.

Punnett Square

The **Punnett square** was developed by Reginald Punnett, with the first published diagrams appearing in 1906. The Punnett square uses the laws of probability to predict and calculate the possible genotypes and phenotypes of offspring produced from a cross between two individuals of known genetic composition. When constructing a Punnett square, each individual allele a female gamete receives is aligned along the top of the diagram, while aligned on the left side of the diagram are the individual alleles a male gamete receives (Figure 8.12). Capital and lowercase letters are used to represent the alleles as being either dominant or recessive. Moving along each square within the diagram, the alleles are combined to form an offspring's predicted genotype. In addition, one can also predict the offspring's phenotype, based on the predicted genotype of the offspring.

Once the Punnett square is complete, the mathematical laws of probability can be applied. The laws of probability use two rules to evaluate the results of a Punnett square, the **product rule**, used to determine the probability of independent events occurring, and the **sum rule**, which determines the probability of the same event occurring in multiple ways.

Using figure 8.5, recall that pea plants have two alleles for seed form, round (R) and wrinkled (r). When Mendel crossed two true-breeding individuals for these traits (RR × rr), the resulting F_1 generation were all heterozygotes (Rr). The genotype and phenotype of the offspring resulting from a cross between two heterozygotes (Rr) can be determined by using a Punnett square and applying the two probability rules. When a heterozygote gamete forms, the probability of the gamete receiving one allele or the other is ½.

To determine the likelihood of the offspring having one genotype or another, the *product rule* (Figure 8.12) is applied by multiplying the probabilities of an egg or sperm having a particular allele (R or r). For example, the probability that a cross between two heterozygotes (Rr) will produce a plant homozygous recessive (rr) for wrinkled seeds is ¼. To calculate this, the probability of a recessive allele (R) coming from the mother (½) and the probability of a recessive alle (r) coming from the father (½) are multiplied together: ½ × ½ = ¼.

The *sum rule* (Figure 8.12) can be applied to the same Punnett square to determine the probability of phenotypes. If the number of offspring exhibiting the dominant allele (R) wants to be known, the probabilities of homozygous dominant (RR) and heterozygous (Rr) offspring produced are added together. The results of the Punnett square show that ¼ of the offspring is homozygous dominant (RR), while two of the offspring are heterozygous (Rr), which is ¼ + ¼, or ½. The total number of offspring exhibiting the dominant allele (R) is then added together, ¼ + ½, or ¾. This indicates that ¾, or 75 percent, of the offspring produced when two heterozygotes (Rr) are crossed exhibit the dominant trait (round seeds).

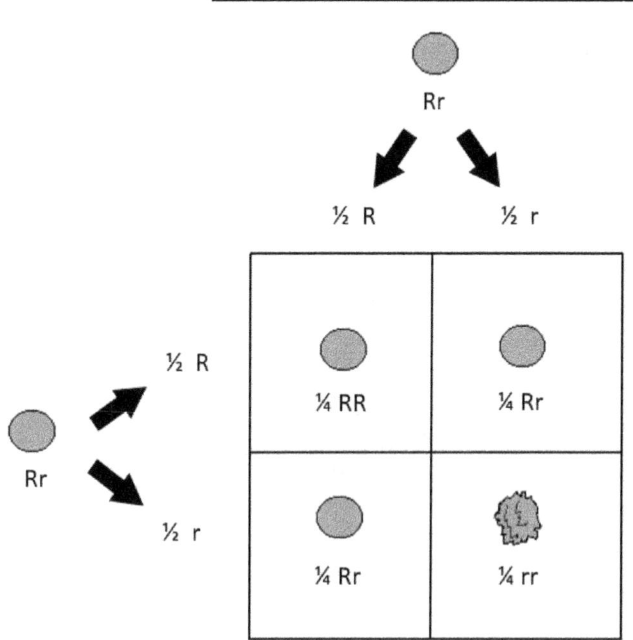

FIGURE 8.12 The laws of probability are used to predict if a chance event will occur. In genetics, a Punnett square is a visual tool that uses the laws of probability to predict and calculate the chances that a particular genotype and phenotype will be produced from a cross between two individuals. When evaluating the results of the Punnett square, two rules are applied, the product rule and the sum rule. The product rule involves multiplying together the probabilities of gametes having certain alleles to determine if an individual will have one genotype or another. The sum rule predicts the chances that certain offspring will exhibit phenotypes by means of adding together the probabilities.

Test Cross

When an organism exhibits a dominant phenotype, it is difficult to determine whether the organism is homozygous dominant or heterozygous. Geneticists use a **test cross**, a tool devised by Mendel that involves crossing a homozygous recessive individual with an individual with the dominant phenotype to determine the latter's genotype (Figure 8.13). During his law of segregation experiments, Mendel used a test cross to determine whether the plants produced in his F_1 generation were heterozygotes. After he bred the F_1 generation plants with a homozygous recessive plant, half of the resulting offspring exhibited the dominant phenotype, while the other half exhibited the recessive phenotype, producing a 1:1 ratio. The results confirmed that the F_1 generation plants were heterozygous for the dominant phenotype because the recessive allele is the only allele produced in the gametes of the homozygous recessive plant. If the individual was homozygous dominant for the phenotype, then a cross with a homozygous recessive individual would produce offspring all exhibiting the dominant phenotype. For example, round seeds are the dominant phenotype for seed form. To determine the genotype of a mystery plant exhibiting round seeds, a homozygous recessive plant (rr) would be crossed with the dominant plant (RR or Rr). If half of the offspring exhibited dominant (round) and half of the offspring exhibited recessive (wrinkled) phenotypes, the mystery plant must be heterozygous for the phenotype (Rr). However, if all the offspring exhibited the dominant (round) phenotype, the mystery plant must be homozygous dominant for the phenotype (RR).

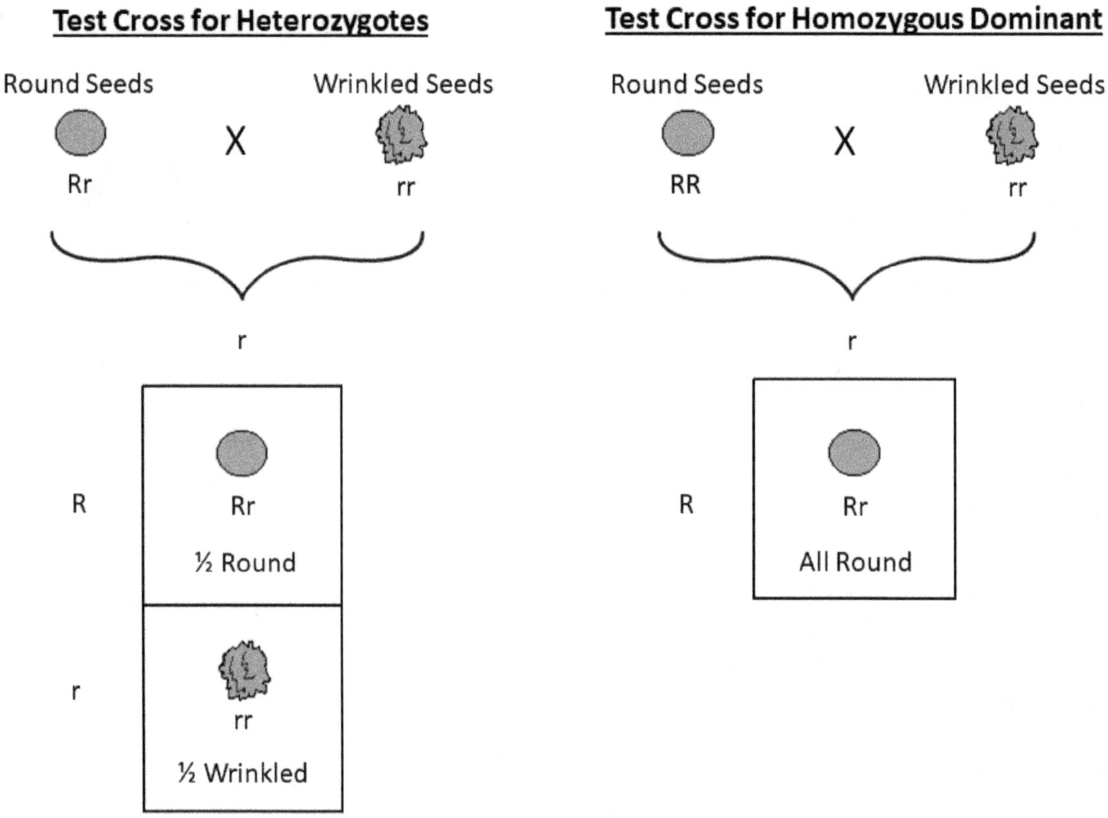

FIGURE 8.13 Testcrosses are used to help determine the genotype of an individual with a dominant phenotype. A dominant phenotype can be heterozygous or homozygous dominant for that particular trait. During a testcross (left), the individual with a dominant phenotype is crossed with a homozygous recessive individual. If half of the offspring are dominant and the other half are recessive for the trait, then the individual with the dominant phenotype is heterozygous for that trait. If all the offspring exhibit the dominant phenotype (right), the individual is homozygous dominant.

8-5. Complex Patterns of Inheritance

Mendelian genetics is not always as simple as the results Mendel observed during his pea experiments. Mendel experimented with traits determined by one gene with two alleles. However, traits may be influenced in more complicated ways or express genotypes and phenotypes that are not as straightforward as one would expect. Despite this, the general laws of inheritance—the law of segregation and the law of independent assortment—still apply to even the most complex patterns of inheritance.

Incomplete Dominance

Incomplete dominance is a situation in which a heterozygote offspring produced when crossing two homozygous individuals has an intermediate phenotype between that of the two homozygotes. This situation results when neither of the alleles express dominance. An example of this phenomenon is seen when red snapdragons are crossed with white snapdragons (Figure 8.14). Both parents are homozygous but when crossed produce a heterozygote offspring that is pink. Hair texture is an example of incomplete dominance observed in humans. When an individual with curly hair (homozygous dominant) mates with an individual with straight hair (homozygous recessive), the resulting offspring has wavy hair (heterozygous), a trait intermediate of both parents.

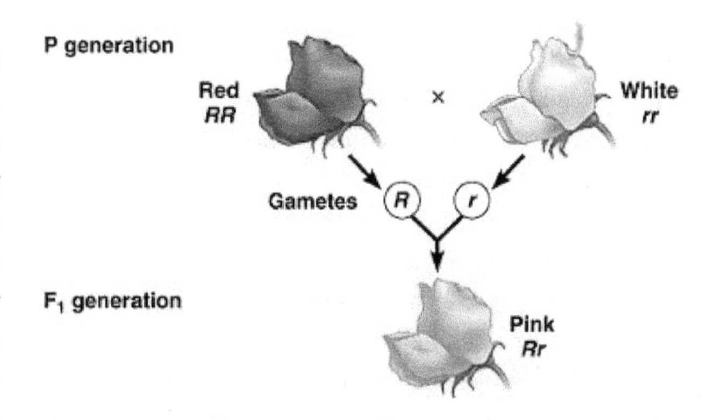

FIGURE 8.14 In incomplete dominance, the resulting heterozygote exhibits a trait that is intermediate of that of two homozygotes.

Codominance

Codominance results when both homozygous alleles inherited by the heterozygote are expressed equally in its phenotype. A common example is evident in the inheritance of feather color in chickens. When a hen that is homozygous for white feathers is crossed with a rooster that is homozygous for black feathers, the resulting heterozygote exhibits a "checkered" pattern of both white and black feathers (Figure 8.15).

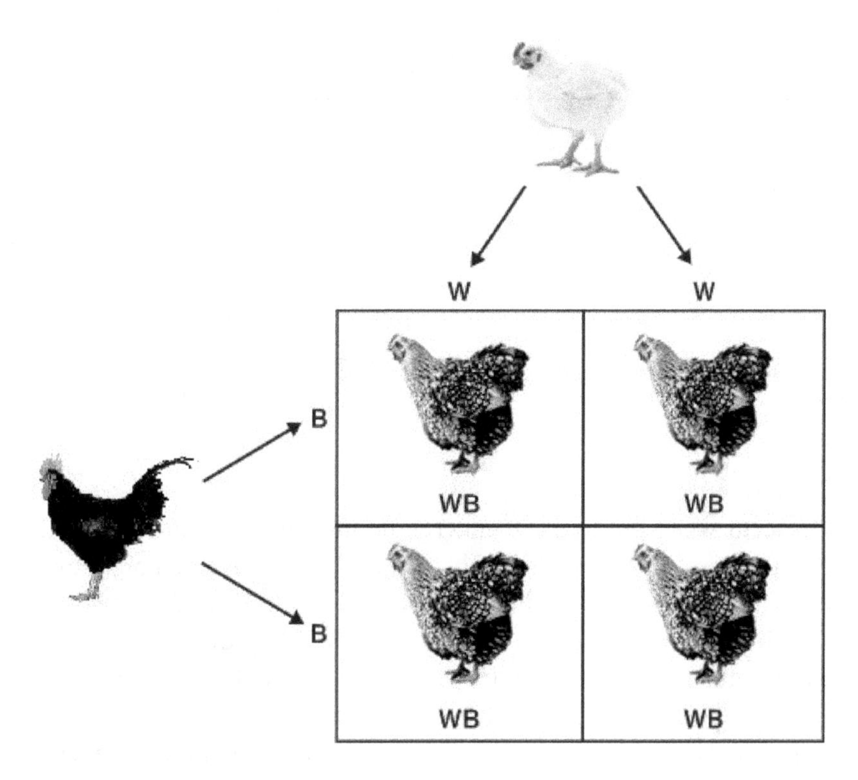

FIGURE 8.15 The "checkered" pattern observed in chickens is an example of codominance. This pattern of inheritance occurs when the phenotypes of both homozygous parents are equally expressed in a heterozygote.

Multiple Allelism

There are situations in which a trait is determined by a single gene containing more than two alleles, a pattern of inheritance known as **multiple allelism**. One of the most common examples is the ABO system of blood groups (Figure 8.16), which is determined by three different alleles located on a gene found on chromosome 9. The three different alleles are I^A, I^B, and i, and each code for an enzyme that catalyzes the transfer of a specific carbohydrate to the surface of red blood cells, called an *antigen*. This group of antigens is responsible for indicating to the body's immune system that the cell belongs to the body and is not a foreign invader. Allele I^A codes for the A antigen, allele I^B codes for the B antigen, alleles I^A and I^B together code for both the A and B antigens, and allele i does not code for either of the two antigens. As a result, the three alleles give rise to four different blood types, based on the type of antigen present on the surface of the red blood cell—A, B, AB, and O.

	A	B	AB	O
Allele	I^A	I^B	I^A *and* I^B	i
Red Blood Cell Antigen	antigen A	antigen B	antigens A and B	No antigens
Red Blood Cell Type	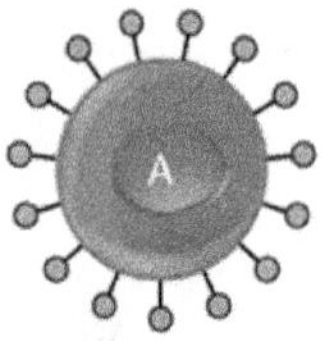 A	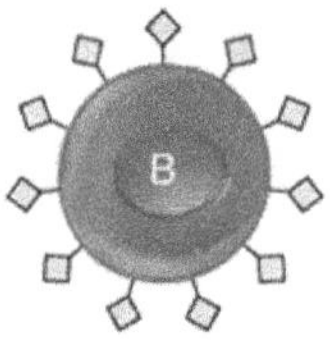 B	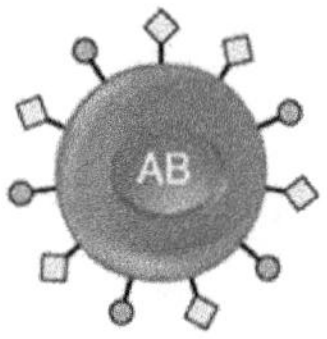 AB	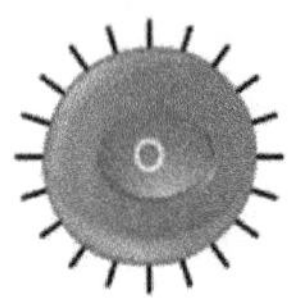 O

FIGURE 8.16 In the ABO system of blood groups, the different blood types are determined by multiple alleles.

Polygenic Traits

Polygenic traits occur when the additive effect of multiple genes influences the inheritance of a single trait. A classic polygenic trait is height. Select studies have demonstrated that roughly 80 percent of the normal variation in height observed in adults is due to the additive effect of multiple genes. In one genomic study of over 250,000 individuals performed in 2014, the researchers identified almost 700 variants occurring along more than 400 different loci. Another classic example of a polygenic trait is skin color (Figure 8.17), which is produced by a variety of skin pigments called melanin. The production of melanin has been linked to over 370 separately inherited genes.

	ABC	*ABc*	*AbC*	*aBC*	*Abc*	*aBc*	*abC*	*abc*
ABC	6	5	5	5	4	4	4	3
ABc	5	4	4	4	3	3	3	2
AbC	5	4	4	4	3	3	3	2
aBC	5	4	4	4	3	3	3	2
Abc	4	3	3	3	2	2	2	1
aBc	4	3	3	3	2	2	2	1
abC	4	3	3	3	2	2	2	1
abc	3	2	2	2	1	1	1	0

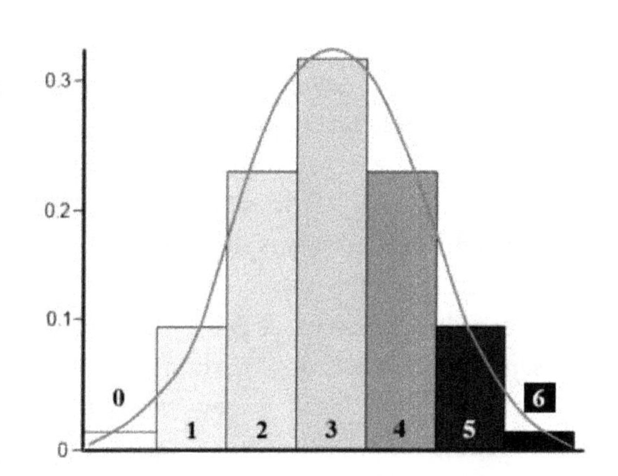

FIGURE 8.17 In this model of polygenic inheritance of skin color, three separate genes are evaluated, each carrying an allele for dark skin (A, B, and C) and an allele for light skin (a, b, and c). Upon mating two heterozygotes (AaBbCc × AaBbCc), both of which have an intermediate skin color, the resulting Punnett square shows seven different phenotypes, with most of the individuals having alleles for an intermediate skin color (see histogram on right). The numbers in each box represent the number of dark-skin alleles the individual has for skin color. For example, the individual with the genotype AABBCC has six dark-skin alleles and would be very dark. The individual with the aabbcc genotype has no dark-skin alleles and would be very light.

Pleiotropy

Pleiotropy is a pattern of inheritance associated with a single gene giving rise to a multitude of different, unrelated phenotypic traits. A common example of pleiotropy is sickle-cell disease, which will be discussed in more detail in section 8-6. Sickle-cell disease is a disorder in which red blood cells become "sickle-shaped" due to abnormal hemoglobin molecules and are unable to effectively distribute oxygen throughout the body. The disease occurs when an individual inherits the two abnormal alleles of the gene that codes for hemoglobin, the protein responsible for carrying oxygen in red blood cells. The allele for normal hemoglobin is *HbA* and the abnormal allele is *HbS*. Three different scenarios arise during the inheritance of these alleles. If an individual inherits both normal alleles (*HbAHbA*) or if an individual inherits both the normal allele and the abnormal allele (*HbAHbS*), neither individual has sickle-cell disease. However, when an individual inherits both abnormal alleles (*HbSHbS*), the individual is diagnosed with sickle-cell disease. The hemoglobin gene has a pleiotropic effect by also providing malarial resistance to individuals that are homozygous (*HbSHbS*) and heterozygous (*HbAHbS*) for sickle-cell disease. Due to the presence of sickle-shaped cells in both individuals, the malarial parasite is unable to effectively reside within the uniquely shaped red blood cells.

Sex-Linked Traits

Recall from chapter 7 that sex chromosomes are chromosomes that determine an offspring's sex. In humans, a female has two X chromosomes (XX) and a male has both an X and a Y chromosome (XY). Like autosomal chromosomes, sex chromosomes contain a specific set of genes that play important roles in the expression of certain traits. Due to the differences in the size of the sex chromosomes, each carries a different number of genes. The Y chromosome is much smaller than the X chromosome; therefore, it contains only a few dozen genes, which are primarily responsible for triggering male embryonic development and male reproductive processes, such as sperm production. The X chromosome is much larger than the Y chromosome and contains over 1,000 genes, which orchestrate female reproductive processes such as egg production and code for such traits as color vision and blood-clotting. Traits expressed by genes located on sex chromosomes are called **sex-linked traits**, a pattern of inheritance discovered in *Drosophila* by a group of researchers led by Thomas Hunt Morgan in 1910.

During the course of his experiments, Morgan discovered that when crossing a white-eyed male *Drosophila* with a homozygous dominant red-eyed female, all the flies had red eyes in the F_1 generation, an expected result, as red eyes are dominant over white eyes. Upon crossing individuals of the F_1 generation, the expected 3:1 phenotypic ratio was revealed; however, he noticed that half of the males had red eyes, and the other half had white-eyes (Figure 8.18). Morgan concluded that the allele for eye color must be present on the X chromosome and not the Y chromosome; therefore, males inherit their white eye color from the female parent.

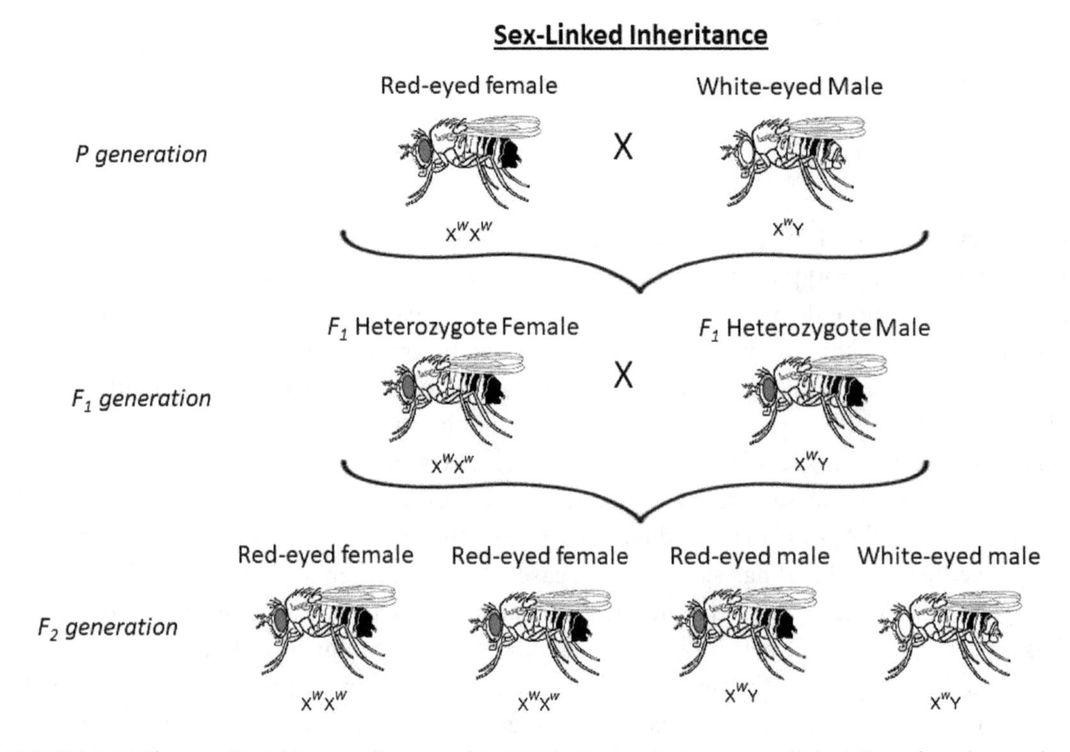

FIGURE 8.18 Thomas Hunt Morgan discovered in 1910 that some traits are sex-linked. Based on the results of his experiments, he concluded that the allele for eye color in *Drosophila* must be located on the X chromosome after observing that half of the males in his F_2 *generation* were white-eyed.

Males inheriting certain traits from their mother can be best explained by the presence of a female's two X chromosomes. The X chromosomes of a female are unique in that they can be homozygous or heterozygous for alleles, therefore expressing both dominant and recessive traits.

Since males only have one X chromosome, the allele inherited on the X chromosome received from the mother is the one expressed, regardless if the allele is dominant or recessive. The most common sex-linked traits occur in males and are recessive in nature, including color blindness and hemophilia. These types of sex-linked traits have a unique pattern of inheritance in that they typically skip a generation, meaning the daughter of an affected male will become a carrier of the recessive trait and potentially pass this trait to her sons.

The gene for color vision is located on the X chromosome and codes for light-sensitive proteins that allow the eyes to differentiate between red and green wavelengths of light. However, the color vision gene may contain an abnormal allele that produces nonfunctioning light-sensitive proteins, which is carried on an X chromosome. When a mother passes the X chromosome containing the abnormal allele to her son, it results in color blindness. Color blindness is a disorder that prevents an individual from distinguishing between certain colors, especially red and green. The disorder typically affects more males than females because males only have one X chromosome, and if the male inherits the X chromosome with the abnormal allele, they inherit the disorder. Color blindness does affect females but usually occurs in less than 1 percent of the population. Since a female has two X chromosomes, the normal, functional allele compensates for the abnormal allele; therefore, females are often called carriers.

The genes responsible for producing the proteins required for blood-clotting are also located on the X chromosome. An individual who is unable to produce the necessary blood-clotting proteins has a disorder known as hemophilia. Individuals with hemophilia either lack the presence of or have low levels of clotting factor VIII or clotting factor IX; therefore, they are more prone to bruising; extensive bleeding from minor injuries; internal bleeding, especially around the joints; and anemia. As with color blindness, hemophilia affects more males than females because of the abnormal allele being inherited on the X chromosome.

Hemophilia was a disorder that greatly affected males of prominent royal families in Europe, especially during the mid-to-late 1800s (Figure 8.19). The prevalence of the disorder in the British royal family was traced back to Queen Victoria, who reigned between 1837 and 1901, as seen in Figure 8.19. The abnormal allele was also introduced into the Russian, German, and Spanish royal families by Queen Victoria's daughters, Alice and Beatrice, who were carriers of the disorder. King Edward VII, Queen Victoria's son, and heir to the throne after her death, did not receive the abnormal allele; therefore, hemophilia is not present in the current British royal family.

8-6. Genetic Disorders

The inheritance of traits, as we have discussed thus far, is genetic in nature and follows the strict guidelines outlined in both Mendelian laws. Unfortunately, some individuals may inherit unwanted traits, traits they had no control over inheriting, because of the Mendelian laws. An example of an unwanted trait that an individual may inherit is a genetic disorder, which may arise due to the random inheritance of a specific group of alleles or due to errors in meiotic division that resulted in sex cells with an abnormal number of chromosomes. Genetic disorders can occur along autosomal or sex chromosomes, be dominant or recessive, or result from the unequal distribution of chromosomes during the development of sex cells, which can all be easily traced using a pedigree.

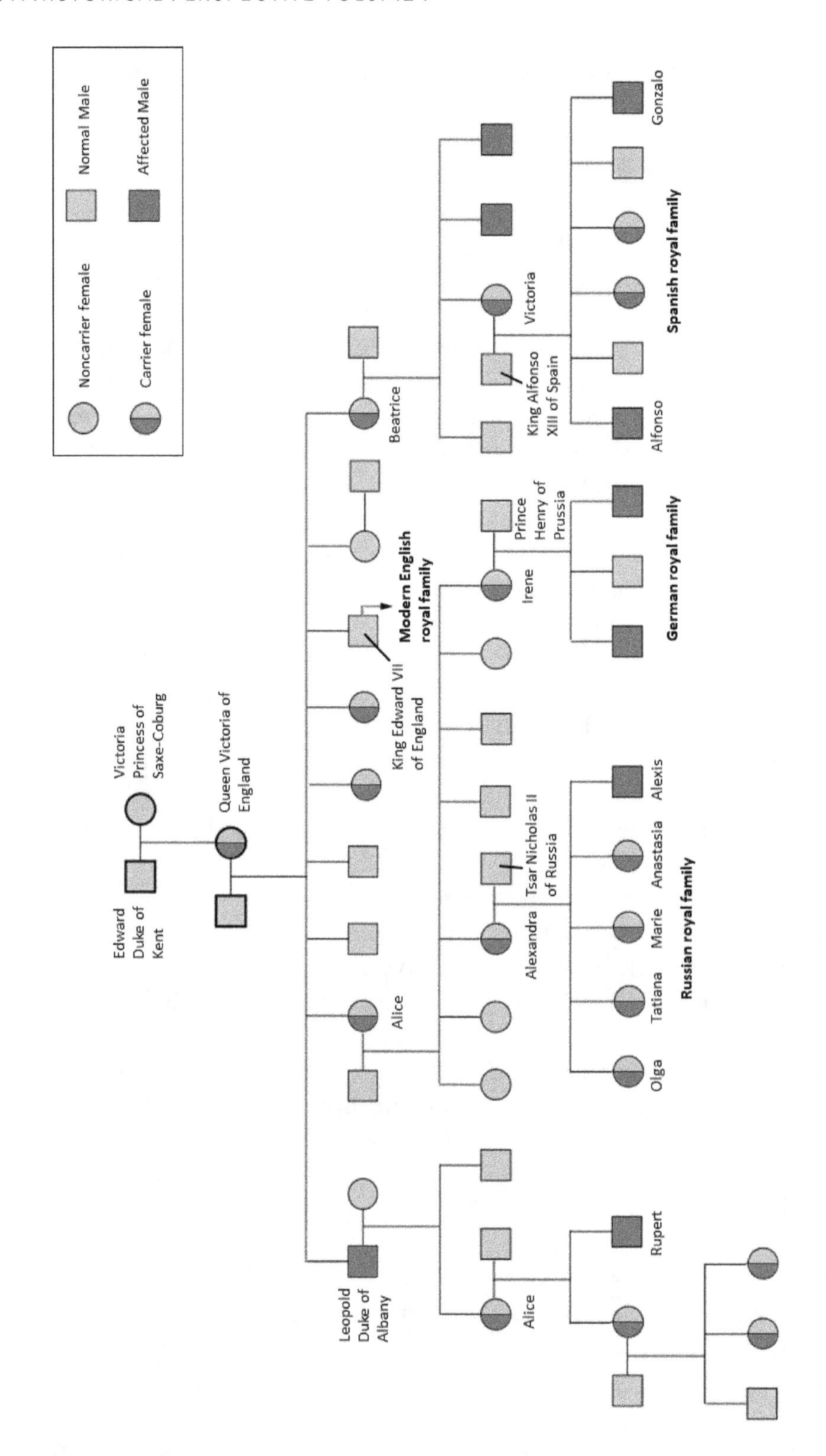

FIGURE 8.19 A pedigree is a useful tool in tracing a pattern of inheritance of a genetic disorder, such as hemophilia in the royal family. Hemophilia is sex-linked trait that affected European royal families and has been traced back to Queen Victoria of England.

Pedigree

A **pedigree** is a chart that traces the occurrence of a trait or genetic disorder in a family over the course of several generations (Figure 8.19). A pedigree is an especially useful tool in genetic counseling, prenatal testing, and understanding and treatment options for genetic disorders. The usefulness of pedigrees includes determining or ruling out the pattern of inheritance of a trait; predicting the probability of offspring having a certain genotype or phenotype; determining whether a trait is sex-linked and dominant or recessive; identifying individuals who may be carrying an abnormal allele but appear normal, which are known as carriers; and determining whether a child poses a risk of inheriting a genetic disorder because of this abnormal allele.

The first step in constructing a pedigree is to obtain as much information as possible from the family, ensuring that it spans several generations. The next step is to convert this information into a variety of shapes and lines that indicate the sexes of each individual, identify unions between individuals, and connect related individuals. Males are represented using squares, and females are represented using circles. A horizontal line between a male and female represents a union, while vertical lines indicate the offspring of these two individuals, and an additional horizontal line is added for any additional offspring produced from this union. If a square or circle is shaded, that individual has the inherited trait of interest or genetic disorder, and if the shape is only half-shaded, the individual is a carrier of the abnormal allele. Once everything has been placed and a pedigree has been constructed, the pattern of inheritance can be traced, and any future inherited traits or genetic disorders can be predicted.

Autosomal Recessive Genetic Disorders

Autosomal recessive genetic disorders occur when two recessive alleles for a particular disorder along an autosomal chromosome are inherited by an individual. There are three common examples, including Tay-Sachs disease, cystic fibrosis, and sickle-cell disease.

Tay-Sachs disease is a prominent disease affecting roughly one in 3,600 individuals born within the Jewish population whose ancestry can be traced back to central and Eastern Europe. The disease occurs when an individual inherits two recessive alleles, resulting in a defective gene located on autosomal chromosome 15, called the *HEXA* gene. The normal-functioning gene produces an active enzyme known as hexosaminidase A (HEXA), which is responsible for digesting lipids within the lysosomes of nerve cells. However, when an individual inherits the defective gene, the body is unable to produce the HEXA enzyme, resulting in the accumulation of lipids in the nerve cells. As the cells continually accumulate lipids, they begin to swell, to the point of burying the nerve cells, which start to die, impairing the brain and its ability to transmit nerve signals to the body. At birth, a child appears normal, but within a few months, the child begins to show symptoms of neurological defects, including the loss of sight, hearing, motor function, and paralysis, eventually dying at age three or four.

Cystic fibrosis is the most common genetic disorder among individuals of European descent, affecting one in 2,500 individuals, with about 4 percent of this population being carriers of the disorder. Individuals diagnosed with cystic fibrosis inherit two recessive alleles that result in a defective *CFTR* gene located on autosomal chromosome 7. A normally functioning *CFTR* gene codes for chloride ion channel proteins, which are located on the cell membrane and function in transporting chloride ions, specifically into and out of the cells found in respiratory and digestive organs, such as the lungs and pancreas. The defective gene produces a chloride ion channel protein that incorrectly folds and is unable to attach properly to the cell membrane. When chloride ions are unable to effectively pass through the defective chloride ion channel, it results in a high concentration of chloride ions within the cell. Typically, when chloride ions leave the cells, water follows; however, water remains in the cell as the concentration of chloride ions increase within the cells. This causes the thin and watery mucus within the lungs and pancreas to become thick and quite sticky, which cannot be removed. The viscous mucus impairs gas exchange within the lungs, causing an individual to have difficulty breathing and increasing the frequency of bacterial infections (Figure 8.20). In addition, the increased viscosity of the mucus in the pancreas prevents the release of digestive enzymes, which affects digestion and nutrient absorption in the small intestine. Individuals diagnosed with cystic fibrosis, if left untreated, have a life expectancy of roughly five years; however, most will survive into their 40s with some form of treatment.

Sickle-cell disease is the most common inherited genetic disorder among Black-Americans, affecting about one in 400. Sickle-cell disease was one of the first genetic disorders whose cause was traced back to a mutation. This mutation occurs along the *hemoglobin-β* gene, located on autosomal chromosome 11, which produces an abnormal hemoglobin molecule that causes the red blood cells to have a sickle shape (Figure 8.21). Functional differences in the hemoglobin of individuals with this disorder were first deduced by Linus Pauling in 1949. However, further studies performed by a team led by German–American biologist Vernon M. Ingram in 1956 located the specific spot along the amino acid sequence where the mutation occurs—the sixth amino acid

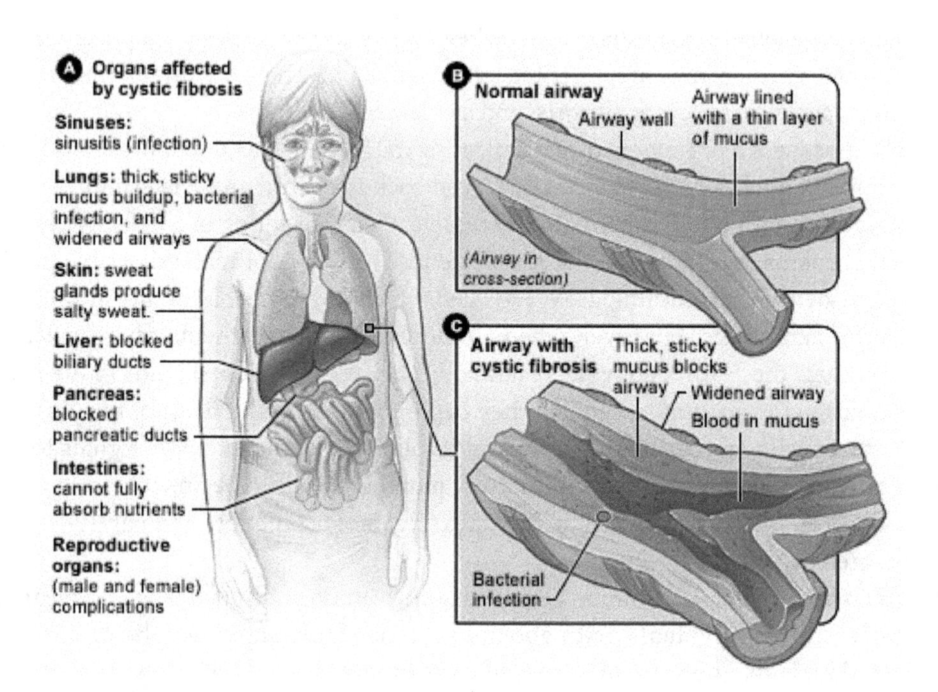

FIGURE 8.20 Cystic fibrosis is an autosomal recessive genetic disorder, which affects calcium channels in the body. A defective calcium channel results in a buildup of mucus in a variety of organs, such as the airways of the lungs, making breathing difficult and increases the risk of bacterial infections.

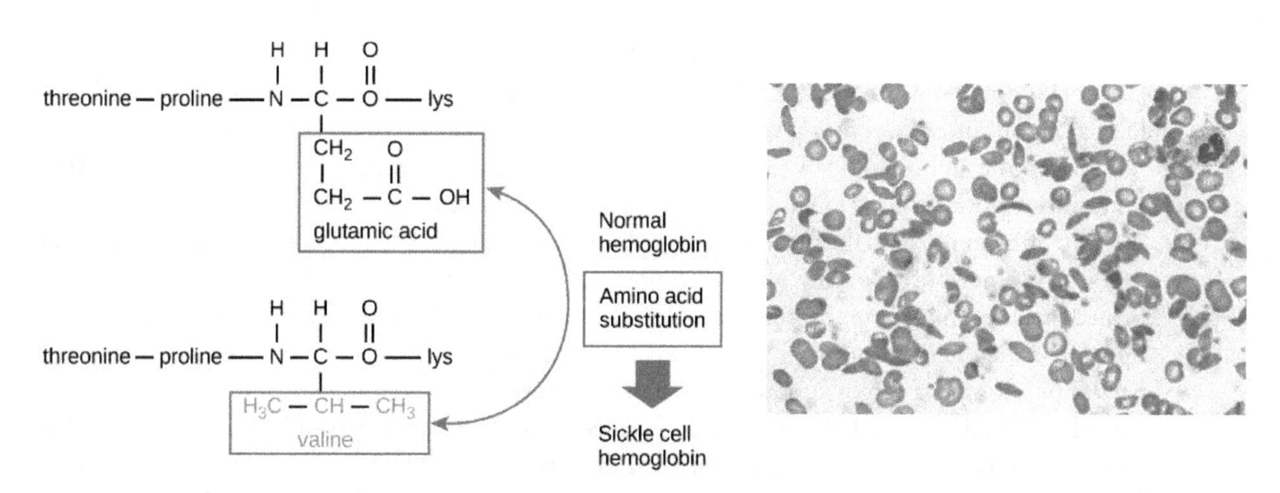

FIGURE 8.21 Sickle-cell disease is a genetic disorder in which a mutation occurs along the hemoglobin-β gene. The mutation causes glutamic acid to replace valine in the amino acid chain (left), which produces abnormal hemoglobin molecules. The abnormal hemoglobin causes the red blood cells to become sickle-shaped (right), making them ineffective in delivering oxygen and leading to other cardiovascular issues, such as blood vessel blockages, poor circulation, and strokes.

where valine was substituted for glutamic acid—resulting in an abnormal hemoglobin molecule. During hypoxic (low-oxygen) events, the abnormal hemoglobin forms long rods within the red blood cells, causing them to become sickle-shaped. The change in shape makes the red blood cells more fragile, breaking apart and dying more quickly, leading to such symptoms as anemia, with small pieces causing blockages in the capillaries. In addition, blockages in narrow blood vessels also occur due to the sickle shape of the cells, which can lead to joint pain, poor circulation, organ damage, physical weakness, and strokes. Recall from section 8-5, individuals with sickle-cell disease are carriers for the disorder and are resistant to malaria, a phenomenon first reported in 1954 by British geneticist Anthony Allison. Allison collected and analyzed blood samples from about 5,000 East Africans and demonstrated that individuals living in coastal regions, where malaria was the most prevalent, had a higher percentage of individuals that were carriers of sickle-cell disease (over 20 percent). In addition, he discovered that in regions where individuals were naturally infected with malaria, individuals who were carriers were more resistant and had a better survival rate.

There are a variety of treatment options for sickle-cell patients, including blood transfusions and bone marrow transplants. However, in 2019, a black woman from Mississippi was the first person to be treated for sickle-cell disease using CRISPR, the gene-editing technique mentioned in chapter 6. The process involved removing her bone marrow cells and editing a gene that would allow her cells to produce fetal hemoglobin, a form of hemoglobin that enables a fetus to obtain oxygen from his or her mother's blood. One year later, the results of the technique have been quite promising. The symptoms she suffered most of her life are basically gone, including the severe pain and physical weakness, along with her frequent hospital visits for blood transfusions. In addition, the researchers discovered that the concentration of fetal hemoglobin in her blood was much higher than expected and has remained high in her red blood cells, while more than 80 percent of her bone marrow cells still contain the CRISPR edit.

Autosomal Dominant Genetic Disorders

Autosomal dominant genetic disorders occur when an individual inherits at least one dominant allele along an autosomal chromosome. Therefore, an individual with the genetic disorder can either be homozygous dominant (AA) or heterozygous (Aa). Many common examples exist, including neurofibromatosis, achondroplasia, and Huntington's disease.

Neurofibromatosis is the most common inherited neurological disorder, affecting approximately one in 3,500 newborns worldwide. Neurofibromatosis occurs because of an *NF1* gene mutation located on autosomal chromosome 17. A normal-functioning *NF1* gene is responsible for producing a protein called neurofibromin in various cells, including nerve cells. The protein regulates cell growth by acting as a tumor suppressor that prevents cells from dividing uncontrollably. When an individual inherits a mutated *NF1* gene, it produces a nonfunctional neurofibromin protein that is unable to regulate cell division. This results in the formation of tumors, known as neurofibromas, that grow along nerves throughout the body, including those in the brain and spinal cord. In addition to tumors, individuals are born with tan to dark skin spots, called café-au-lait spots, which grow and increase in size as the individual ages. Although most individuals have mild symptoms and live a normal life, some symptoms can be quite severe, including high blood pressure, skeletal deformities such as scoliosis (spine curvature), and, at its worst, hearing and sight loss.

Achondroplasia is a form of short-limbed dwarfism that can affect up to one in 40,000 newborns. Individuals born with achondroplasia have inherited one copy of a mutated *FGFR3* gene, located on autosomal chromosome 4. The *FGFR3* gene produces a protein called fibroblast growth factor receptor 3 that is responsible for bone development during a process known as ossification, in which cartilage is converted to bone. However, a mutated version of the *FGFR3* gene produces an overactive protein that interferes with the ossification process. As a result, individuals with achondroplasia are small in stature, averaging no more than four feet tall; have short limbs; and have an accompanying sway when they walk. In addition, achondroplasia individuals are characterized by macrocephaly (large head), with a prominent forehead, short fingers, and frequent health problems such as ear infections and sleep apnea. In cases in which an individual inherits both dominant alleles for the mutated *FGFR3* gene, the individual dies shortly before or after birth due to an underdeveloped rib cage, which leads to respiratory failure.

Huntington's disease is a progressive, irreversible, degenerative nervous system disorder that affects an estimated one in 10,000 individuals, especially those of European descent. Huntington's disease is caused when an individual inherits a defective *HTT* gene, which is located on autosomal chromosome 4. The *HTT* gene produces a protein called huntingtin, a protein whose function is relatively unknown but has been associated with normal brain development before birth and other important cellular functions, such as signaling and transport. The mutation occurs along a region of trinucleotide repeats of cytosine, adenine, and guanine (CAG) that normally repeat ten to 35 times (Figure 8.22). In the mutated version of the *HTT* gene, the trinucleotide region repeats more than 40 times, producing a longer-than-normal huntingtin protein. As the protein increases in length, it changes shape and begins to accumulate within neurons, disrupting their normal function and resulting in their eventual death. Many individuals with Huntington's disease are asymptomatic until the phenotypic effects of the disorder begin, around 35 to 40 years of age. Currently, individuals begin to experience such symptoms as personality changes, poor coordination, and muscle spasms. Upon the onset of symptoms, an individual may only live another 15 to 20 years.

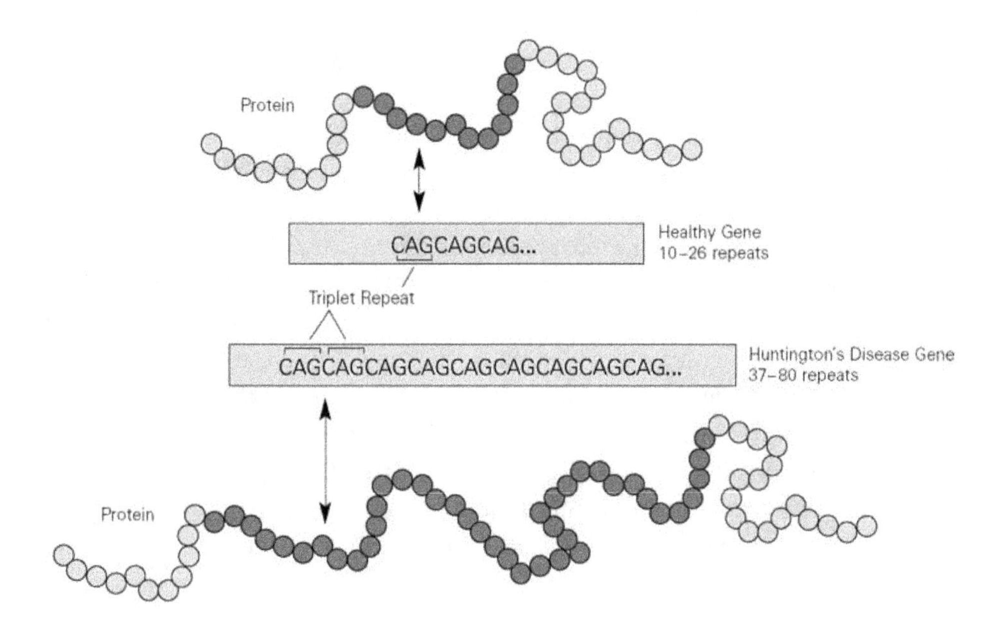

FIGURE 8.22 Huntington's disease is an autosomal dominant disorder that produces a mutated version of the *HTT* gene. The mutated gene increases the trinucleotide region of cytosine, adenine, and guanine (CAG) of the huntingtin protein. The resulting extra-long protein accumulates in the neurons affecting neurological function and eventual death later in life.

Nondisjunction

Nondisjunction occurs when chromosomes fail to equally distribute during meiosis, resulting in gametes with an abnormal number of chromosomes (Figure 8.23). Nondisjunction affects both autosomal and sex chromosomes, leading to different types of genetic disorders.

Autosomal Genetic Disorders Due to Nondisjunction

Patau syndrome (trisomy 13) occurs when an individual inherits an extra copy of chromosome 13, instead of the normal two. This genetic disorder occurs in about one in 16,000 newborns, with

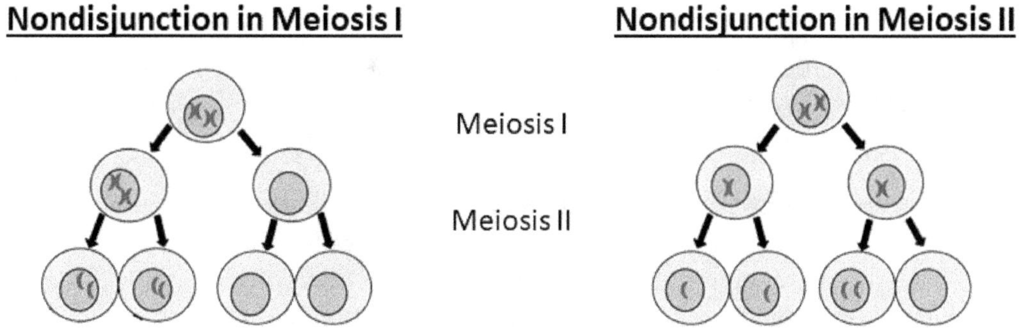

Nondisjunction in Meiosis I

Nondisjunction in Meiosis II

Meiosis I

Meiosis II

FIGURE 8.23 Nondisjunction can occur during meiosis I when the homologous chromosomes fail to separate or during meiosis II if the sister chromatids fail to separate. This error during meiosis results in gametes with an abnormal number of chromosomes and can lead to different genetic disorders.

most dying soon after birth and only 5 to 10 percent living longer than one year. Individuals with Patau syndrome have many life-threatening complications, including heart defects, nervous system abnormalities, and a cleft lip, which may or may not be accompanied by a cleft palate. For those individuals that do survive, they have severe intellectual disabilities, small adrenal glands, and an underdeveloped face.

Down syndrome (trisomy 21) is the most observed chromosomal disorder in humans, occurring in approximately one in 800 newborns. The disorder results when an individual receives an extra copy of chromosome 21 (Figure 8.24). Roughly half of those born with the disorder die within the first year due to organ defects or infections. For those that do survive, the extra chromosome interferes with the normal development of the individual, leading to many different, distinct physical characteristics. The most common are the distinct facial features, including a flat face, slanted eyes, and small mouth held partially open due to a large, protruding tongue (Figure 8.24). Another common feature of Down syndrome is learning disabilities, generally associated with varying degrees of mental retardation. Other characteristics include short stature, weak muscle tone, and an increased risk of developing leukemia and Alzheimer's later in life.

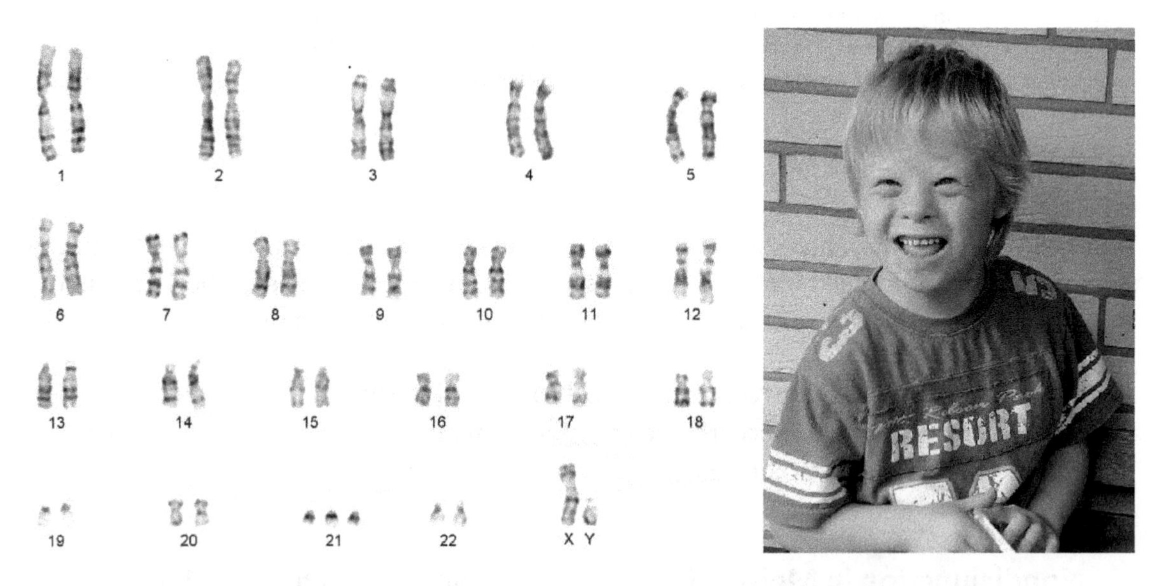

FIGURE 8.24 Nondisjunction resulting in additional autosomal chromosomes can lead to different types of genetic disorders such as Down syndrome (trisomy 21). Individuals with Down syndrome have an extra copy of chromosome 21 (left) and have distinct facial features, such as a flat face and slanted eyes (right).

Sex Chromosome Genetic Disorders Due to Nondisjunction

When gametes are formed during meiosis, half of the sperm cells will carry an X chromosome and the other half a Y chromosome, while in females, all four cells will carry an X chromosome. However, during nondisjunction, a sperm cell may have both sex chromosomes, such as XX, XY, YY, or a sperm cell lacking a sex chromosome, known as an "O" sperm. In females, nondisjunction may produce an egg with two X chromosomes, XX, or an "O" egg, which does not contain either X chromosome. After fertilization, the resulting zygote will have an abnormal number of sex chromosomes and will typically result in a miscarriage, especially a zygote without at least one X chromosome, as these chromosomes contain several genes important for the proper development and survival of the offspring. However, some zygotes do survive, although the individuals born have an abnormal number of sex chromosomes, which can lead to variety of different genetic disorders, such as Turner syndrome, trisomy X, Klinefelter syndrome, and Jacob's syndrome.

Turner syndrome is a unique genetic disorder because it is the only disorder in which an individual can survive without having two sex chromosomes, denoted XO (Figure 8.25a). Although 98 percent of fetuses do not survive to term, roughly one in 2,500 females suffer from Turner syndrome. The disorder is most characterized by varying degrees of intelligence, short height (as seen by age five), and a web of skin located between the neck and shoulders, which only occurs in about 30 percent of females. Another common characteristic of Turner syndrome is the lack of knowledge that they have the disorder until puberty. Upon reaching pubescence, deficiencies in hormone production prevent both normal sexual development and menstruation from occurring. However, with hormone replacement therapy, the female secondary sex characteristics begin to develop. Despite hormone replacement therapy, Turner females are sterile because they lack mature eggs, although

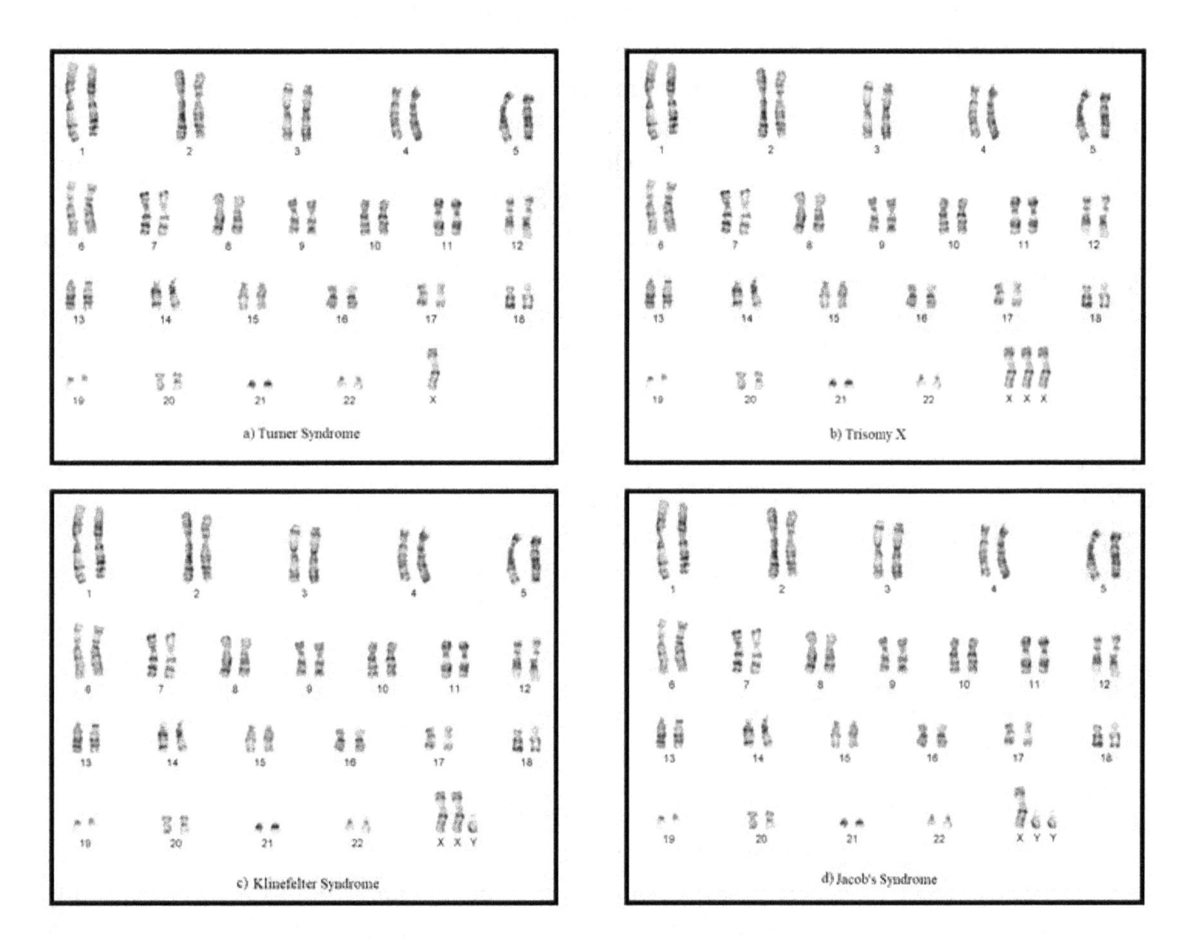

FIGURE 8.25 The karyotypes show individuals who are born with an abnormal number of sex chromosomes. The genetic disorders that arise from an abnormal number of sex chromosomes include a) Turner Syndrome (XO), b) Trisomy X (XXX), c) Klinefelter Syndrome (XXY), and d) Jacob's Syndrome (XYY).

many have given birth with the help of donor eggs and in vitro fertilization. Some females with Turner syndrome have life-threatening cardiovascular issues, such as aortic narrowing or aortic valve abnormalities, and are more prone to X-linked recessive disorders like hemophilia.

Trisomy X, denoted XXX, occurs when a female inherits an additional X chromosome (Figure 8.25b). Trisomy X affects approximately one in 1,000 females but does not typically present any abnormal physical features, except for being taller than normal and thin. Females with Trisomy X have varying degrees of intelligence, ranging from normal IQ to some slight learning disabilities, resulting in delayed speech and language skills. Despite some menstrual irregularities, sexual development is normal in Trisomy X females. Most females can conceive offspring with a regular set of sex chromosomes, although there is an increased risk of conceiving offspring with an abnormal set, such as an XXX daughter or an XXY son.

Klinefelter syndrome is one of the most common sex chromosome genetic disorders, affecting one in 650 newborn males, and is denoted XXY (Figure 8.25c). Males with Klinefelter display minimal physical and mental characteristics, such as being taller than average and demonstrating some learning disabilities. As with Turner syndrome, males who have Klinefelter syndrome may not realize they have the genetic disorder until pubescence, when both female and male features begin to develop, even though about 75 percent are never diagnosed. One of the most common female features is breast development, along with minimal to no facial hair, while low levels of testosterone are produced due to underdeveloped testes, which results in sterility. Most Klinefelter males utilize testosterone replacement therapy to help stimulate muscle strength, promote libido, and reduce aspects of female development, but it has no effect on testicular development or sterility. In comparison to unaffected males, about 50 percent of males with Klinefelter syndrome will develop high blood pressure, type 2 diabetes, or rheumatoid arthritis later in life.

Jacob's syndrome affects one in 1,000 newborn males and occurs when a male inherits an additional Y chromosome (Figure 8.25d). Denoted as XYY, males suffering from Jacob's syndrome do not typically present abnormal physical features, except for being quite tall, with over 60 percent being taller than six feet. However, males with Jacob's syndrome have slightly lower intelligence, which contributes to developmental delays in language and speech skills. As with most sex chromosome genetic disorders, males with Jacob's do not show any signs of having an additional Y chromosome. Despite having higher than normal levels of testosterone, males with Jacob's syndrome develop secondary sex characteristics as normal and are usually capable of bearing offspring.

Chapter Summary

8-1. Genetics

- Genetics is the scientific study of the inheritance of traits and variation of inherited characteristics. The term *genetic* was first used in a series of papers published in 1819 to resolve arguments among sheep breeders in Brno (Czech Republic) about inbreeding and the quality of wool. The published papers of Imre Festetics were in response to the "physiological laws of nature" of inheritance argument proposed by J. M. Ehrenfels, suggesting the inheritance of traits occurred as a result of the "genetic laws of nature."

- Mendelian genetics began in 1854 in the experimental garden of the Augustinian St. Thomas's Abbey monastery. Gregor Mendel would spend two years evaluating different varieties of pea plants before choosing *Pisum sativum*, the common garden pea, to perform experiments on trait inheritance with. Mendel chose *Pisum sativum* for several reasons. The peas were easy to grow, developed quickly, were easy to fertilize manually, produced hundreds of offspring, and each character had two easily distinguishable forms. Mendel chose to focus on seven different characters, including seed form, seed color, seed coat color, pod form, pod color, flower position, and stem length.
- Prior to performing his experiments, Mendel created a distinct population of true-breeding plants. True-breeding plants were plants that, when crossed, would produce offspring exhibiting the same traits as the parents, such as round-seed plants producing round-seed offspring, in every generation. Mendel would perform his inheritance experiments over the course of eight years, studying several thousand *Pisum sativum* plants. Mendel kept a detailed catalogue of his experiments, including the traits he observed in every generation produced. Mendel was the first to apply a quantitative and mathematical approach to analyzing his data. This approach enabled Mendel to propose his particulate theory of inheritance, which stated that heredity units called *elementen* were exchanged and reshuffled and responsible for the inheritance of traits between generations. The particulate theory of inheritance consisted of two fundamental laws, the law of segregation and the law of independent assortment. Mendel detailed the results of his experiments in his publication "Experiments on Plant Hybridization" in 1866. After he received and distributed 40 copies of his paper, many in the scientific community discredited his work, but copies of the *Proceedings of the Science Research Society of Brünn*, in which it was published, reached other parts of the world. Between 1869 and 1899, Mendel's paper and name were rarely referenced, fading into obscurity.
- Mendelian genetics was rediscovered by four botanists in the early 1900s, each working independently of one another in different countries. Carl Correns was a German botanist whose work on corn and peas led to the 1900 publication "G. Mendel's Law Concerning the Behavior of the Progeny of Varietal Hybrids," in which he referenced Mendel's previous work. Erich von Tschermak, an Austrian botanist who, like Mendel, was experimenting with peas, published his paper "On Deliberate Cross-Fertilization in the Garden Pea" in 1900, outlining the results of his experiments and referencing Mendel in a footnote. Recent discoveries in unpublished memoires and family letters have cast doubt regarding von Tschermak's role in being considered a rediscoverer of Mendelian genetics. Dutch botanist Hugo de Vries proposed a particulate theory of inheritance in his 1868 publication, *Intracellular Pangenesis*, suggesting that hereditary units called *pangenes* were responsible for the inheritance of traits. After experimenting with corn, peas, and poppies, de Vries published *The Law of Segregation of Hybrids* in 1900, which detailed his observations of segregation and patterns of inheritance. William Jasper Spillman was an American botanist whose work with wheat varieties in the state of Washington demonstrated that traits separated and recombined. Spillman would publish "Quantitative Studies on the Transmission of Parental Characteristics to Hybrid Offspring" in 1902; however, whether he had any previous knowledge of Mendel or Mendel's work is uncertain. Upon his paper being reprinted in 1903 in the *Journal of the Royal Horticultural Society*, it was suggested Spillman be added to the list of rediscoverers.

8-2. Law of Segregation

- Mendel's first group of experiments focused on single characters with two distinct traits. Mendel began his experiment by crossing two different true-breeding parental plants, referencing them as the parental (P) generation. Mendel observed that the resulting offspring from the parental plants, the filial (F_1) generation, did not exhibit intermediate traits, as the blending concept of inheritance theory would suggest. All the F_1 generation offspring exhibited a trait resembling one of the parents, while the other trait had "disappeared." Mendel performed a monohybrid cross to generate a second generation, known as the F_2 generation. A monohybrid cross is performed to trace the occurrence of a single character by crossing parental hybrids. The resulting F_2 generation offspring produced traits in a 3:1 ratio, with 75 percent of the offspring exhibiting one trait and 25 percent exhibiting the "disappeared" trait from the F_1 generation. Mendel would conclude that the "disappeared" trait was hidden or masked by the other trait. Mendel described the trait masking the other trait as dominant, while he designated the trait being masked as recessive. Upon analysis, Mendel would formulate three main conclusions, which would become the law of segregation. Each plant contains two hereditary units (*elementen*) for each trait. *Elementen* segregate in the formation of gametes, with each gamete containing one unit or the other for each trait. When gametes are fertilized, *elementen* recombine, giving the resulting individual a new combination of two units for each trait.
- Modern genetics has introduced a series of terms that are important in helping understand the law of segregation. Chromosomes were first described in 1879 by Walther Flemming and named in 1888 by Wilhelm von Waldeyer-Hartz. Researchers in the 20th century began to see similarities between Mendel's *elementen* and chromosomes, especially in how they separate during cell division. The chromosomal theory of inheritance was introduced by Theodor Boveri and Walter Sutton in 1902, which stated that chromosomes were responsible for the inheritance of traits. The term *gene* was introduced in 1909 by Danish geneticist Wilhelm Johannsen in his publication, *The Elements of an Exact Theory of Heredity.* Although Johannsen shortened the term *pangene*, used by Hugo de Vries to describe the hereditary units of inheritance, Mendel's *elementen* now had an official name. Thomas Hunt Morgan's *Drosophila* experiments of 1910 demonstrated that genes were located on chromosomes. Genes are located on a specific region of a chromosome, now known as a locus. William Bateson and Edith Rebecca Saunders introduced the term *allelomorph* in 1902. The term was shortened to *allele*, which is the hereditary unit of a gene. Each gene has two alleles for every trait. Alleles for traits are located on homologous chromosomes. When the terms *dominant* and *recessive* are used to describe the different alleles, each allele is represented using letters of the alphabet. Dominant alleles are represented using capital letters, and recessive alleles are represented using lowercase letters. The term homozygous refers to two identical alleles of a gene present on both chromosomes. Homozygous dominant is used to describe a gene having two dominant alleles that code for the dominant trait and are represented by capital letters, such as RR. Homozygous recessive is used to describe a gene having two recessive alleles that code for the recessive trait and are represented by lowercase letters, such as rr. The term heterozygous refers to two different alleles occupying a gene's position on a chromosome and

represented by a capital letter (dominant allele) and a lowercase letter (recessive allele), such as Rr. Wilhelm Johannsen's 1903 publication, "On Heredity in Society and in Pure Lines," introduced the terms *genotype* and *phenotype*. Genotype is the specific combination of alleles an organism inherits. An organism's genotype is represented using letters and descriptive phrases, such as homozygous dominant (RR), heterozygous (Rr), and homozygous recessive (rr). Phenotype is the expression of an organism's genotype and refers to an organism's outward appearance and its physiological and psychological traits. The phenotype of an organism can also be influenced by the environment.

8-3. Law of Independent Assortment

- Mendel's next group of experiments focused on the inheritance of two character traits. Mendel crossed true-breeding parental plants having two different traits to produce his P generation. Mendel observed that the F_1 generation were all heterozygotes exhibiting the dominant traits. Mendel performed a dihybrid cross to produce an F_2 generation. A dihybrid cross is performed to trace the occurrence of two characters by crossing parental hybrids. The F_2 generation produced four different seed varieties, with traits being exhibited in a 9:3:3:1 ratio. Nine were dominant for both traits; three were dominant for one trait, recessive for the other; three were recessive for one trait, dominant for the other; and one was recessive for both traits. Upon analysis, Mendel formulated his law of independent assortment, which states that *elementen* assort independently of one another during gamete formation; therefore, traits are inherited independently of each other.
- Linked genes are genes that are located close to each other on a chromosome and are likely to be inherited together, which violates Mendel's law of independent assortment. The pattern of inheritance associated with linked genes was first observed and described in 1905 by William Bateson, Edith Rebecca Saunders, and Reginald Punnett while examining flower color and pollen shape in *Lathyrus odoratus* (sweet pea plants). After performing a cross with true-breeding plants in their P generation, the researchers crossed individuals from the F_1 generation with the expectation that a 9:3:3:1 would result in the F_2 generation. The observed ratio deviated from their expected, with more of the offspring exhibiting parental traits. The researchers concluded that the alleles must be linked on the same chromosome, do not independently assort, and are inherited together. Experiments performed in 1912 by Clara J. Lynch and Thomas Hunt Morgan on *Drosophila* provided more insight on the pattern of inheritance observed with linked genes. The experiments involved examining body color and wing size, two traits located on autosomal chromosomes. In their P generation, the researchers crossed two true-breeding parents, a wild-type fly (gray body and normal wings) with a mutant fly (black body and vestigial wings). The resulting offspring in the F_1 generation all exhibited heterozygous traits (gray body and normal wings). To determine whether the genes were linked, Morgan and Lynch crossed a heterozygous fly (gray body and normal wings) with a homozygous recessive fly (black body and vestigial wings). Most of the resulting offspring expressed traits of the parents in the P generation, named parental phenotypes. The researchers concluded that the traits for body color and wing size do not assort independently, are linked, and are inherited together. The remaining offspring produced

during this cross exhibited traits different from those of the P generation and were named nonparental phenotypes. The researchers concluded that the linked alleles must be broken and recombined during gamete formation. Morgan concluded with subsequent experiments that genetic material must be exchanged during a process known as crossing over. Crossing over is a process that occurs during meiosis, in which genetic material is exchanged between homologous chromosomes. Crossing over can occur along any portion of a chromosome but typically occurs in regions where the genes are farther apart. This idea was first proposed by Alfred H. Sturtevant in 1911. Genes that are close to one another on a chromosome are less likely to separate during crossing over, resulting in them being inherited together. Bateson, Saunders, Punnett, Morgan, and Lynch all observed the presence of nonparental phenotypes, suggesting that linked genes do occasionally separate during meiosis and crossing over.

8-4. Laws of Probability

- The laws of probability predict the chance a particular event will occur. The chance an event will occur is measured using a scale ranging from 0 to 1, with 0 representing that the event will not occur and 1 representing that the event will occur. The laws of probability state that the occurrence of one event is not dictated by the occurrence of another. Upon applying the laws of probability to a particular event, the sum of all probabilities must equal 1. The laws of probability played a central role in Mendel's formulating his laws of inheritance. In the law of segregation, it is impossible to know which gamete will receive which allele, and each resulting gamete with a specific group of alleles has an equal chance of being fertilized. In the law of independent assortment, the allele a gamete receives is an independent event and not influenced by other alleles.
- There are two tools geneticists use to calculate and determine the probability of offspring having certain genotypes and phenotypes, the Punnett square and test cross The Punnett square was developed by Reginald Punnett in 1906. The Punnett square is used to trace possible outcomes of a cross between two individuals of known genetic composition. All possible types of sperm alleles are lined up vertically, and all possible egg alleles are lined up horizontally. Allele combinations are placed in each corresponding square to form an offspring's predicted genotype. The laws of probability use two rules to evaluate the resulting genotypes and phenotypes of the Punnett square. The product rule is used to determine the probability of independent events giving rise to one genotype or another. The application of the product rule involves multiplying the probabilities of an egg or sperm having a particular allele. The sum rule is used to determine the probability of a particular phenotype occurring in multiple ways. The application of the sum rule involves adding together the resulting probabilities of certain phenotypes. The test cross was a tool developed by Mendel to confirm his law of segregation. Geneticists use a test cross to determine the genotype of an organism exhibiting a dominant phenotype. When applying a test cross, a homozygous recessive individual is crossed with an individual either homozygous dominant or heterozygous for the dominant trait. If the individual is homozygous dominant for the trait, all the resulting offspring will be heterozygous. If the individual is heterozygous for the trait, half of the offspring will be heterozygous and the other half homozygous recessive.

8-5. Complex Patterns of Inheritance

- Mendel observed a simple, straightforward approach to the inheritance of traits. The inheritance of traits may be influenced by more complicated means than observed by Mendel, expressing unexpected genotypes and phenotypes.
- There are a variety of different complex patterns of inheritance that are governed by the general laws of inheritance.
    - Incomplete dominance is when a heterozygote has a phenotype intermediate between those of two homozygotes. Examples include pink snapdragons (results from a cross between a homozygous white snapdragon and homozygous red snapdragon) and wavy hair texture in humans (results from a cross between a homozygous dominant individual with curly hair and a homozygous recessive individual with straight hair).
    - Codominance results when a heterozygote displays characteristics of both alleles. An example includes feather color in chickens: when a white hen is crossed with a black rooster, the heterozygote offspring exhibit both white and black feathers.
    - Multiple allelism is a pattern of inheritance in which a single gene has more than two alleles for a gene within a population. The most common example is the ABO system of blood groups, which is determined by three alleles (I^A, I^B, and i) coding for antigens that give rise to four different blood types (A, B, AB, and O).
    - A polygenic trait is when a single trait is influenced by the additive effect of multiple genes. Examples of polygenic traits include height and skin color.
    - Pleiotropy occurs when a single gene influences multiple unrelated phenotypic traits. An example of pleiotropy is sickle-cell disease, a fatal condition in which defective red blood cells are produced due to abnormal hemoglobin molecules and are ineffective in carrying oxygen within the body.
    - Sex-linked traits are traits coded for by a gene on a sex chromosome. Sex-linked inheritance was first observed during Thomas Hunt Morgan's *Drosophila* experiments of 1910. Sex chromosomes carry different types of genes associated with both male and female reproductive processes. The X chromosome is larger than the Y chromosome, carrying over 1,000 genes. Sex-linked traits are more common in men than women, because men only inherit one X chromosome, and the alleles on that chromosome are expressed. The most common examples of sex-linked traits include color blindness and hemophilia. Color blindness occurs when the color-vision gene has an abnormal allele that cannot produce the light-sensitive proteins responsible for seeing red and green wavelengths of light. Hemophilia is a blood-clotting disorder that occurs when the body is unable to produce the necessary blood-clotting proteins responsible for clotting the blood after an injury. Hemophilia has been found in many prominent royal European families, and its occurrence has been traced back to the British royal family and Queen Victoria. Sex-linked traits usually skip a generation, with the daughter of an affected male being a carrier for the recessive trait, which she may pass to her son.

8-6. Genetic Disorders

- Genetic disorders arise when a specific group of dominant or recessive alleles are inherited or when gametes contain an abnormal number of chromosomes due to errors in meiotic division. Genetic disorders are nonspecific in nature, occurring on either autosomal or sex chromosomes. A pedigree is a useful tool that can trace the inheritance of a trait or the familial occurrence of a genetic disorder over many generations.
- The pedigree can be used to determine or rule out a pattern of inheritance, predict the chances of offspring having a particular genotype or phenotype, determine whether a trait is carried on sex chromosomes and whether the trait is dominant or recessive, or determine individuals who may be carriers of a particular trait. A pedigree is constructed by obtaining information from the family over several generations. The information is converted into a variety of shapes and lines. Males are designated by squares, females by circles. Shaded circles and squares are individuals with the trait or genetic disorder of interest. Circles or squares that are half-shaded represent carriers of the trait or genetic disorder. A horizontal line between a square and circle represents a union. A vertical line identifies the offspring produced by the union. Additional horizontal lines are added for each additional offspring.
- Autosomal recessive genetic disorders occur when an individual inherits two recessive alleles for a particular disorder.
    - Tay-Sachs: Individual inherits two recessive alleles, resulting in a defective *HEXA* gene, located on chromosome 15. Affects one in 3,600 individuals of the Jewish population who are of central and Eastern European descent. Individuals are unable to produce the enzyme HEXA, causing lipids to accumulate in the lysosomes of nerve cells, burying the nerve cells in lipids. Characteristics include impaired nerve signaling in the brain; neurological defects such as loss of sight, hearing, and motor function; paralysis; and early death, by age three or four.
    - Cystic fibrosis: Individual inherits two recessive alleles, resulting in a defective *CFTR* gene, located on chromosome 7. Affects one in 2,500 individuals of European descent, with 4 percent being carriers of the disorder. A defective *CFTR* gene produces an incorrectly folded chloride ion channel that does not allow chloride ions to exit the cell, resulting in an increased concentration of chloride and water in the cell. Characteristics include thick, sticky mucus in the lungs and pancreas, impairing gas exchange in the lungs, as well as digestion and absorption in the small intestine; with treatment, individuals can survive into their 40s.
    - Sickle-cell disease: Individual inherits two recessive alleles, resulting in a defective *hemoglobin-β* gene, located on chromosome 11. Affects one in 400 individuals of African American descent. A defective *hemoglobin-β* gene produces an abnormal hemoglobin molecule, which results in sickle-shaped red blood cells, which are fragile and die shortly after breaking apart as they pass through the blood vessels. Characteristics include anemia, blood vessel blockages, joint pain, poor circulation, organ damage, physical weakness, and strokes. Individuals with sickle-cell disease are resistant to malaria. Treatment options include blood transfusions, bone marrow transplants, and CRISPR technology.

- Autosomal dominant genetic disorders occur when an individual inherits at least one dominant allele for a particular disorder.
  - Neurofibromatosis: Individual inherits at least one dominant allele, resulting in a defective *NF1* gene, located on chromosome 17. Affects one in 3,500 newborns. A defective *NF1* gene is unable to produce neurofibromin, a protein that regulates cell growth. Characteristics include the formation of tumors that grow along the brain and spinal cord, café-au-lait spots, and, in the most severe cases, high blood pressure, scoliosis, and hearing and sight loss.
  - Achondroplasia: Individual inherits at least one dominant allele, resulting in a defective *FGFR3* gene, located on chromosome 4. Affects one in 40,000 newborns. A defective *FGFR3* gene produces an overactive fibroblast growth factor receptor 3 protein, which interferes with the ossification of cartilage to bone. An individual born inheriting both dominant alleles dies prior to or shortly after birth due to respiratory failure. Characteristics include short stature (dwarfism), short limbs, walking sway, and macrocephaly with prominent forehead, short fingers, ear infections, and sleep apnea.
  - Huntington's disease: Individual inherits at least one dominant allele, resulting in a defective *HTT* gene, located on chromosome 4. Affects one in 10,000 individuals of European descent. A defective *HTT* gene produces a longer-than-normal huntintin protein, consisting of more than 40 trinucleotide repeats of CAG, which build up in the brain. The function of the huntintin protein is relatively unknown but may be associated with normal brain development and cellular signaling and transport. Characteristics include neurological disorders later in life, starting around 35 to 40 years of age, such as personality changes, poor coordination, and muscle spasms, with imminent death within 15 to 20 years after onset.
- Nondisjunction results in gametes having an abnormal number of chromosomes due to unequal distribution during meiosis. Autosomal genetic disorders due to nondisjunction occur when an individual inherits additional autosomal chromosomes.
  - Patau syndrome (trisomy 13): Individual has an additional copy of chromosome 13. Affects one in 16,000 newborns. Most individuals die after birth, with only 5 to 10 percent living more than one year. Characteristics include heart defects, nervous system abnormalities, cleft lip and/or cleft palate, severe intellectual disabilities, small adrenal glands, and underdeveloped face.
  - Down syndrome (trisomy 21): Individual has an additional copy of chromosome 21. Affects one in 800 newborns. Most observed chromosomal disorder in humans. About 50 percent die within the first year. Characteristics include distinct facial features, such as a flat face, slanted eyes, and small mouth with a large tongue; learning disabilities; short stature; weak muscle tone; and increased risk of leukemia and Alzheimer's.
- Sex chromosome genetic disorders due to nondisjunction result when a gamete has either no copies or two copies of a sex chromosome, rather than a single copy.
  - Turner syndrome: Females have only one X chromosome, denoted as XO. Affects one in 2,500 females. Only disorder in which a person can survive without one of a pair of chromosomes. Characteristics include varying degrees of intelligence, short height, web

of skin between neck and shoulders, hormone deficiencies, underdeveloped sex characteristics, sterility, and cardiovascular issues. Hormone replacement therapy is an option, allowing sex characteristics to develop.

- Trisomy X: Females have three X chromosomes, denoted as XXX. Affects one in 1,000 females. Characteristics include being taller than normal, thin, varying degrees of intelligence, and normal sexual development despite irregular menstrual cycles; most conceive offspring.
- Klinefelter syndrome: Males have two X chromosomes and one Y chromosome, denoted as XXY. Affects one in 650 males. Most common genetic abnormality in humans. Characteristics include being taller than average, some learning disabilities, hormone deficiencies, underdeveloped testes, sterility, development of female features, high blood pressure, type 2 diabetes, rheumatoid arthritis. Hormone replacement therapy is an option but only stimulates muscle strength, libido, and reduction of female development and has no effect on testicular development or sterility.
- Jacob's syndrome: Males have one X chromosome and two Y chromosomes, denoted as XYY. Affects in one in 1,000 males. Characteristics include being tall (some six feet or taller), slightly lower intelligence, high levels of testosterone, and development of male secondary sex characteristics as normal; most conceive offspring.

End-of-Chapter Activities and Questions

Directions: Please apply what you learned in this chapter to complete the following activities.

Define Each Term in Your Own Words

1. Pangene
2. Monohybrid Cross
3. Parental Phenotype
4. Multiplication Rule
5. Pedigree

Chapter Review

1. Compare and describe the origin of the following terms: *elementen*, *pangene*, and *gene*.
2. Two mice with brown hair are crossed and produce a white mouse. Use the following terms to provide a detailed description of this cross (please note that not all the terms will be used): *genotype, phenotype, heterozygous, homozygous dominant, homozygous recessive, dominant allele(s)*, and *recessive allele(s)*.
3. Explain how the experiments performed by Bateson et al. and Morgan and Lynch demonstrate how the presence of linked genes violates Mendel's law of independent assortment.
4. Define a test cross and explain how Mendel used this tool to support his law of segregation.
5. How do autosomal genetic disorders occur? Give an example and a short description of an autosomal genetic disorder.

Multiple Choice

1. Which of the following plants did Gregor Mendel use during the course of experiments?
    - **a.** *Phaseolus vulgaris*
    - **b.** *Pisum sativum*
    - **c.** *Hordeum vulgare*
    - **d.** *Lathyrus odoratus*

2. Which of the following conclusions from Mendel's experiments laid the foundation for his law of segregation?
    - **a.** alleles segregate during gamete formation
    - **b.** there are two distinct alleles for every trait
    - **c.** alleles assort independently of one another
    - **d.** alleles recombine during fertilization

3. Why are males more likely than females to inherit sex-linked traits?
    - **a.** X chromosome is smaller than the Y chromosome
    - **b.** males only have one X chromosome
    - **c.** females do not inherit sex-linked traits
    - **d.** all of the choices are correct

4. If two individuals with wavy hair mate, what is the probability their offspring will have curly hair? (Hint: this is an example of incomplete dominance.)
    - **a.** 50%
    - **b.** 75%
    - **c.** 25%
    - **d.** impossible

5. As a result of nondisjunction, an egg without either X chromosome (an "O" egg) was fertilized by a sperm cell carrying a Y chromosome. Which of the following genetic disorders would result upon fertilization?
    - **a.** Turner syndrome
    - **b.** Klinefelter syndrome
    - **c.** Jacob's syndrome
    - **d.** none of the choices are correct

Reference

Carlson, L. M. (2005). *William J. Spillman and the birth of agricultural economics.* University of Missouri Press.

Image Credits

Fig. 8.1: Source: https://commons.wikimedia.org/wiki/File:Gregor_Mendel_with_cross.jpg.

Fig. 8.2: Copyright © by Szabó T. Attila (CC BY-SA 3.0) at https://commons.wikimedia.org/wiki/File:Grof_Festetics_Imre.PNG.

Fig. 8.3a: Copyright © by Jamain (CC BY-SA 3.0) at https://commons.wikimedia.org/wiki/File:Pisum_sativum_flowers_J1.jpg.

Fig 8.3b: Source: https://commons.wikimedia.org/wiki/File:Mendel_seven_characters.svg.

Fig. 8.4a: Source: https://commons.wikimedia.org/wiki/File:Carl_Correns_1910s.jpg.

Fig. 8.4b: Source: https://commons.wikimedia.org/wiki/File:Acta_Horti_berg._-_1905_-_tafl._124._-_Erich_Tschermak.jpg.

Fig. 8.4c: Source: https://commons.wikimedia.org/wiki/File:Hugo_de_Vries.jpg.

Fig. 8.4d: Source: https://commons.wikimedia.org/wiki/File:Farm_grasses_of_the_United_States;_a_practical_treatise_on_the_grass_crop,_seeding_and_management_of_meadows_and_pastures,_descriptions_of_the_best_varieties,_the_seed_and_its_impurities,_grasses_for_(14800013853).jpg.

Fig. 8.5a: Source: https://incois.gov.in/Tutor/science+society/lectures/illustrations/lecture34/law3.html.

Fig. 8.7a: Source: https://commons.wikimedia.org/wiki/File:Wilhelm_Johannsen_1857-1927.jpg.

Fig. 8.7b: Source: https://commons.wikimedia.org/wiki/File:Thomas_Hunt_Morgan.jpg.

Fig. 8.7c: Source: https://commons.wikimedia.org/wiki/File:William_Bateson.jpg.

Fig. 8.7d: Source: https://commons.wikimedia.org/wiki/File:Edith_Rebecca_Saunders,_1919.jpg.

Fig. 8.8a: Source: https://incois.gov.in/Tutor/science+society/lectures/illustrations/lecture34/law3.html.

Fig. 8.9: Source: https://commons.wikimedia.org/wiki/File:Independent_assortment_%26_segregation.svg.

Fig. 8.10a: Copyright © by OpenStax (CC BY 4.0) at https://openstax.org/books/biology-2e/pages/12-2-characteristics-and-traits#fig-ch12_02_10.

Fig. 8.11: Copyright © by OpenStax (CC BY 4.0) at https://commons.wikimedia.org/wiki/File:Figure_12_03_04.jpg.

Fig. 8.12a: Source: https://incois.gov.in/Tutor/science+society/lectures/illustrations/lecture34/law3.html.

Fig. 8.13a: Source: https://incois.gov.in/Tutor/science+society/lectures/illustrations/lecture34/law3.html.

Fig. 8.14: Copyright © by RudLus02 (CC BY-SA 4.0) at https://commons.wikimedia.org/wiki/File:09_11aIncompleteDominance-L.jpg.

Fig. 8.15a: Copyright © 2014 Depositphotos/Farinosa.

Fig. 8.15b: Copyright © 2019 Depositphotos/Olhastock.

Fig. 8.15c: Copyright © 2010 Depositphotos/stefan1234.

Fig. 8.16: Copyright © by OpenStax (CC BY 3.0) at https://commons.wikimedia.org/wiki/File:1913_ABO_Blood_Groups.jpg.

Fig. 8.17: Copyright © by CKRobinson (CC BY-SA 4.0) at https://commons.wikimedia.org/wiki/File:Human_skin_colour_chart_%26_histogram.PNG.

Fig. 8.18: Copyright © by OpenStax (CC BY 4.0) at https://commons.wikimedia.org/wiki/File:Figure_12_02_09.jpg.

Fig. 8.20: Source: https://commons.wikimedia.org/wiki/File:Cysticfibrosis01.jpg.

Fig. 8.21a: Copyright © by OpenStax (CC BY 4.0) at https://commons.wikimedia.org/wiki/File:Figure_03_04_05.jpg.

Fig. 8.21b: Copyright © by Ed Uthman (CC BY 2.0) at https://commons.wikimedia.org/wiki/File:Sickle_Cell_Anemia_(5610746554).jpg.

Fig. 8.22: Source: https://commons.wikimedia.org/wiki/File:Huntington%27s_disease_(5880985560).jpg.

Fig. 8.24a: Source: https://commons.wikimedia.org/wiki/File:Human_chromosomesXXY01.png.

Fig. 8.24b: Copyright © by Vanellus Foto (CC BY-SA 3.0) at https://commons.wikimedia.org/wiki/File:Boy_with_Down_Syndrome.JPG.

Fig. 8.25: Source: https://commons.wikimedia.org/wiki/File:Human_chromosomesXXY01.png.

CHAPTER

9

The Origins of the Theory of Evolution

PROFILES IN SCIENCE

FIGURE 9.1 Alfred Russel Wallace.

Alfred Russel Wallace, co-discoverer of the theory of evolution, was born on January 8, 1823, in Llanbadoc, Wales, and died in Dorset, England, in 1913. His father's unprofitable business ventures resulted in the Wallace family moving from Wales to Hertford, England, in 1828. After being withdrawn from Hertford Grammar School in 1836 because of the family's waning finances, Wallace traveled around England, living in London, Bedfordshire, and Leicester, before returning to Wales in 1845. During this time, he was learning, practicing, and teaching the land-surveying trade. As a practicing land surveyor, Wallace spent most days outdoors, allowing him to remain committed to his interest in natural history. Additionally, Wallace's free time was used remaining current in the field by reading scientific works on the topic. As Wallace's interest shifted, he wanted to become a traveling naturalist and explore life's evolutionary history. In 1848, Wallace, accompanied by William Henry Bates, a famed entomologist he met during his time in Leicester, launched an expedition to South America. Wallace spent the next four years mapping and surveying unexplored regions of the Amazon River basin, writing about the culture of the locals, and studying the region's natural history. Upon experiencing failing health in 1852, Wallace returned to England, where he was met with disaster as the ship caught fire and sank, along with a majority of his collected specimens and research. Eighteen months after his return to England, Wallace began a near-eight-year endeavor exploring the Malay Archipelago, collecting over 125,000 specimens, including several thousand new species unknown to Western science, and identifying a transitional faunal boundary between Australia and Asia (the Wallace Line). While on his trip, Wallace articulated ideas about evolution and natural selection in an essay, sending it to Charles Darwin in 1858. Wallace returned to England in 1862, spending the remaining years of his life corresponding with Darwin and writing on topics such as zoogeography, island biogeography, and natural selection, as well as his expeditions, writing one of his most popular books, *The Malay Archipelago* (1869), and giving lectures, including in the United States, defending the ideas of evolution and natural selection. Wallace would win multiple awards, including the Royal Medal (1868), Darwin Medal (1890), and Copley Medal (1908).

Introduction to the Chapter

Many realms of scientific thought prior to the 18th century were influenced by Greek philosophy, including the **fixity of species**, an idea expressed in the writings of Aristotle that all species were divinely created beings permanently fixed in nature. Aristotle believed that each species demonstrated differing levels of complexity but was created with the innate ability to become a more perfect version of itself. Recognizing this, Aristotle arranged all species in sequential order from the simplest to the most complex on a ladder of life, known as *scala naturae*, with humans occupying the top of the ladder. Each species had its permanent place on Aristotle's *scala naturae*, solidifying the idea of the fixity of species, a view held by many for nearly 2,000 years. By the 18th century, many expeditions around the world included naturalists, scientists responsible for collecting and categorizing the flora and fauna of newly explored regions. Upon examining their collections, the naturalists began to observe unique patterns of species distribution in similar and different geographical regions, bringing into question the idea of species fixation. This chapter will introduce the scientists who began to question the fixity of species, along with their thoughts on evolutionary change, including Charles Darwin, whose ideas would eventually lay the foundation for the theory of evolution.

Chapter Objectives

In this chapter, students will learn the following:

9-1. Several scientists suggested early thoughts on evolutionary change, which were instrumental in providing the foundation on which Charles Darwin built his ideas about species change.

9-2. Charles Darwin's voyage on the HMS *Beagle* provided the evidence he needed to publish *On the Origin of Species,* outlining natural selection as the primary means of evolutionary change.

9-1. Pre-Darwinian Science

The teachings of Aristotle, along with those of other Greek philosophers, greatly influenced the scientific thoughts of the Western world. Many of the teachings were consistent with the theological beliefs of the Judeo-Christian religion introduced in *Genesis* and other books of the Old Testament. The books of the Old Testament described the origination of living species as a single act of creation by God, which were therefore special creations that remained unaltered through time. While analyzing the works of the Old Testament, Archbishop James Ussher of Ireland, a prolific biblical scholar of his time, developed a chronology of the origin of the world. Ussher published in 1654 that the earth was created around 4004 BC, suggesting the planet was no more than 6,000 years old, and the single act of creation must have occurred around this time period. In addition, with the belief that each species was a special creation, any idea that God's creations would become extinct was firmly dismissed. Although fossils were discovered as early as the sixth century BC, many remained defiant to believe that fossils represented extinct life. Some scientists argued that the organismal remains were organisms that did not survive the biblical flood described in *Genesis*, while others suggested the discoveries were nothing more than rocks shaped by the coincidental results of nature, resulting in the resemblance of once-living organisms.

However, as more fossils were being discovered by the mid-1700s, the unique impressions of extinct plants and animals began to shed light on possible connections between fossils and currently living organisms. As paleontology began to develop more in the 18th and 19th centuries, scientists began to discover that the fossilized remains represented unobserved organisms that must have gone extinct, as there was no evidence that members of the species remained on Earth. This suggested that different groups of organisms must have inhabited Earth during various periods of time in Earth's history. The distribution of fossils in the rock layers provided paleontologists further evidence of a connection between extinct and modern species, because fossils discovered in younger layers of rock more closely resembled modern species than fossils found in older rock layers. As scientists began to observe more of these unique patterns in the fossil record, it began to ignite conflict and division between contemporaries of the time, as these discoveries were inconsistent with theological ideology. The scientists responsible for bringing into question the theological ideas of the time included Georges-Louis Leclerc, James Hutton, Georges Cuvier, Erasmus Darwin, Jean-Baptiste Lamarck, and Charles Lyell (Figure 9.2).

Georges-Louis Leclerc (1707–1788)

Georges-Louis Leclerc (Figure 9.2a), better known as Comte de Buffon, was an 18th-century French naturalist. His greatest work, *Histoire Naturelle*, was written over the span of 50 years and described several aspects of natural history, with a primary focus on minerals, birds, and quadrupeds (animals with four feet). Comte de Buffon was one of the first scientists to suggest that species change by means of descent with modification, the idea that descendants obtain traits from their ancestors. He speculated that the species environment, migration, or the species struggle for existence were possible mechanisms that could influence modification. Comte de Buffon also asserted that the Earth was considerably older than 6,000 years, suggesting it was about 75,000 years old. He countered the theological idea by stating that when the planets formed because of a comet striking the sun, it took Earth at least 75,000 years for the planet's molten state to cool. Comte de Buffon also paid close attention to the internal anatomy of organisms, describing it as "the foundation of nature's design," and comparing the morphological characteristics of different species. This was evident in his description of the morphological similarities between apes and humans, suggesting an ape's lack of cognitive ability was the main difference between the two species. However, due to Comte

FIGURE 9.2 The scientists who shaped the earliest thoughts on evolutionary change included a) Georges Louis Leclerc, b) James Hutton, c) Georges Cuvier, d) Erasmus Darwin, e) Jean-Baptiste Lamarck, and f) Charles Lyell.

de Buffon's belief in the fixity of species and special creation, he remained indecisive on professing evolutionary descent, in addition offering no causative mechanism by which the process would occur.

James Hutton (1726–1797)

James Hutton (Figure 9.2b) was a Scottish geologist, whose famous work *Theory of the Earth*, first published in 1788, established the science of modern geology. After studying the coastlines and landscapes of the Scottish lowlands for roughly 25 years, Hutton proposed that gradual, natural forces were responsible for the formation of Earth's different geological features. Hutton suggested these forces acted in a slow yet continuous manner, and the cumulative effects would produce profound geological changes over time. Although Hutton's ideas were not universally recognized at the time, his ideas would later became the foundation of the theory of uniformitarianism proposed by Charles Lyell, discussed below.

Georges Cuvier (1769–1832)

Georges Cuvier (Figure 9.2c) was a French zoologist whose work on animal anatomy and the study of fossils was paramount in the development of new fields of biology, such as comparative anatomy and vertebrate paleontology. Cuvier was one of the first scientists to use comparative anatomy of living species and the fossil record to expand the Linnaean classification of animals. Cuvier's examination of strata in coal and slate mines in France revealed fossilized forms of once-living organisms, which he observed were quite different from current living species. Upon examining each successive strata layer more carefully, Cuvier observed that the more recent fossils were in the upper layers, while the deeper layers consisted of much older fossils. As a believer in special creation, Cuvier suggested that an initial mass of species was created, with the belief that the fossilized species he observed within the strata layers were species that had gone extinct by a series of catastrophic events, such as floods or volcanic eruptions, that had occurred throughout Earth's history. He speculated that each strata layer represented a different catastrophic event, and the more recent fossils were remnants of newer species that had repopulated the area after the previous species had gone extinct. In 1813, Cuvier compiled his ideas and proposed the theory of **catastrophism** in his publication entitled "Essay on the Theory of the Earth" and expanded upon his theory in *A Discourse on the Revolutions of the Surface of the Globe*, published in 1825. Within each publication, Cuvier speculated that catastrophic events throughout the history of Earth were responsible for producing different geological changes and the succession of animal fossils discovered within the strata.

Erasmus Darwin (1731–1802)

Erasmus Darwin (Figure 9.2d) was a prominent English physician and naturalist and the grandfather of Charles Darwin. In his most famous scientific work, *Zoonomia* (1794), Darwin wrote, "all warm-blooded animals have arisen from one living filament," an idea suggesting evolutionary descent from a common ancestor. Although Darwin would not propose a causative mechanism by which evolutionary descent might occur, his ideas were based solely on his observations of comparative anatomy and embryology. Darwin would suggest another evolutionary idea in *Zoonomia*—sexual selection—writing that when males compete for a mate, "the strongest and most active animal should propagate the species, which should thence become improved." Finally, in a poem published after his death, "The Temple of Nature" (1803), Darwin would describe nature as "a great slaughter-house," suggesting nature's struggle for existence, an idea that would become prominent in Charles Darwin's theory of evolution.

Jean-Baptiste Lamarck (1744–1829)

Jean-Baptiste Lamarck (Figure 9.2e) was a French naturalist and one of the first scientists to propose a potential mechanism through which organisms might change. Lamarck hypothesized that evolutionary change and extinction were directly related to how an organism interacted with and adapted to its environment. He believed organisms well-suited to their environment would experience gradual change, while those that were not would become extinct. The basis of Lamarck's hypothesis was a unique pattern he observed within the fossil record, noting how younger fossils, which were complex, revealed more recent lines of descent to current living species than older fossils. In 1809, Lamarck published his hypothesis in *Zoological Philosophy: An Exposition with Regard to the Natural History of Animals*, proposing two laws suggesting how the environment would influence evolutionary change. The first law Lamarck proposed was the **use and disuse law**, stating that the extensive use of a body part would increase an organism's chance of survival and success in its environment, while the disuse of a body part would eventually cause it to deteriorate and lose function. Lamarck suggested that the length of a giraffe's neck was the result of ancestral giraffes stretching their necks to reach leaves in tall trees, while mole rats, who spend most of their time underground, were blind because of the disuse of the eyes. Lamarck's second law was the **inheritance of acquired characteristics law**, which stated that organisms who acquired characteristics making them better adapted and more successful in their environment would pass these characteristics on to their offspring. Lamarck suggested that as ancestral giraffes continued stretching their necks, the resulting long necks were inherited in subsequent generations. Unfortunately, the premise of Lamarck's hypothesis was wrong, as experimentation has never validated the inheritance of a trait based on the use and disuse law. In other words, changes in an organism's phenotype do not influence its genotype. However, Lamarck's insight about the inheritance of traits was instrumental in setting the stage for the development of the theory of evolution.

Charles Lyell (1797–1875)

Charles Lyell (Figure 9.2f) was a Scottish geologist who challenged the ideas of catastrophism by proposing in his most famous work, *Principles of Geology* (1830), a theory of uniformitarianism. Lyell, supporting an idea first proposed by James Hutton in 1788, argued that gradual yet relentless natural forces responsible for the formation of Earth's geological features were uniform through time. Greatly influenced by his travels around Europe observing many of Earth's features, such as mountains, valleys, and canyons, Lyell argued that wind, rain, volcanic eruptions, and earthquakes were responsible for producing these features, and these forces were continuing to modify them. Lyell believed that the formation and modification of these features occurred gradually over the course of many millions of years, therefore concluding that the earth was quite old. Charles Darwin, who was greatly influenced by reading Lyell's *Principles of Geology* while aboard the HMS *Beagle*, developed a similar view regarding how organisms undergo gradual, constant change over time.

9-2. Charles Darwin: The Man, the Voyage, the Theory

Charles Darwin (Figure 9.3) was born in 1809 in Shrewsbury, England, into a wealthy family, which included two prominent physicians: his father, Robert Darwin, and grandfather, Erasmus Darwin. As a child, Darwin never excelled in school, finding it unappealing and being therefore uninterested in completing his work. Instead, Darwin preferred more intriguing subjects, such as natural history, and spending his time outdoors exploring the countryside; collecting and cataloguing insects, plants, and animals; and fishing and hunting, with the hopes of becoming a naturalist. However, Darwin's father, believing that his son's interest in being a naturalist would be a disgrace to the family, sent Darwin to the University of Edinburgh to study medicine. As a medical student, Darwin was miserable, finding the subject boring and dull, but most of all, the practice of surgery distressing, for at that time, the procedure was performed without the use of anesthesia. Due to Darwin's waning interest in the field, he neglected his medical studies, which was met with great annoyance from his father. After two years, Darwin dropped out of medical school, prompting his father to enroll him in Christ College, a divinity school at Cambridge University, with hopes that Darwin would pursue theology to become a clergyman. Although Darwin found the idea of studying theology uninspiring, he remained at Christ College, satisfying his interest in natural history by attending biology and geology courses and performing fieldwork with many renowned geologists, including Adam Sedgwick, who was known for his work with Welsh rock strata and proposing two of Earth's different geological periods, the Cambrian and Devonian. Through his second cousin, William D. Fox, who was also in attendance at Christ College at the time, Darwin was introduced to botany professor John Henslow, who would become Darwin's mentor and close friend.

FIGURE 9.3 Charles Darwin has been identified as the primary contributor to the theory of evolution and one of the first scientists to propose a mechanism of species change through natural selection.

Shortly after Darwin graduated with a theology degree from Christ College in 1831, Henslow would recommend Darwin to Robert FitzRoy, captain of the British surveying ship the HMS *Beagle*, to accompany him and his crew as a "gentleman companion" and naturalist on their upcoming surveying expedition around the world. Captain FitzRoy accepted Henslow's recommendation, and the crew, with Darwin on board, would leave England in December of 1831. The main purpose of the expedition was to survey and chart unexplored regions of the South American coastline. Darwin's job on the voyage was to observe, describe, and collect biological specimens, as well describe the main geological features observed along the coastline. Unfortunately, the trip, originally scheduled to take two years, lasted almost five years, which was not good for Darwin, who despised sea travel because of the continual bouts of seasickness he experienced. Within days of the *Beagle*'s departure, Henslow presented Darwin with the first edition of Charles Lyell's *Principles of Geology*, which Darwin found helpful in battling his seasickness, along with the shore time he would spend performing fieldwork.

As the *Beagle* made its way around the world (Figure 9.4, top), most of the voyage was spent along the South American coast. Here, Darwin was astonished at the abundance and diversity of plant and animal species he observed in the rainforests, along the beaches, and high in the mountains, taking note of how certain features of these species made them suitable to these unique environments. Darwin was also intrigued by the various fossils he discovered in South America, noting their resemblance to extant species. For example, Darwin discovered a group of fossils from an extinct group of organisms known as glyptodonts in Argentina. Darwin observed how the armadillo-like features of glyptodonts closely resembled armadillos currently living in the same region, suggesting that glyptodonts may have once been living ancestors of armadillos. Darwin also noted the unique pattern in which fossils were distributed. Darwin discovered fossils of marine life high in the Andes Mountains and plant fossils buried deep in sea sediment. Darwin concluded that these fossils, miles away from where they originated, must have been placed there because of geological change due to earthquakes, volcanoes, or erosion. This conclusion was reinforced after Darwin observed firsthand how rocks along the Chilean coastline were uplifted after a violent earthquake shook the region while he was there. Darwin's observations of geological change in South America made him an advocate for the idea of uniformitarianism, as proposed by James Hutton and Charles Lyell.

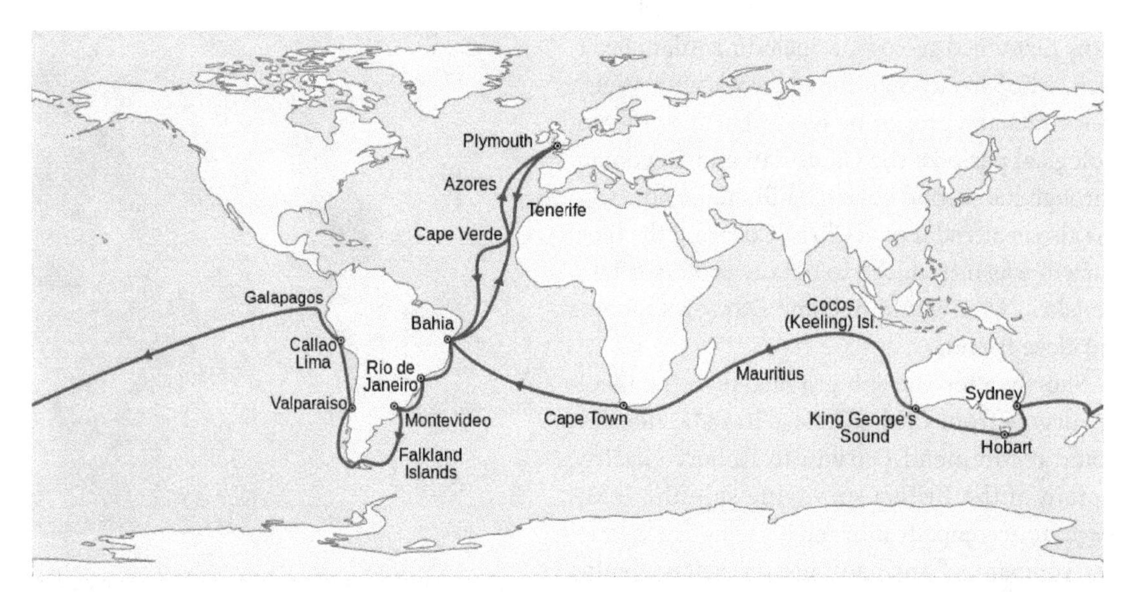

FIGURE 9.4 Charles Darwin's five-year voyage around the world on the HMS *Beagle* (top) was instrumental in shaping his ideas for species change, especially his stop along the volcanic islands of the Galápagos Islands (bottom).

One of the most influential and significant stops on the *Beagle*'s route around the world was a five-week visit along a group of volcanic islands located off the northwest coast of South America called the Galápagos Islands (Figure 9.4, bottom). The observations and notes Darwin would make on these groups of islands would become the foundation on which he built his theory of evolution. The Galápagos Islands are home to some of the most diverse and unusual species of life Darwin had ever witnessed, including iguanas, huge tortoises, and variable groups of finches, which greatly interested Darwin. Darwin made note of the differences he observed in the features of the tortoises, such as the length of their necks and shell size, and the varying beak sizes of the finches and how these features were variable among the species on different islands. Darwin also made detailed geological notes about the volcanic landscape of the Galápagos Islands, describing craters, their formation, and the remnants of lava flow that surrounded them. Darwin also observed the presence of marine shells in volcanic deposits, suggesting that the islands had uplifted from the sea during their formation from volcanic eruptions. While anchored in the Galápagos Islands, Darwin collected various species of flowering plants, animals, and fossils, making detailed notes on which island he obtained his specimens from, along with important details of the island's habitat and its geological features.

After five long years traveling the world, Darwin and the crew returned to England in 1836. Upon regaining his "land legs," Darwin began to compile his notes about the natural history of the regions he visited, relying on the expertise of other naturalists and geologists for analysis and interpretation. One of these experts was John Gould, a leading ornithologist at the Zoological Society of London, who Darwin had analyze his ornithological specimens, which included bird skins, whole birds, and parts of birds. Of Darwin's collection, Gould was most excited about the finches of the Galápagos Islands. Due to Darwin's limited knowledge of ornithology, he had identified the finches as a mix of gross-beaks, finches, blackbirds, and wrens. Upon examining the gross-beaks, finches, and blackbirds more closely, Gould recognized that the beak structures of the birds were unique, each adapted to a specific food source (different-sized seeds, fruit, or insects) found on the island they inhabited. Gould concluded that Darwin's group of birds were "a series of ground finches which are so peculiar as to form an entirely new group, containing 12 species" (Desmond & Moore 1991, 209), with Gould adding that the features were quite similar to a finch species from the South American mainland. Meeting with Darwin again a few months later, Gould would reveal that another bird collected from the islands, the Galápagos "wren," was another species of finch, totaling 13 distinct species of finches isolated to the Galápagos Islands.

Gould's analysis of the different species of Galápagos finches puzzled Darwin. Why would the Galápagos Islands, a group of islands volcanic in nature, millions of years old, which were never connected to the mainland, have slightly different finches on each island and have a "close affinity of most of these birds to American species in every character, in their habits, gestures, and tones of voice" (Darwin 1872, 354)? Darwin speculated that the Galápagos finches must be descendants of the South American mainland species, which must have flown or been windblown to the islands. Upon reaching the various Galápagos Islands, Darwin suggested the species diverged as they experienced uninhabited habitats with different food sources, began to flourish in these habitats, and over time diversified into new, separate species. This idea began to lay the foundation for Darwin's early thoughts on the evolutionary process. Today, the finches of the Galápagos Islands, now known as "Darwin's finches," have been extensively studied and have become an iconic model for the study of evolution (Figure 9.5, left). Studies suggest that the mainland species arrived on the Galápagos Islands approximately two million years ago and, according to mitochondrial DNA (mtDNA) studies, are closely related to a group of birds of the genus *Tiaris*, with the dull-colored grassquit, *Tiaris obscura*, as the closest living species related to "Darwin's finches." (Figure 9.5, right)

FIGURE 9.5 The finches of the Galápagos Islands, generally known as "Darwin's finches," are an iconic representation of the study of evolution (left). Based on genetic studies, the dull-colored grassquit (*Tiaris obscura*) (right) is considered the closest living relative of "Darwin's finches."

In 1837, Darwin was asked to compile his natural history notes at the request of Captain FitzRoy, who was in the process of putting together a captain's account of the voyage. By 1839, Darwin published *Journal and Remarks, 1832–1835*, the third volume in Captain FitzRoy's four-volume set entitled *Narrative of the Surveying Voyages of His Majesty's Ships Adventure and Beagle.* Within months of its release, Darwin's volume was so popular that the publisher rereleased Darwin's account, changing the title to *Journal of Researches into the Geology and Natural History of the Various Countries Visited by H.M.S. Beagle.* Since its publication in 1839, the title of Darwin's book has changed several times and is universally recognized today as *The Voyage of the Beagle.*

FIGURE 9.6 Thomas Malthus's *Essay on the Principle of Population* suggested that population growth was influenced by the availability of resources, a principle Darwin felt impacted both animal and plant populations as well.

During his note compilation between 1837 and 1839, Darwin continued to wonder how an isolated population of finches traveling from the South American mainland could result in different species of finches inhabiting the Galápagos Islands. An idea of a mechanistic cause came in 1838, while Darwin read "for amusement" a book entitled *Essay on the Principle of Population*, published in 1798 by political economist Thomas Malthus (van Whye 2007, 187). Within the confines of the text, Malthus (Figure 9.6) described that the growth of the human population correlated to increases in agricultural production, leading to a much more improved life for the population. However, as the human population continued to grow at an increased rate, it would eventually outpace agricultural production, thereby resulting in famine, disease, and death within the population. Malthus concluded an "increase of population is necessarily limited by the means of subsistence," or in other words, the size of a population is limited by the resources available to that population. Therefore, a limited availability of resources would not be sufficient for all members of the population to survive, and these individuals would die.

Upon reading this, Darwin applied this principle to both animal and plant populations, including the finches of the Galápagos Islands. Darwin speculated that once the South American mainland finch species arrived in the Galápagos, the population increased to a point at which the availability of a specific food source was limited, resulting in the death of those who could not eat anything else to survive. Those that did survive, Darwin thought, must have inherited a slight modification in their beak structure that enabled them to become better adapted to eating other types of foods, like larger seeds or fruit. With a more bountiful food source available to these groups of finches, these adaptations enhanced their survival, enabling the population to reproduce at higher rates, flourish, and grow. Darwin would write that under the struggle for existence, "favorable variations would tend to be preserved, and unfavorable ones to be destroyed. The results of this would be the formation of a new species" (Darwin 1958, 120). This idea became the foundation Darwin needed to formulate a hypothesis for evolutionary change. Darwin proposed that evolutionary change occurs because of natural selection, a mechanism that increases the survival and reproductive rates of organisms with favorable adaptations to a specific environment.

Darwin would spend most of the 1840s patiently gathering and assembling more information, facts, and research on the aspects of his natural selection hypothesis. In 1842, Darwin "drew up some short notes" (van Whye 2007, 178) to organize his thoughts and coherently describe his hypothesis in a 35-page sketch, written in pencil, coining the term *natural selection* for the first time. Within the "Sketch of 1842," Darwin compares natural selection to **artificial selection**, a practice used by humans for thousands of years to specifically breed and produce individuals with desirable traits, evident in the several hundreds of different breeds of dogs produced from a domesticated descendant of wolves. Darwin also integrated thoughts on variations in nature and the geographic distribution of plants and animals into his "sketch." Darwin expanded on his ideas of natural selection in 1844 with a 230-page essay, again referencing artificial selection, suggesting that variations in organisms were the result of environmental effects, and if the environment produced a desirable variation within these organisms, the breeder would breed these individuals for that specific variation. Within the essay, Darwin describes natural selection as an intelligent, compassionate force capable of producing slight variations in nature as the environment slowly changes, maintaining that individuals without adaptable traits would be eliminated.

Despite providing a clear definition of his hypothesis in his "Essay of 1844," Darwin would write he had "no intention of publication in its present form," because he needed more time for "correcting and enlarging and altering" (van Whye 2007, 188). Darwin would ask his wife, Emma, to see it improved, enlarged, and published in the event of his death, due to his recurrent bouts of illness, which would debilitate him for months to years on end. After compiling a list of potential editors for Emma to contact, Darwin decided that his close friend Joseph D. Hooker, the botanist who analyzed the plant specimens collected on the *Beagle* voyage, would be the recipient of his portfolios of notes and books, as well as the primary editor of his work. Although it was never published by Darwin himself, his son Francis published the essay in 1909.

Between the years of 1839 and 1859, a period known as "Darwin's delay," Darwin never published his work. Some scholars believe Darwin avoided publication because of fear stemming from upsetting those with strong religious convictions, especially those of his wife; fear of tainting his reputation; or fear of receiving negative reactions and disapproval from his closest scientific colleagues. However, no evidence exists within any of his writings to suggest he avoided publishing because of these reasons. During those 20 years, Darwin "steadily pursued the same object," writing later that he "gained much by my delay in publishing ... and lost nothing by it" (van Whye 2007, 178; 185–6). This suggests that Darwin used this time to collect additional facts and make relevant observations to develop and prepare his case. In addition, the delay of conceiving and publishing his idea on natural selection was not of priority to Darwin at the time. Darwin was concentrating his time on writing and publishing his scientific results from the *Beagle* voyage, reading (which included Malthus's *Essay on the Principle of Population*), and working on other natural history projects, such as publishing his observations and descriptions of the anatomy of invertebrates, specifically barnacles.

Other scholars believe that Darwin kept his hypothesis a secret during the delay because Darwin viewed his ideas as a "social crime," and he would be seen as a traitor among his contemporaries. Contrary to this idea, Darwin did not maintain his hypothesis a secret, as his correspondences and other writings suggest he was open to discussing and describing it among his family and other close friends, including Charles Lyell and Joseph D. Hooker. Darwin wrote, "I cared in the highest degree for the approbation of such men as Lyell and Hooker, who were my friends, I did not care much about the general public" (van Whye 2007, 184). Although Darwin discussed his hypothesis with the likes of Lyell and Hooker, he was unsuccessful in convincing them of natural selection at the time. However, Lyell pushed Darwin to publish his ideas, as he did not want Darwin to be surpassed by someone else.

Darwin began gathering and sorting his notes on what he called his "species theory" as early as 1854. By 1856, he "began by Lyell's advice" working exclusively on the project (van Whye 2007, 193). In June of 1858, with more than ten chapters completed of his "species theory," Darwin received a 4,000-word essay from British naturalist Alfred Russel Wallace entitled "On the Tendency of Varieties to Depart Indefinitely from the Original Type". Wallace, who was exploring the Malay Archipelago at the time, sent his essay to Darwin for review, asking Darwin to send the essay to Charles Lyell if he felt it deserved merit. Upon reading Wallace's essay, Darwin was startled to read that Wallace, much like himself, had articulated a mechanistic idea of species divergence in response to environmental change. Per Wallace's request, Darwin forwarded the essay to Lyell, enclosed with a letter dated 18 June 1858 stating, "I never saw a more striking coincidence" to his own theory, and that Wallace "could not have made a better short abstract." With knowledge of Darwin's "Essay of 1844," Charles Lyell and Joseph D. Hooker presented abstracts of this essay along with Wallace's essay to the Linnaean Society of London in July 1858. The presentation of the papers was met with little exuberance, as suggested by the societal secretary writing within the annual report that nothing of value happened that year.

Prompted by the receipt of Wallace's essay, Darwin began putting together an abstract of his "species theory" shortly after the presentation of their papers in July 1858. Darwin completed the abstract by March 1859, and roughly nine months later, *On the Origin of Species by Means of Natural Selection, or the Preservation of Favoured Races in the Struggle for Life* (shortened to *On the Origin of Species* hereafter) was published. Unlike the reception Darwin's abstract received during the Linnaean Society presentation one year earlier, his book was an instant bestseller, as all 1,250 copies printed sold out within a day.

Within *On the Origin of Species,* Darwin dedicates the first five chapters to describing his theory and providing arguments for the process of natural selection. In chapter 1, Darwin discusses artificial selection and how the process is used to isolate desirable variations for human benefit in domesticated organisms, while in chapter 2 he explains that variations in species are selected by nature and occur slowly as they are inherited within wild animal populations. Darwin uses chapter 3 to present his Malthus-inspired argument that the struggle for existence is the mechanism that drives natural selection in wild animal populations and keeps these populations in check. He outlines different methods that are responsible for keeping a population in check, such as the ability to obtain food, potential of early death in juveniles, and effects of climatic change on survival. In chapter 4, Darwin describes natural selection as the "preservation of favorable individual differences and variations, and the destruction of those which are injurious," resulting in large-scale changes that act "with extreme slowness" within a population (Darwin 1872, 63; 84). Also, within the chapter, Darwin argues that the variability of species arises because of environmental change. Darwin uses chapter 5 to theorize as to the mechanism responsible for species variations that result from natural selection, presenting different examples to explain some basic laws. Unfortunately, Gregor Mendel had yet to publish the results of his pea plant hybridization experiments (1865); therefore, Darwin was unaware of the fundamental laws of genetic inheritance at the time of his publication.

Darwin presents expected objections to his theory in chapter 6, as in the case of transitional forms or structures, which are lacking either due to the slowness at which natural selection occurs or due to the incompleteness of the geological record. Darwin also addresses the emergence of unique structures such as the wings of bats and how natural selection may have influenced the development of complex organs like the eye. In the remaining chapters, Darwin provides observations that can only be explained by natural selection, including the presence of instinctual traits; the effects of hybridism (two species reproducing with each other); although imperfect, how the geological record connects extinct with extant species; and the geographical distribution of species. Of these chapters, it is within chapter 14 that Darwin argues that all organisms have descended from a common ancestor, with each generation having slightly different characteristics than the generation prior, and that the accumulation of these characteristics provides the diversification seen among different species. This idea is known as **descent with modification**. Darwin suggests that species variation is the result of the accumulation of differences over a long period of time from a common ancestor, which has equated to the diversity of life seen on Earth today.

Over the course of the next 13 years, Darwin would revise *On the Origin of Species* multiple times, with the sixth and final edition published in 1872 (Figure 9.7). Over the course of his revisions, Darwin used the term "survival of the fittest" for the first time in the fifth edition, published in 1869. The term, however, was not coined by Darwin but by Herbert Spencer, a 19th-century British philosopher, in 1864. Darwin uses the term in the fifth edition to not only suggest individual survival but also the survival of many individuals capable of reproducing to ensure the survival of the population.

FIGURE 9.7 Darwin published six editions of *On the Origin of Species* within a 13-year span, with the first published in 1859 and the final edition in 1872.

Darwin's book was instrumental in changing the scientific and social culture of thought regarding species change, sparking discussion and debate among many who were indoctrinated in believing in the fixity of species, an idea that had remained unchanged for thousands of years. *On the Origin of Species* provided an evolutionary view of the world, suggesting a mechanistic cause of species change resulting from extinction and species variation to produce the diversity of life, supplemented not by opinion but by facts supported by logic and evidence. Although some of Darwin's contemporaries and closest friends failed to embrace his view of the world at the time, many changed their view and recognized the relevance of this scientific contribution over the course of time. Darwin's idea, however, has been continually met with disdain by others and viewed as heretical for many years because it disrupts the common core of religious beliefs, including the belief that humans are special creations unrelated to other species, distinguished by their distinct place within the *scala naturae* as Aristotle intended. Darwin belied this idea, suggesting that humans, like other species, were in a constant battle with one another for resources, struggling to exist to ensure their survival in a nonharmonious world. Despite these views and its many opponents, the "species theory" has survived years of experimental and observational testing and will always remain the most important and enduring contribution ever made in science.

Chapter Summary

9-1. Pre-Darwinian Science

- Many early theological beliefs, including the idea that species were special creations that remain fixed over time and did not go extinct, were influenced by the teachings of Greek philosophers, such as Aristotle, and based off the book of *Genesis*. Archbishop James Ussher's publication of his analysis of the Old Testament in 1654 suggested that species creation occurred in 4004 BC, making Earth about 6,000 years old. Despite the discovery of fossils in the sixth century, many believed them to be species that did not survive the biblical flood or rocks shaped by nature. Connections between fossils and resemblance to modern-day organisms began in the mid-1700s, and this pattern became more prevalent in the 18th and 19th century as fossils of extinct organisms were uncovered, sparking conflict between theology and science.

- As more patterns became apparent in the fossil record, the theological beliefs that had remained steadfast for thousands of years came under scrutiny by Georges-Louis Leclerc, James Hutton, Georges Cuvier, Erasmus Darwin, Jean-Baptiste Lamarck, and Charles Lyell.
  - Georges-Louis Leclerc (Comte de Buffon) suggested the idea of descent with modification, which he claimed was influenced by environment, migration, or struggle for existence. Leclerc proposed that Earth was 75,000 years old as he compared morphological differences between species, although he believed in fixity of species and special creation.
  - James Hutton established the science of geology and proposed that the formation of Earth's geological formations occurred because of slow, gradual, and continuous forces, which laid the foundation for the theory of uniformitarianism.
  - Georges Cuvier expanded animal classification through comparative anatomy and fossils, which he believed were the remains of specially created species that went extinct due to catastrophic events. He proposed the theory of catastrophism, the idea that geological changes and the unique pattern of fossils found within Earth's rock layers were produced by catastrophic events such as volcanoes or floods.
  - Erasmus Darwin was one of the first scientists to suggest descent from a common ancestor. Using his observations of comparative anatomy and embryology, Darwin promoted the idea of sexual selection and described nature's struggle for existence.
  - Jean-Baptiste Lamarck proposed a mechanism of species change outlining two laws, the use and disuse law and the inheritance of acquired characteristics law. The use and disuse law stated that extensive use of a body part would increase survival, while another part not used would deteriorate and become nonfunctional. The inheritance of acquired characteristics law stated that traits making an organism better adapted to its environment would be passed to its offspring.
  - Charles Lyell renewed the idea that Earth's geological features formed uniformly through slow, gradual, and continuous forces over the course of many millions of years, an idea known as the theory of uniformitarianism.

9-2. Charles Darwin: The Man, the Voyage, the theory

- Charles Darwin was the son and grandson of physicians; however, he found school uninteresting and medical school distressing. Darwin attended divinity school at Christ College to study theology but found biology and geology courses and fieldwork more intriguing. Upon graduating from Christ College with his theology degree, Darwin joined the crew of the HMS *Beagle* as a naturalist. The five-year voyage of the *Beagle* included a stop along the South American coast, where Darwin collected fossils from the coast to the mountains, taking note of how the fossils closely resembled modern-day organisms while also witnessing geological uplifting occurring along the Chilean coast following an earthquake. Darwin's most famous account of the *Beagle* voyage is entitled *The Voyage of the Beagle.*
- The stop along the Galápagos Islands, a group of volcanic islands off the coast of South America, laid the foundation for Darwin's theory of evolution. On the Galápagos Islands,

Darwin observed a diversity of species, most notably the tortoises and finches, in addition to collecting different plant and animal species and fossils and taking detailed notes on the island's geological formations. Darwin's collection of Galápagos finches was identified shortly after his return to England by John Gould, who concluded that Darwin had collected 13 different finch species with unique beaks specific to a particular food, whose features were like a species found on the mainland of South America. Darwin speculated the different finch species were descendants of the South American species, which diverged after flying to or being blown to the islands. The closest living species related to the Galápagos finches is the dull-colored grassquit (*Tiaris obscura*).

- After reading Thomas Malthus's *Essay on the Principle of Population*, which described the effects that limited resources would have on human population growth, such as death, disease, and famine, Darwin applied this idea to animals and plants, including the Galápagos finches. Darwin suspected the divergence of the different finch species on the Galápagos Islands resulted when the population of South American mainland species outgrew its resource availability. Darwin believed the lack of resources led many finches to die, but those that survived inherited modifications in their beak structures, enabling them to eat other types of foods, thereby enhancing their survival and increasing reproductive rates. Darwin called the mechanism by which an adaptation enhances an organism's survival, leading to higher reproductive rates, *natural selection*, an idea he compared to the practice of artificial selection. Artificial selection is the process by which humans breed species for specific traits.
- Darwin spent the next 20 years, known as "Darwin's delay," focusing on other projects, in addition to collecting more facts, observing, and compiling notes on natural selection for a much larger project. Darwin wrote out his thoughts about natural selection in two short pieces, the "Sketch of 1842" and the "Essay of 1844," prior to beginning major work on what he called his "species theory" in 1858. Darwin received an essay from Alfred Russel Wallace in 1858 and was asked to review the sketch and pass it along to Charles Lyell, which Darwin agreed to do. Upon reading the essay, Darwin realized it was a comprehensive sketch of his own "species theory." Lyell, along with Joseph D. Hooker presented Darwin's "Essay of 1844" and Wallace's essay to the Linnaean Society of London but were met with disinterest.
- Darwin published his "species theory" in 1859 under the title *On the Origin of Species*. The 14 chapters of *On the Origin of Species* describe and provide arguments and observations for the process of natural selection and suggest that all organisms are descendants of a common ancestor, an idea known as descent with modification. Darwin published six revised editions of *On the Origin of Species*, using the term *survival of the fittest* for the first time in the fifth edition. Although *On the Origin of Species* has continually prompted favorable scientific and social discussions about natural selection and species change since its first publication, its acceptance has also been countered by opposition from individuals with staunch religious beliefs. Despite the opposition, many years of experimentation and observations have made the theory of evolution by means of natural selection one of science's most important contributions.

End-of-Chapter Activities and Questions

Directions: Please refer back to what you learned in this chapter to complete the following activities.

Define Each Term in Your Own Words

1. Uniformitarianism
2. Galápagos Islands
3. Thomas Malthus
4. Essay of 1844
5. Natural Selection

Chapter Review

1. Identify the early thinkers of evolutionary change and describe why their observations and hypotheses conflicted with the theological ideas of the time.
2. Which ideas proposed by the early thinkers of evolutionary change were instrumental in helping Darwin formulate his theory of evolution? Which idea(s) do you believe may have been the most influential and why?
3. The HMS *Beagle* spanned the globe, with stops along the coasts of South America, Africa, and Australia. Although a majority of the trip was spent along the coastline of South America, Darwin collected many different unique fossils during the ship's five-year voyage. Perform an online search to determine what types of fossils Darwin collected along the coasts of Africa and Australia. Compare and contrast these fossils with those he collected in South America.
4. What are some differences between natural selection and artificial selection? Explain why Darwin would reference artificial selection in the first sketch of his natural selection hypothesis in 1842.
5. Why is the period between 1839 and 1859 called "Darwin's delay," and what are the two main misconceptions scholars have suggested as to why Darwin delayed publishing his "species theory"?

Multiple Choice

1. Which pre-Darwinian scientist is correctly paired with his early thoughts on evolutionary change?
   - **a.** Charles Lyell—catastrophism
   - **b.** Georges-Louis Leclerc—descent with modification
   - **c.** Erasmus Darwin—use and disuse law
   - **d.** Georges Cuvier—uniformitarianism

2. Which of the following observations made by Darwin during the HMS *Beagle* voyage is best paired with an idea proposed by an early thinker of evolutionary change?
   - **a.** distribution of marine fossils in the Andes—descent with modification
   - **b.** formation of the Galápagos Islands due to volcanic eruptions—common ancestry
   - **c.** modified beak structures in finches—inheritance of acquired characteristics
   - **d.** glyptodonts' features closely resembling modern-day armadillos—uniformitarianism

3. The species considered the closest living relative of the Galápagos finches is __________.
   a. *Certhidea fusca*
   b. *Tiaris olivaceus*
   c. *Geospiza fortis*
   d. *Tiaris obscura*

4. What was the name of Alfred Russel Wallace's essay that prompted Charles Darwin to publish his "species theory"?
   a. "On the Tendency of Varieties to Depart Indefinitely from the Original Type"
   b. "Essay on the Principle of Population"
   c. "A Discourse on the Revolutions of the Surface of the Globe"
   d. "Zoological Philosophy: An Exposition with Regard to the Natural History of Animals"

5. Who is credited with coining the term *survival of the fittest*?
   a. Herbert Spencer
   b. Charles Darwin
   c. John Gould
   d. Thomas Malthus

References

Darwin, C. (1872). *The origin of species by means of natural selection, or the preservation of favoured races in the struggle for life* (2nd ed.). John Murray.

Darwin, C., & Barlow, N. D. (1958). *The autobiography of Charles Darwin, 1809–1882: With original omissions restored; ed. with appendix and notes by his grand-daughter Nora Barlow.* Collins.

Desmond, A., & Moore, J. (1991). *Darwin.* London: Michael Joseph, Penguin Group.

van Wyhe, J. (2007). Mind the gap: Did Darwin avoid publishing his theory for many years? *Notes and Records of the Royal Society, 61*: 177–205.

Image Credits

Fig. 9.1: Source: https://commons.wikimedia.org/wiki/File:Alfred-Russel-Wallace-c1895.jpg.

Fig. 9.2a: Source: https://commons.wikimedia.org/wiki/File:Buffon_1707-1788.jpg.

Fig. 9.2b: Source: https://commons.wikimedia.org/wiki/File:Hutton_James_portrait_Raeburn.jpg.

Fig. 9.2c: Source: https://commons.wikimedia.org/wiki/File:-Georges_Cuvier_large.jpg.

Fig. 9.2d: Source: https://commons.wikimedia.org/wiki/File:Portrait_of_Erasmus_Darwin_by_Joseph_Wright_of_Derby_(1792).jpg.

Fig. 9.2e: Source: https://commons.wikimedia.org/wiki/File:-Jean-Baptiste_de_Lamarck.jpg.

Fig. 9.2f: Source: https://commons.wikimedia.org/wiki/File:Sir_Charles_Lyell,_1st_Bt.jpg.

Fig. 9.3: Source: https://commons.wikimedia.org/wiki/File:Charles_Darwin_seated_crop.jpg.

Fig. 9.4a: Copyright © by Sémhur (CC BY-SA 4.0) at https://commons.wikimedia.org/wiki/File:Voyage_of_the_Beagle-en.svg.

Fig. 9.4b: Copyright © by Pete from USA (CC BY-SA 2.0) at https://commons.wikimedia.org/wiki/File:Bartoleme_Island.jpg.

Fig. 9.5a: Source: https://commons.wikimedia.org/wiki/File:F-MIB_47321_Finches_from_Galapagos_Archipelago.jpeg.

Fig. 9.5b: Copyright © by Dominic Sherony (CC BY-SA 2.0) at https://commons.wikimedia.org/wiki/File:Dull-colored_Grassquit_(Tiaris_obscurus)_(cropped).jpg.

Fig. 9.6: Source: https://commons.wikimedia.org/wiki/File:Thomas_Robert_Malthus_Wellcome_L0069037_-crop.jpg.

Fig. 9.7: Copyright © by Wellcome Collection (CC BY 4.0) at https://commons.wikimedia.org/wiki/File:6_editions_of_%27The_Origin_of_Species%27_by_C._Darwin,_Wellcome_L0051092.jpg.

CHAPTER 10

Mechanisms of and Evidence for Evolution

PROFILES IN SCIENCE

Wilhelm Weinberg was a German physician born in Stuttgart on December 25, 1862, and died in Tübingen in November 1937, a few years after retiring there from private practice. Weinberg was known for his intellect and careful attention to details, especially during his mathematics courses at his local high school. Weinberg pursued medicine at the universities of Tübingen, Berlin, and Munich, graduating with his degree in 1886. Weinberg spent the next three years gaining invaluable clinical experience in Berlin, Vienna, and Frankfurt. In 1889, Weinberg returned to Stuttgart to operate a private practice specializing in obstetrics and gynecology. For over 40 years, Weinberg provided medical care to the Stuttgart community, along with delivering several thousand babies, including more than 100 sets of twins. In his spare time, Weinberg maintained interest in the independent study and careful analysis of data he collected associated with human and medical genetics. Most of his 160 publications focused on inheritance in twins, which eventually led him to recognize the problem of ascertainment bias (the deviation of results due to factors unaccounted for during a study) in genetics. One of Weinberg's most prominent works was presented in 1908 during a lecture given before the Society for the Natural History of the Fatherland in Württemberg, in which he outlined his genetic equilibrium principle, the idea that genotypic frequencies within a population remained unchanged from generation to generation. When the study was published in the society's annals that same year, the article was predominantly disregarded by the scientific community. Months

FIGURE 10.1 Wilhelm Weinberg.

following Weinberg's initial lecture and subsequent publication, British mathematician G. H. Hardy independently presented a similar premise when prompted by British geneticist Reginald Punnett into the matter of whether a recessive genotype could be eliminated by the dominant genotype in a population. Hardy's proposal gained immediate recognition upon publication. Conversely, Weinberg's genetic equilibrium principle remained obscure until German geneticist Curt Stern recognized his contribution in 1943, suggesting that Weinberg's outline of the principle was more inclusive and extensive than Hardy's. Although his contribution to the genetic equilibrium principle was not recognized until well after his death, his name is now forever associated with what population geneticists and evolutionary biologists call the Hardy-Weinberg principle.

Introduction to the Chapter

Comte de Buffon, James Hutton, Georges Cuvier, Erasmus Darwin, Jean-Baptiste Lamarck, and Charles Lyell proposed various forms of evolutionary change as an alternate explanation to the idea of species fixation. For example, Comte de Buffon suggested descent with modification, Darwin proposed common ancestry, and Lamarck outlined the law of acquired inheritance. While some failed to offer causative mechanisms to substantiate their suggestive form of evolutionary change, Lamarck proposed a mechanism of evolutionary change through two laws: use and disuse and the inheritance of acquired characteristics. Although some lacked causative mechanisms, all of these scientists supported their alternate ideas through various types of evidence, mainly comparative anatomy and/or fossils. When Charles Darwin published *On the Origin of Species* in 1859, he presented an all-encompassing narrative of arguments, observations, and evidence to support evolutionary change through natural selection and descent with modification from a common ancestor. Over the past century and a half, scientists have continually tested Darwin's hypothesis, discovering different mechanisms that drive evolutionary change, all of which they have supported with empirical evidence. This chapter presents these specific mechanisms, along with a mathematical model known as the Hardy-Weinberg equilibrium. This model states that when evolutionary mechanisms are not at work, the allele frequencies within a population from generation to generation remain in equilibrium, making evolution unlikely to occur. Finally, the chapter concludes with different lines of evidence that scientists have collected supporting the theory of evolution.

Chapter Objectives

In this chapter, students will learn the following:

10-1. Mutations, genetic drift, gene flow, and natural selection are the mechanisms of evolutionary change and are responsible for influencing the allele frequencies of traits in a population over time.

10-2. The Hardy-Weinberg principle states that allele frequencies of traits in a population remain in constant equilibrium from one generation to the next, given that none of the mechanisms of evolutionary change occur.

10-3. Darwin's theory of evolution is supported by different lines of evidence including the fossil record, biogeography, comparative anatomy and embryology, molecular biology, and laboratory and field studies.

10-1. Mechanisms of Evolutionary Change

Evolution

Evolution occurs when allele frequencies of traits within a population change over time. Recall from chapter 8 that alleles are located on genes and a specific combination of alleles determines how a particular trait will be expressed within an individual. However, evolution does not influence genetic change within an individual; instead, it changes the genetics and physical features of the entire population, making the population better suited to a particular environment. A **population** is a group of interbreeding individuals that live in the same area and can produce fertile offspring. Darwin recognized that evolution was a population-level process, suggesting that change within an individual will be of "benefit of the whole community; if the community profits by the selected change" (Darwin 1872, 67).

At the time of publishing *On the Origin of Species*, Darwin was unaware of the laws surrounding genetic inheritance and how these laws influence species variation; therefore, he was only able to theorize on the specific mechanism. Today, evolutionary biologists understand how species variation occurs within a population using a branch of study known as **population genetics**. Population genetics is a useful tool in determining the frequency of all the alleles of the genes within a given population, known as that population's gene pool, and how these alleles are inherited among the descendants of the population. To determine the allele frequencies of a particular gene in a population, all of the alleles for that gene are added together. For example, using Mendel's pea plants as a model, a population of 25 plants contains a gene that codes for seed form, as discussed in chapter 8. The gene for seed form contains two alleles, one for round and one for wrinkled, which means the pea plant population has 50 total alleles for that gene (25 x 2). If 30 of those 50 alleles are responsible for producing round seeds, then the frequency of the round allele in the pea plant population is 30/50, or 60 percent. Conversely, the remaining twenty alleles in the pea plant population produce wrinkled seeds; therefore, the allele frequency for wrinkled seeds in the population is 20/50 or 40 percent. For evolution to occur within this pea plant population, these allele frequencies will have to change. A change in allele frequencies resulting in the evolution of a population can occur in one of four ways: mutation, genetic drift, gene flow, or natural selection.

Mutation

A **mutation** is an alteration in the base pair sequence of DNA nucleotides for a particular gene. The resulting mutation can convert one allele to another or introduce new alleles into a population, changing the population's overall allele frequency. Mutations are rare, randomly occurring events induced through exposure to environmental factors, such as radiation or chemical mutagens, or spontaneous errors during DNA replication. Regardless of the means through which a mutation occurs, the affected allele is introduced into the gene pool of the population through gamete-producing cells. As the mutated allele is passed from generation to generation, it produces genetic

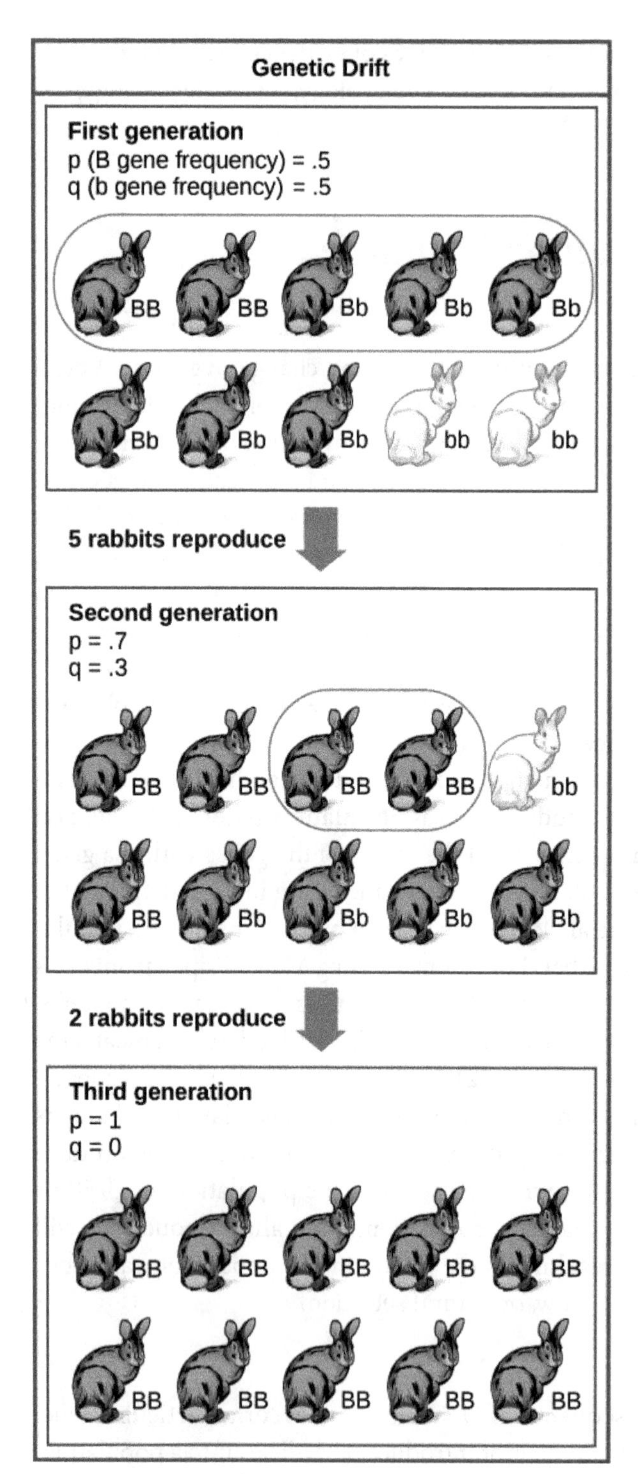

FIGURE 10.2 Genetic drift is a change in allele frequencies of population through the occurrence of chance events, such as a random set of individuals mating. The resulting offspring have the same allele frequencies as those of the parents, which leads to allele fixation.

variance and introduces novel phenotypes into the population. Whether a novel phenotype is beneficial, neutral, or harmful to the population is dependent on the environment and other evolutionary processes, such as natural selection, which can increase its frequency or completely remove it from the gene pool. Due to the rarity in which mutations occur, the overall effect a single mutation has on the allele frequency of a population is relatively small. However, an accumulation of mutations within a large gene pool is more essential in generating evolutionary change, because it gives evolutionary processes more variety with which to work.

Genetic Drift

Genetic drift is another mechanism of evolution influencing a change in allele frequencies of a population due to random, chance events (Figure 10.2). Although genetic drift has the potential to affect any population size, changes in allele frequencies tend to be more problematic in smaller populations, especially those consisting of endangered species, than in larger populations. The reason is that in a smaller population, if a specific group of individuals carrying particular alleles does not survive to reproduce or only a minimum number of these individuals are reproducing, these alleles are lost from the gene pool, affecting the allele frequency of the population. However, which alleles are lost is merely random. In other words, genetic drift does not differentiate what type of allele is eliminated, whether it is beneficial, neutral, or harmful, but it does influence the frequency of the remaining alleles in the population after the event. As the remaining alleles are subjected to genetic drift, some alleles can become fixed in the population, meaning they have reached a frequency of 100 percent. If certain alleles reach 100-percent frequency, it greatly reduces genetic variation, affecting the adaptability of the population if their environment begins to change.

Genetic drift can strongly influence the allele frequencies of a population in two ways: the bottleneck effect and the founder effect.

A **bottleneck effect** occurs when a population experiences a significant reduction in size because of either a randomly occurring natural event that changes the population's environment, such as an earthquake, or a human event, like habitat destruction or overhunting (Figure 10.3). When such an event occurs, the genetic diversity of the remaining individuals of the population is drastically reduced due to a significant change in the allele frequencies, which are now considerably different from those of the ancestral population. Although the population can restore their numbers with the few individuals remaining, the devastating impact it has on the genetic diversity of the population will remain apparent for multiple generations.

One of the most common examples of a population that has experienced bottleneck events is the cheetah population. The first bottleneck event occurred roughly 100,000 years ago, when North American cheetahs began expanding their range into Europe, Africa, and Asia. During this rapid dispersal event, the cheetahs were spread over such a large geographical range, their ability to mate and exchange genes was restricted. About 10,000 to 12,000 years ago, around the end of the last ice age, the cheetah population experienced their second bottleneck event with the extinction of the North American and European populations, leaving only the Asian and African populations. Another bottleneck event that has greatly affected the cheetah population is human influence beginning in the 19th century, with overhunting and, more recently, habitat destruction. All these events have greatly decimated the cheetah population, with only several thousand individuals remaining across Asia and Africa.

Although their numbers are steadily rebounding, genetic analyses have revealed that the current cheetah population can trace its ancestry back to several individuals who survived the various bottleneck events. To increase their numbers, the cheetahs began **inbreeding**, or mating with close relatives. Inbreeding has greatly reduced the gene pool of the cheetah population, leading to genetic uniformity. Studies have suggested several lines of evidence that support the idea that inbreeding has probably occurred among the cheetah population for many

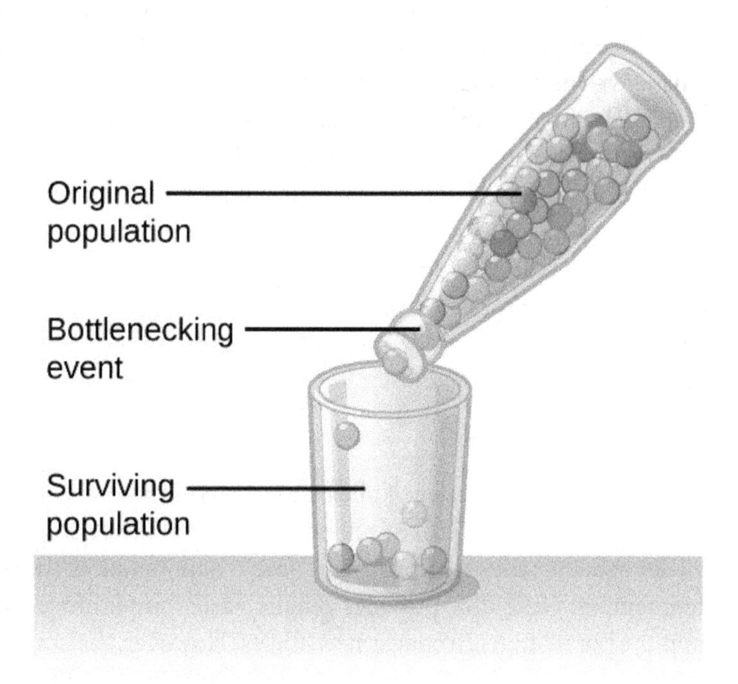

FIGURE 10.3 The bottleneck effect is a form of genetic drift, which causes allele frequencies to change when the size of a population is drastically reduced because of a random, natural event or human event.

generations. Skin grafting, a process that involves transplanting skin from one individual to another, has revealed that in inbred cheetah populations, a skin graft fails to be immunologically rejected by unrelated individuals about 50 percent of the time. This is a unique phenomenon, as most skin grafts are rejected when accepted from another individual, except in the case of identical twins, who are genetically similar. Another line of evidence suggesting inbreeding can be found in the asymmetrical shape of cheetah skulls when compared to ancestral skull specimens found in museums around the world. Scientists have discovered that the right and left sides of the skulls of the current cheetah population are not mirror images of one another, a common occurrence of inbreeding. Third, it is typical for no two individuals, especially those within the same population, to have enzymes that are genetically identical, unless these individuals are the byproduct of inbreeding. Enzyme studies have revealed that the enzymes of cheetahs are about 97-percent identical, a strong, compelling piece of evidence to support inbreeding. Finally, genomic studies have revealed that cheetahs have a high level of *homozygosity*, or genetic similarity, reaching above 90 percent. All these pieces of evidence suggest that genetic uniformity has occurred due to inbreeding among individuals within the cheetah population. Unfortunately, this puts the population at a much greater risk of experiencing another extinction event, especially if the environment unexpectedly changes.

The **founder effect** is another example of genetic drift and occurs when a small group of individuals leaves or is geographically isolated from the original population (Figure 10.4). Since the gene pool of the founding members of the new population is only a subset of the total gene pool of the original population, the resulting gene pool will be quite different, because the alleles represented in the new population occur by chance and will occur at much higher frequencies.

The founder effect occurs commonly in the human population as small groups of individuals leave the original population to establish new settlements elsewhere. A common example is the Amish, a group of individuals who established residence in the United States in the early 18th century to escape religious persecution in Europe. After leaving Europe, the newly founded population possessed allele frequencies quite different from the original Amish population. Unfortunately, some of the founding members of the new colony in the United States carried alleles for a rare genetic disorder known as Ellis-van Creveld, characterized by dwarfism and polydactyly (extra fingers or toes). As individuals of the community began to marry and have children, the prevalence of this disorder, rare in the original population, began to increase in frequency in the new population.

Gene Flow

Gene flow is the transfer of alleles between individuals as they migrate into and out of different populations (Figure 10.5). As an individual moves from one population to another, it introduces its alleles, whether advantageous or not, into that population, therefore changing that population's allele frequencies. Gene flow provides genetic variation within the population, if the individual moving into the population introduces new alleles. However, the continual transfer of alleles into and out of the population makes the gene pools less diversified, ultimately making the populations more like one another. The more similar the gene pools become, the likelihood of one of these populations becoming a new species is greatly reduced. Geographical barriers, like a mountain or river, are important structures that can limit the migration of individuals between populations. This has the potential to make both populations genetically different enough for a new species to arise. Most

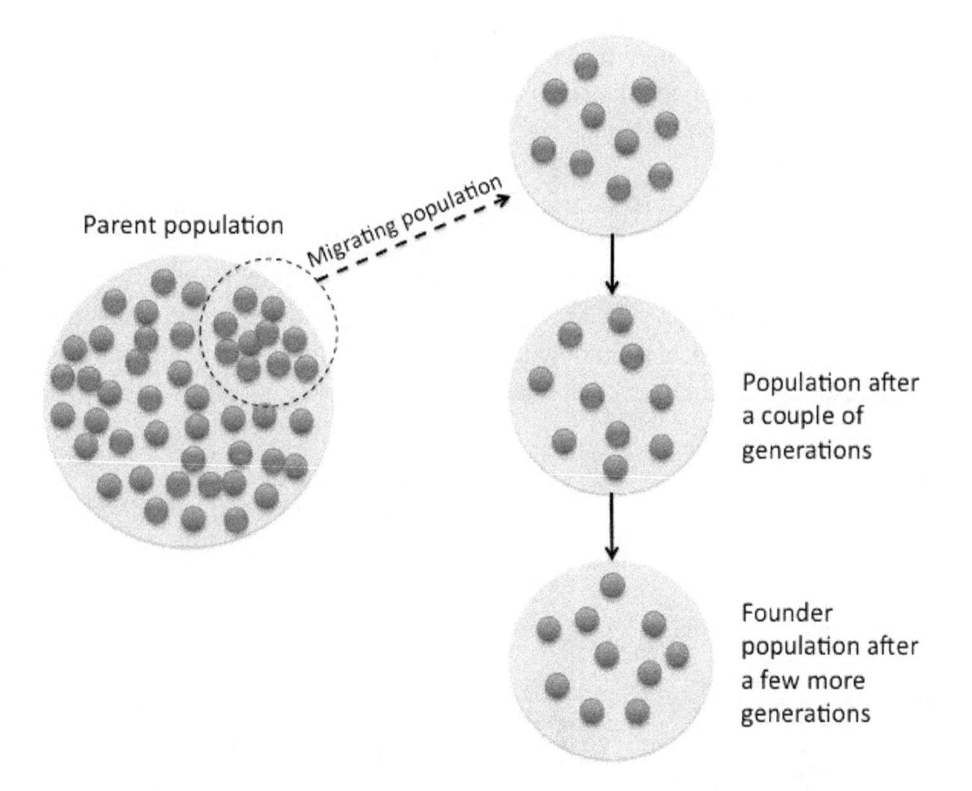

FIGURE 10.4 The founder effect is another form of genetic drift and causes allele frequencies to change when individuals leave a population or are geographically isolated. The resulting population of founding individuals only have a subset of the total genetic variability as the original population.

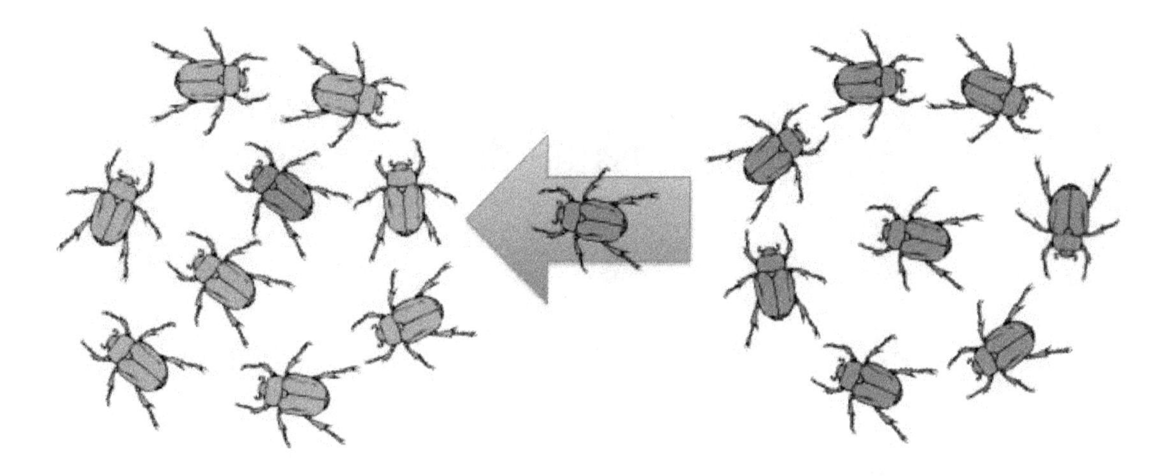

FIGURE 10.5 The migration of individuals from one population to another is called gene flow, causing allele frequencies to change as different or new alleles are introduced.

plants are an ideal example of gene flow, as factors such as wind and water can carry seeds with rare and novel alleles from one population to another.

Natural Selection

Natural selection is a strong evolutionary force that favors individuals with phenotypes better suited for environmental change. Individuals with favored phenotypes tend to have higher reproductive rates relative to other individuals in the population with another phenotype, causing allele frequencies to fluctuate, resulting in evolutionary change. Both Darwin and Wallace independently observed and proposed the mechanism of natural selection, arguing that its effect on a population was the direct result of four different components (Figure 10.6):

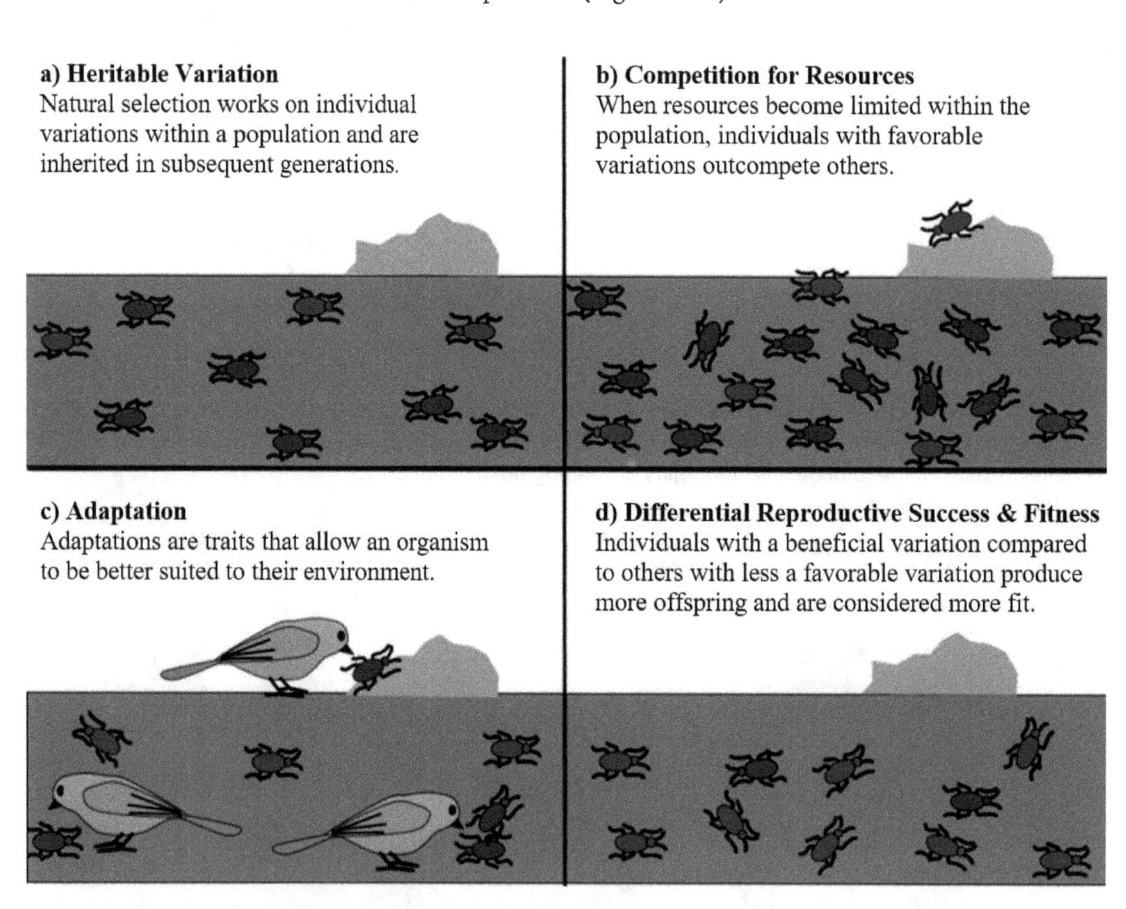

FIGURE 10.6 The four components of natural selection include a) heritable variation, b) competition for resources, c) adaptation, and d) differential reproductive success and fitness.

Heritable Variation

The common belief among pre-Darwinian contemporaries was that individual variations were imperfections, merely used as a means of distraction, and were to be ignored by an observer. However, Darwin and Wallace believed that individual variations, whether physical or behavioral, would arise randomly in nature and were the necessary instruments by which natural selection works. At the time, Darwin and Wallace were unaware of genetic inheritance; therefore, they did not understand how variations arise and why these variations would be observable in subsequent generations. Because of Mendel's publication in 1866, it is now understood that variation arises because of mutations occurring within the existent gene pool, and these variations are passed from parents to offspring by means of inheritance (Figure 10.6a).

Competition for Resources

Recall that in the 1789 essay entitled "Essay on the Principle of Population," Thomas Malthus wrote an "increase of population is necessarily limited by the means of subsistence." This statement greatly influenced Darwin and Wallace because it suggested a population might outgrow the availability of resources. Both scientists suggested that a limited availability of resources would lead to intense **competition** between individuals within the population, influencing their overall survival and ability to reproduce. Darwin and Wallace concluded that individuals with specific variations that helped increased their ability to outcompete a competitor for limited resources would ultimately survive and have more energy to devote to reproduction (Figure 10.6b).

Adaptation

Darwin and Wallace both observed how individuals with fit traits are better matched to their environment and are more likely to survive and reproduce, passing these same traits to their offspring. A trait that enables an organism to be better matched to its environment is an **adaptation** (Figure 10.6c). Although many other mechanisms of evolution have been discussed, natural selection is the only evolutionary force capable of generating an adaptation. Natural selection generates adaptations by increasing the frequencies of favorable alleles necessary for survival and reproduction related to how an individual's environment changes. When the environment changes, natural selection favors the more adaptable alleles and propagates these through the population, increasing the number of individuals better matched to these current environmental conditions. However, if the current environmental conditions change, natural selection will once again operate on alleles that are more favorable, producing organisms best suited to the new environmental conditions. Therefore, natural selection is a fluid, efficient process, responsible for continually changing the dynamics of a population by selecting adaptations that are beneficial to a particular environment (Figure 10.7).

FIGURE 10.7 The Arctic hare (left) and the stoat (right) have specific adaptations, such as their fur coats, that enable them to be better suited to their environment.

Phenotypic variation in a population is unique because it produces a range of phenotypes that fit perfectly into a bell-shaped curve. As the environment changes, natural selection affects which phenotypes are favored, influencing the curve in one of three ways.

Directional selection occurs when an extreme phenotype is favored, shifting the frequency of individuals with that phenotype in one direction or another (Figure 10.8a). Beginning in 1973, a team of researchers led by American biologists Peter and Rosemary Grant began studying how natural selection affects the finch populations of Daphne Major, one of the Galápagos Islands. In 1977, a long period of drought greatly affected the vegetation of Daphne Major, resulting in a drastic decline in the availability of seeds on the island. Small-beaked birds, whose primary food source are small, soft seeds, began to die as most other finches on the island ate these seeds first. When the small, soft seeds were gone, the large, hard seeds became more prevalent, and finches with larger, deeper beaks, such as the medium ground finch (*Geospiza fortis*), were more successful at cracking and eating these seeds. The medium ground finches that survived the drought reproduced the following year, and upon measuring the beak sizes of the next generation, the Grants discovered the beak sizes were slightly larger, suggesting a directional change toward this phenotype.

Diversifying (disruptive) selection occurs when the frequency of individuals with extreme phenotypes are favored over those with an intermediate phenotype (Figure 10.8b). According to studies performed by American biologist Thomas Bates Smith, diversifying selection is responsible for maintaining two different bill sizes in the black-bellied seedcracker finches (*Pyrenestes ostrinus*) of Cameroon. Smith's studies revealed that the two distinct bill sizes prevalent in the population are related to the overall hardness of two different types of sedge seeds. Smith discovered large-billed birds are more efficient at feeding on hard-seeded sedge (*Scleria verrucosa*), while small-billed birds are better at feeding on soft-seeded sedge (*Scleria goossensii*). Individuals with an intermediate beak size were found to be less effective in cracking either seed type; therefore, the frequency of individuals within the population is considerably less due to a lower survival rate.

Stabilizing selection occurs when individuals with an intermediate phenotype are favored over those with extreme phenotypes, increasing the frequency of the intermediate phenotype (Figure 10.8c). An example of stabilizing selection occurs in the size of galls made by the goldenrod gall fly (*Eurosta solidaginis*). The goldenrod gall fly parasitizes the tall goldenrod plant (*Solidago altissima*) by injecting its eggs into the stem of the plant, and when the larvae hatches, it bores into the stem, producing a spherical

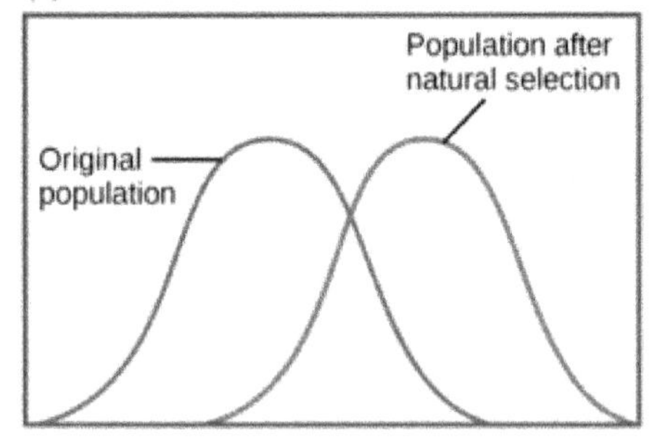

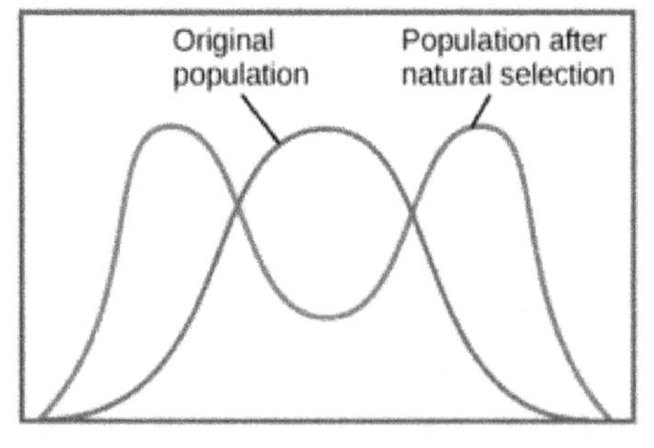

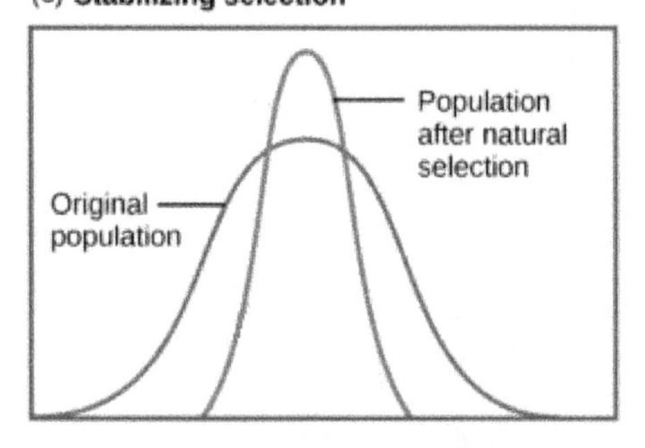

FIGURE 10.8 There are three patterns of natural selection, including a) directional selection (one extreme phenotype or the other is favored), b) diversifying (disruptive) selection (extreme phenotypes are favored over an intermediate phenotype), and c) stabilizing selection (intermediate phenotype is favored over extreme phenotypes).

swelling known as a gall. In a study performed by American biologists Warren Abrahamson and Arthur Weis (1986), the researchers discovered that when the larvae produce small galls, they are more likely to come under attack by parasitoid wasps, such as *Eurytoma gigantea* and *Eurytoma obtusiventris*, and the inquiline beetle (*Mordellistena unicolor*). When larvae produce large galls, they are primarily targeted by the downy woodpecker (*Poicoides pubsecens*) and the black-capped chickadee (*Parus atricapillus*). Larvae that produce galls of intermediate size avoid predation from both the parasitoid wasps and avian predators, therefore making this phenotype favored in the population.

Although natural selection is a strong evolutionary force capable of producing profound change, it does not, however, always lead to perfectly adapted organisms, for several different reasons. One reason is that natural selection only operates on the phenotypic variations currently present within the gene pool of the population. New alleles that may provide a more adaptive advantage to the population are not produced upon demand, but rather through mutations or through gene flow. Due to the random nature in which these situations may occur, the required alleles for a population to be truly adapted to its environment may never exist in the population. Another reason is the environment is a dynamic entity, at times unpredictably changing much faster than natural selection can operate to produce perfectly adapted organisms. A third reason are the historical constraints placed on an organism's existing anatomical structures because of descent with modification. Recall that Darwin described descent with modification as the accumulation of modifications in subsequent generations upon descending from a common ancestor. Therefore, evolution progressively modifies existing anatomical structures based on a required adaptation. For example, birds and bats evolved from nonflying ancestors, but the preexisting ancestral limbs were modified into wings to enable flight. Finally, the adaptation of one feature may require a trade-off with another, restricting evolutionary change. In other words, an adaptation of one feature that provides benefit to the organism might change another feature for the worse. For example, seals may benefit from having legs rather than flippers, enabling them to move more efficiently along a rocky shore. However, their ability to swim would be greatly diminished, as legs are not as effective in the water as flippers.

Differential Reproductive Success and Fitness

Fitness is the measure of the reproductive success of an individual with a favorable phenotype compared to an individual with an alternative phenotype (Figure 10.6d). An individual with a variation making it better suited to its environment, that can produce more offspring, is said to have greater reproductive success and is therefore considered the most fit in the population. Although individuals with advantageous variations will survive the longest and produce the most offspring, Darwin and Wallace recognized that these individuals typically produce far more offspring than necessary, but the size of the population remains relatively constant. The scientists attributed this to individuals dying young, while others fail to reproduce, produce very few offspring, or produce offspring that have disadvantageous variations that result in their inability to survive or reproduce. Darwin and Wallace also recognized that competition among individuals for resources was a contributing factor to survival, thereby helping maintain the relative size of a population.

A commonly studied and documented model of fitness and natural selection in the wild is crypsis in *Peromyscus* mice. Extensive studies along the beaches of Florida, led by American evolutionary biologist Hopi Hoekstra, have revealed a strong selective advantage and higher fitness for different subspecies of *Peromyscus polionotus* mice whose coat color complements its local soil color. These subspecies include the dark-colored oldfield mice (*P. p. subgriseus*), found in abandoned agricultural fields (Figure 10.9, left), and the light-colored Santa Rosa Island beach mice (*P. p. leucocephalus*), which occupy coastal sand dunes (Figure 10.9, right). When the mice are introduced into an environment in which they do not typically occupy, such as light-colored Santa Rosa Island beach mice placed in abandoned agricultural fields, they exhibit a higher selective disadvantage and lower fitness, as their coat color is more conspicuous to avian and mammalian predators. A corresponding effect occurs when dark-colored oldfield mice are introduced to the coastal sand dunes environment. The ability of each subspecies of mice to survive in their own specific environment increases their individual fitness, resulting in changes in their allele frequencies. A perfect example of this has been documented in two studies performed in the Sand Hills of Nebraska, which have revealed that fitness can arise through mutations regardless of timescale.

FIGURE 10.9 The fitness of an organism is dependent on how it best adapts to its environment. Crypsis, an adaptation in which the coat color of *Peromyscus polionotus* mice is matched to its environment, is one of the most common studied examples of fitness and natural selection.

In 2013, a team of researchers, led by Hoekstra, used light and dark plasticine models of deer mice (*Peromyscus maniculatus*) to determine whether the light-colored coat of the mice was a recent adaptation to the light-colored Sand Hills of Nebraska, a landscape modified in the last 10,000 years. The team discovered that predators frequently attacked the dark plasticine models, which were more conspicuous on the light-colored soil. The results of the experiment led the researchers to take a closer look at *agouti*, a gene Hoekstra and other scientists implicated in 2009 as being responsible for deer mouse coat color, to determine whether the adaptation was the result of a single mutation or an accumulation of smaller mutations over time. The researchers discovered that the accumulative effect of nine different mutations along the *agouti* gene over the course of 8,000 years led to the adaptation of light-colored crypsis in the Sand Hills species of deer mice.

In a 2019 study, researchers again used the Sand Hills of Nebraska as a crypsis study site, discovering that within three months, mortality rates were higher among mice whose coat color was more conspicuous against an unmatched soil type. Upon collecting the surviving mice, the researchers observed the mice's coat colors had changed during the experiment, becoming either lighter or darker than the mice first introduced. The researchers expanded their study by sequencing the *agouti* gene of all mice introduced at the beginning of the experiment and discovered the allele frequencies of the gene had changed, supporting a selection event. In addition, the researchers identified several mutations occurring along the *agouti* gene in the light-colored deer mice, but none more prominent than the *delta-Ser* mutation. This mutation resulted in the deletion of serine in the amino acid sequence of the agouti protein, disrupting its required interaction with another protein responsible for pigment production in the deer mice. The reduced interaction between the two proteins lowered pigment production, leading to a lighter coat color within three months.

Mate selection, or **sexual selection**, is another contributing factor to the reproductive success and fitness of an individual, a concept introduced by Darwin in *The Descent of Man and Selection in Relation to Sex* (1871). Sexual selection is a form of natural selection in which certain, favorable traits influence the ability of an individual to obtain a mating partner. There are two forms of sexual selection, intrasexual selection and intersexual selection. **Intrasexual selection** involves direct competition between individuals of the same sex for a mate, typically males. Males compete with one another in a variety of ways. One way is through direct combat, in which a larger, stronger male defends his right to mate with a particular female by defeating a much smaller, weaker opponent. An example includes male elks (bulls) sparring for the right to mate with an estrous female elk (cow) (Figure 10.10, top). Another form of intrasexual selection is through a choreographed, ritualized behavioral display in which one male, usually with ornate features, discourages other potential males from mating with females by means of physiological defeat. For example, competing male peacocks visually assess their competitors by focusing attention on one another's display of brightly colored plumages (trains) or the shaking of their wings. In the event of aggression between the competitors, the male peacocks chase one another and engage in fights that involve jumping and the use of their sharp, powerful spurs. The second form of sexual selection is **intersexual selection**, which involves a female selecting a mate based on characteristics that specify a male's health and fitness. As in the case of the male peacock, a brighter and more elaborate train may signify to a potential female better health and fitness (Figure 10.10, bottom). Elaborate

FIGURE 10.10 Sexual selection includes intrasexual selection (top), which involves direct competition between males for a mate (elk sparring) or intersexual selection (bottom), where a female chooses her mate based on elaborate phenotypic characteristics (female peahen).

displays, as in birds like the peacock, intrigued Darwin, because he could not understand how a display could have an adaptive benefit, especially if it brought greater attention from predators. However, if the display confers a mate, the benefit of reproductive success outweighs the potential risk of predation.

10-2. The Hardy-Weinberg Principle

Gregor Mendel's pea plant experiments were instrumental in establishing the principle laws of inheritance, but their significance in the scientific community did not become apparent until the early 1900s. When Mendel's work was rediscovered, scientists discovered that his laws implied discontinuous variation, in which a trait within a population exists in two or more distinct phenotypic forms, an idea supported by the likes of William Bateson and Reginald Punnett. However, this notion drew the ire of other biologists who maintained the belief that species variation was continuous, and characteristics existed within a range of gradually different forms. Other opponents argued that Mendel's laws were not applicable to all species, while others added that dominant alleles would become more apparent in a population over time, a point argued by British statistician Udny Yule. Referencing a lecture given by Punnett in 1908 on his publication "Mendelism in Relation to Disease," Yule questioned Punnett as to why a dominant allele would not continually increase to the point of displacing a recessive allele, thereby resulting in dominant allele fixation within the population after multiple generations. As an example, how would blue eyes (recessive trait) continue to exist within the human population if and when brown eyes (dominant trait) becomes fixed? Unable to suggest a counter argument, Punnett "sought out G. H. Hardy with whom I was then very friendly" (Punnett 1950, 9).

G. H. Hardy (Figure 10.11, left), considered one of the leading British mathematicians of the day, was tasked to formulate an answer to this seemingly perplexing yet "simple issue" posed to Punnett. Within a day, Hardy provided Punnett with a proposed mathematical model, which established a theoretical state of genetic equilibrium where allele frequencies remain unchanged from one generation to the next within a large, randomly mating population. In July 1908, Hardy's suggested model was published in *Science* magazine, and within the correspondence, he stated, "In a word, there is not the slightest foundation for the idea that a dominant character should show a tendency to spread over a whole population, or that a recessive should tend to die out."

FIGURE 10.11 G. H. Hardy (left) and Wilhelm Weinberg (right) independently proposed the mathematical model used to establish genetic equilibrium in allele frequencies in a population from generation to generation.

However, what Hardy did not realize was that in January 1908, six months prior to his publication, German physician Wilhelm Weinberg (Figure 10.11, right) had presented his paper entitled "About the Evidence of Heredity in Humans," in Stuttgart, Germany, proposing a similar, yet more extensive, mathematical model that "derived the general equilibrium principle for a single locus with two alleles." Unfortunately, for 35 years, Weinberg's contribution failed to be recognized by many in the scientific community. One author suggested this failure may have stemmed from ignorance of work written in other languages (Weinberg's publication was in German), when most genetic studies of the time were dominated by English-speaking scientists. Another author suggested Weinberg's publication in an obscure journal, unlike Hardy's in a more prominent journal, along with lack of name recognition, may have been the culprits. Regardless, the mathematical model, which was named "Hardy's Law" for 35 years was renamed and is now recognized as the Hardy-Weinberg Principle, one of the most important mathematical models and theoretical frameworks in the study of population genetics and evolutionary biology.

The Hardy-Weinberg Principle states that for a gene in which two alleles exist, the allele frequencies are expressed using the following formula: $p + q = 1$—with p representing all the dominant alleles and q representing all of the recessive alleles for that specific gene (Figure 10.12). If the frequency of one allele is known, the frequency of the other allele is calculated by subtracting the known frequency from 1. For example, a population of mice have either brown fur color, which is dominant (B), or the recessive white fur color (b). If the frequency of the dominant brown fur color allele is $p = 0.6$, then one can easily determine the frequency of the recessive allele by using the formula $1 - p = q$ or $q = 0.4$. Therefore, the white fur color allele has a frequency of 0.4 in the mouse population. This also applies if the frequency of the recessive allele is known, using the formula $1 - q = p$. When added together, the two allele frequencies of the fur color gene equal 1.

Upon knowing the allele frequencies of a particular gene, the Hardy-Weinberg Principle states that the proportion of genotype frequencies within the population can be calculated using the following formula:

$$p^2 + 2pq + q^2 = 1$$

Using the mice as an example, since p represents the frequency of the dominant allele, the frequency of the homozygous dominant genotype (DD) in the population is p^2, which equals 0.36 (0.6×0.6). Individuals of the population that are homozygous recessive (dd) is q^2 or $0.4 \times 0.4 = 0.16$. Hardy and Weinberg represented heterozygous individuals as $2pq$ to reference these individuals having one dominant and one recessive allele and receiving one allele or the other in two ways, either from the mother or the father. In the mouse example, $2pq$ equals $2 \times 0.6 \times 0.4$ or 0.48. When the frequencies of the genotypes are added together, they equal 1 ($0.36 + 0.48 + 0.16 = 1$). In addition, the genotype frequencies also represent the percentage of individuals within the population having a particular genotype. For the mouse population, 36 percent of the individuals are homozygous dominant (DD), 48 percent are heterozygous (Dd), and the remaining 16 percent are homozygous recessive (dd).

For the population to remain in genetic equilibrium, or **Hardy-Weinberg equilibrium**, the proposed mathematical model suggests that the allele frequencies, and the resulting genotype frequencies, will remain the same in subsequent generations. Therefore, the allele frequencies

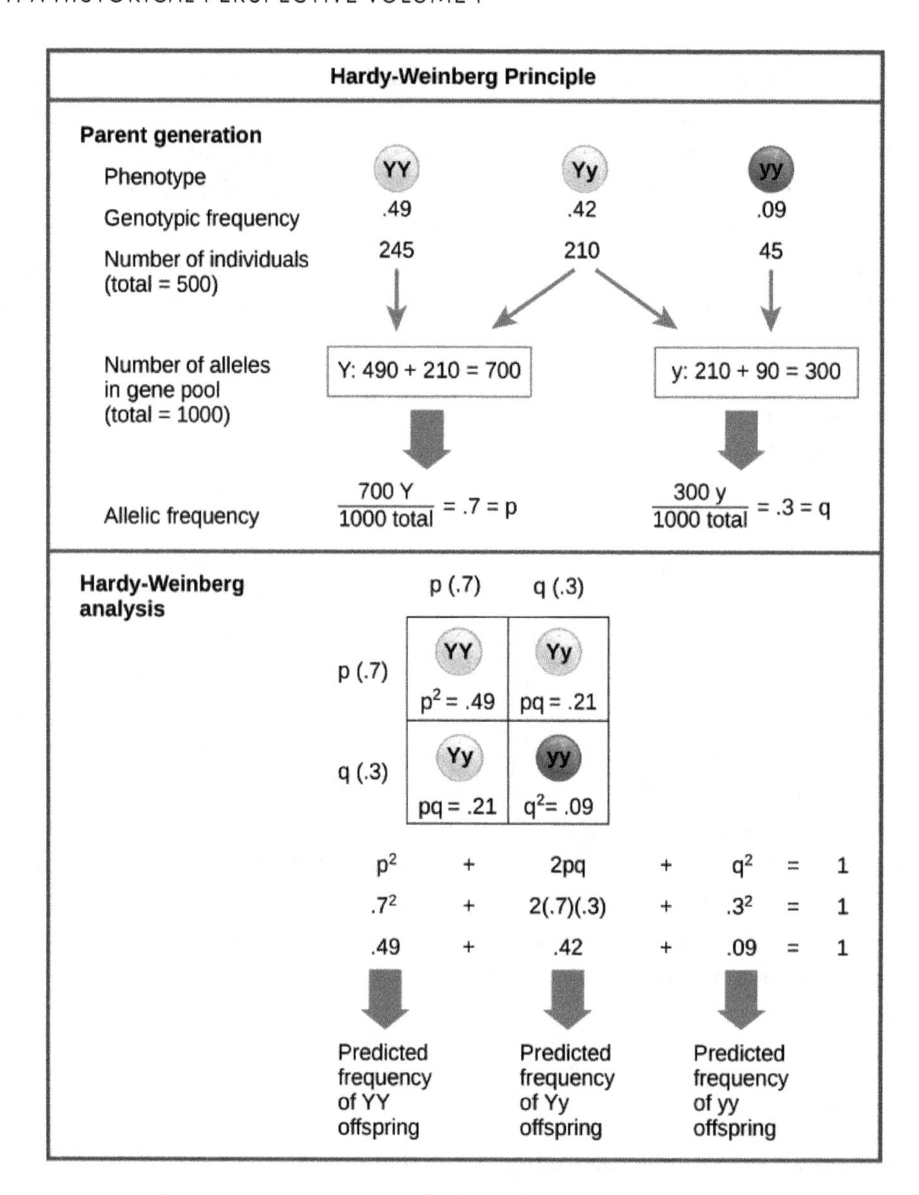

FIGURE 10.12 The Hardy-Weinberg Principle is a useful mathematical tool to calculate the allele frequencies and genotype frequencies of a population. This is especially helpful to population geneticists and evolutionary biologists to determine if evolution is occurring within the population.

of the dominant and recessive alleles in the mouse population must remain 0.6 and 0.4, respectively, in the next generation, which causes the genotype frequencies to also remain the same—36 percent homozygous dominant (DD), 48 percent heterozygous (Dd), and 16 percent homozygous recessive (dd). However, for this theoretical state of allele and genotype frequencies to remain in Hardy-Weinberg equilibrium in subsequent generations, five conditions have been established, and all must be met:

1. **No mutations.** The lack of mutations does not convert one allele to another or introduce new alleles into a population.

2. **Random mating.** All mating occurs by chance and is not influenced by particular genotypes or phenotypes.
3. **No gene flow.** Individuals do not migrate and transfer alleles between populations.
4. **Large population size.** Random changes in allele frequencies due to genetic drift would appear insignificant in a large population.
5. **No natural selection.** Natural selection does not favor alleles that confer selective and reproductive success for a particular phenotype.

If any one of these conditions is violated and allele frequencies within the population change, it disrupts Hardy-Weinberg equilibrium and suggests that the population is evolving. Natural populations rarely ever satisfy all five conditions to maintain Hardy-Weinberg equilibrium, as allele frequencies and genotype frequencies are constantly changing from generation to generation due to at least one of these conditions. Therefore, the conditions are merely a set of guidelines by which population geneticists and evolutionary biologists can track the inevitability of evolution (allele frequency change) occurring in a population.

10-3. Lines of Evidence Supporting Evolution

Since the publication of *On the Origin of Species* in 1859, countless years of observations and experimentation have brought forth various lines of evidence that strongly support Darwin's theory of evolution. These lines of evidence include the fossil record, biogeography, comparative anatomy and embryology, molecular biology, and laboratory and field studies.

Fossil Record

Fossils are remnants of once-living organisms preserved in Earth's stratified rock layers for over thousands of millions of years. Fossils may exist as the skeletal or dental remains of an animal that died whose soft parts are quickly consumed by either scavengers or decay, while other fossils of plants may exist as leaf or flower imprints or seeds. Fossils are an important piece of evidence supporting the theory of evolution because they document the history of an organism's life. Paleontologists, the scientists who study fossils, often refer to this documented evidence as the fossil record. The fossil record is important to evolutionary biologists because it provides a glimpse into an organism's past by providing details of its basic structure, its relatedness to other groups of organisms, the environment or location in which it lived, and certain events such as global changes that may have influenced its extinction or evolution.

Fossil formation is a relatively rare event, because only a limited number of structures can produce a fossil after an organism dies, while some organisms leave limited to no fossils, resulting in evolutionary gaps within the fossil record. Therefore, the fossil record is considered incomplete, which Darwin considered an issue, stating it as "the most obvious and gravest objection which can be urged against my theory" (Darwin 1859, 280). Despite the fossil record being incomplete, many clues have revealed the progressive, intermediate stages in which an ancestral organism accumulated changes over the course of time, supporting the idea of descent with modification.

Transitional fossils are groups of fossils that help provide the missing links necessary to support the descent with modification hypothesis. Transitional fossils demonstrate the intermediate stages of evolutionary change by exhibiting traits shared by a common ancestral group with its descendants. One of the most common examples of a transitional fossil is *Archaeopteryx* (Figure 10.13, left), which links dinosaurs and birds. The initial discovery of an *Archaeopteryx* feather occurred in 1860 by German paleontologist Hermann von Meyer, who described and named the fossil in 1861. The first skeletal remains of *Archaeopteryx* were discovered in Germany in 1861 and sold to the Natural History Museum in London. Despite the fossil missing most of its head and neck, British anatomist and paleontologist Richard Owen described the *Archaeopteryx* specimen in 1863, outlining distinct reptilian characteristics, such as its long, lizard-like tail and two free claws on its wings, along with its unique avian characteristics, its body completely covered in feathers. Later fossils discovered, which were more complete, revealed that *Archaeopteryx* had teeth, another distinct reptilian characteristic. Other transitional fossils discovered that have revealed links between other groups of organisms include *Tiktaalik*, (fishes and land animals) (Figure 10.13, middle), *Seymouria* (amphibians and reptiles) (Figure 10.13, right), and therapsids (reptiles and mammals).

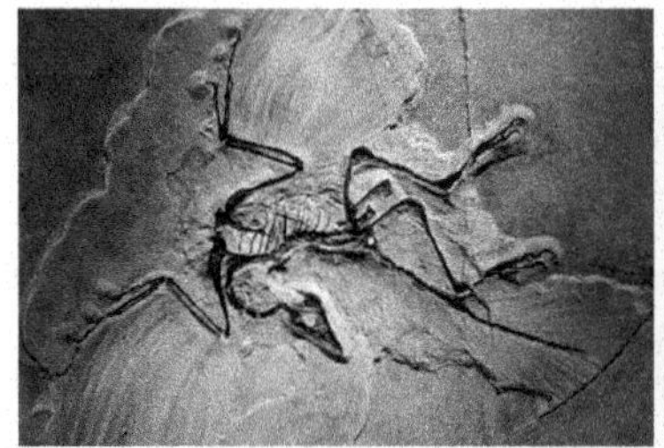
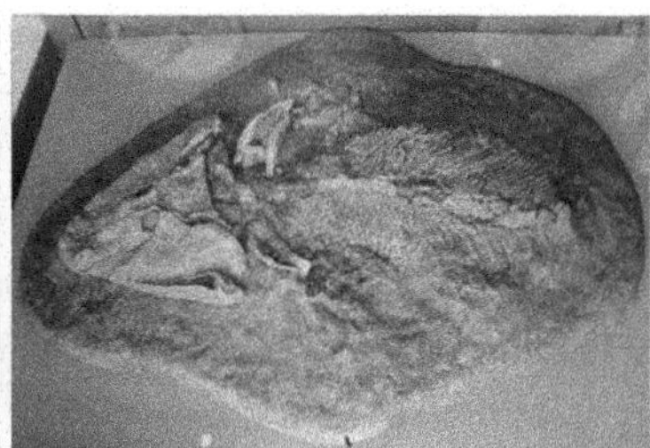

FIGURE 10.13 Transitional fossils provide the missing links to support the idea of descent with modification. Some of the most important transitional fossils include *Archaeopteryx* (left), *Tiktaalik* (middle), and *Seymouria* (right).

One of the most complete, well-preserved fossil records, consisting of many different transitional fossils, is of the modern-day horse (*Equus*). The fossil record of *Equus* (Figure 10.14) begins with *Hyracotherium*, a doglike ancestor that flourished in regions of North America, Europe, and Africa roughly 55 million years ago. The *Hyracotherium* fossil, first discovered and described by Richard Owen in 1839, revealed that the ancestral horse had four hoofed toes on its front feet and three hoofed toes on its hind feet, along with low-crowned molars, characteristics suggestive of a browsing, forest-dwelling herbivore. As habitats began to change to grasslands, slight gradations of anatomical change were revealed in the fossil record, suggestive of adaptations to this new environment. The most evident adaptations included modified teeth for grazing, adapted foot and leg anatomy for grazing and escaping predators, and an increase in body size. As individuals of the various evolutionary branches beginning with *Hyracotherium* went extinct, a single evolutionary branch emerged about 5 million years ago, giving rise to the genus *Equus*, which includes the modern-day horse.

Biogeography

Another line of evidence that supports the theory of evolution is **biogeography**, defined as the study of the unique patterns of geographical distribution of organisms living around the world. One of

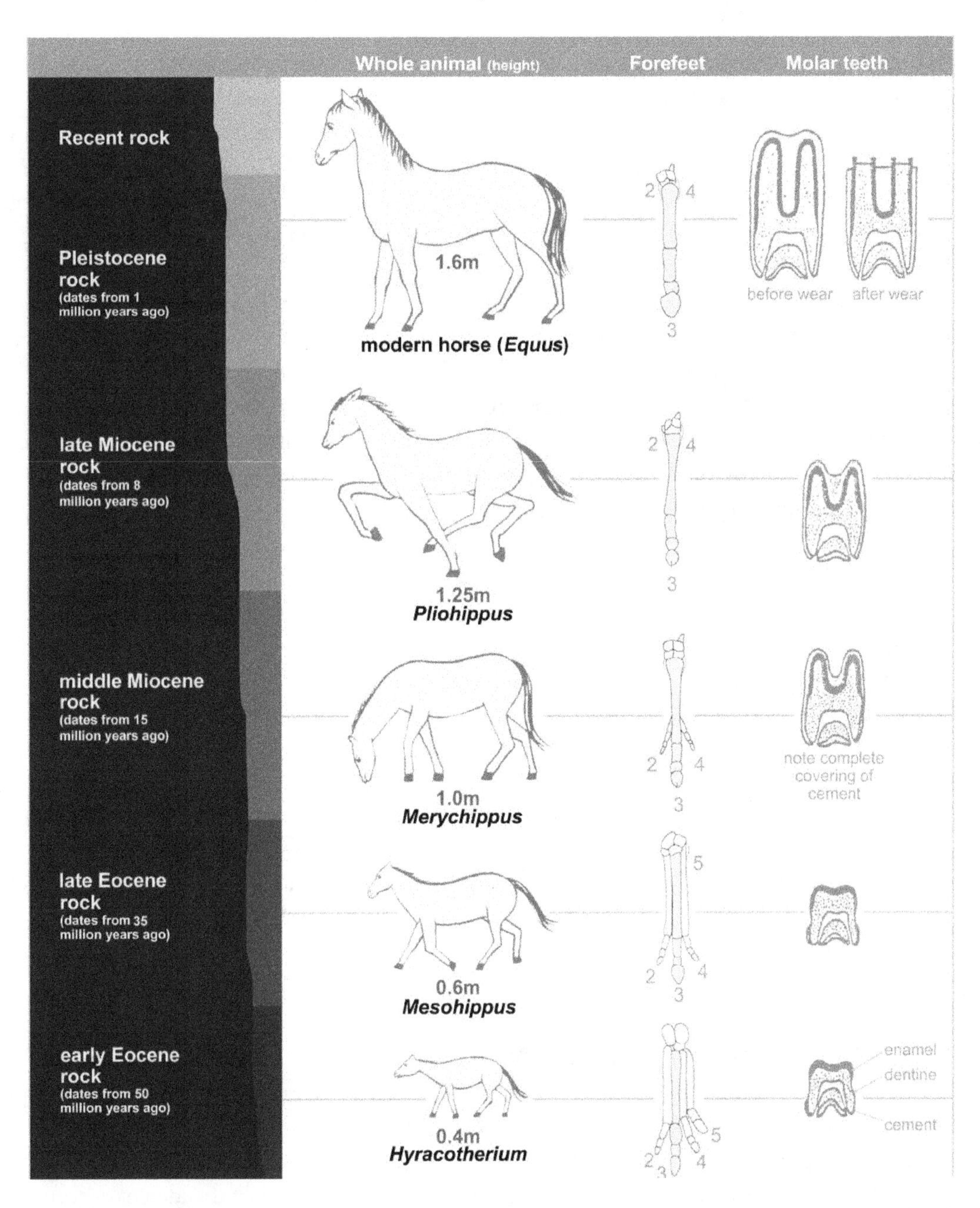

FIGURE 10.14 The evolutionary history of the horse is well documented due to its detailed fossil record and numerous transitional fossils outlining important anatomical gradations in their teeth and foot and leg anatomy as habitats began to shift.

the most common explanations as to why living organisms are distributed in such patterns is due to the continental shift of Earth's landmasses that occurred about 200 million years ago, known as the theory of continental drift, first proposed by German scientist Alfred Wegener in his 1912 publication, "The Origin of Continents." All the continents once existed as a large land mass called Pangaea, which allowed organisms to travel among its different regions. When Pangaea broke apart, the continents began shifting into their current locations, preventing migration, and therefore isolating certain organisms to specific continents or separating and confining populations of others to different continents. Darwin and Wallace both made note of species distribution during their travels, observing how species living in similar habitats in different regions of the world lacked resemblance, while species occupying different habitats that were in closer proximity to one another shared more resemblance. For example, Darwin observed that species living in the temperate regions of South America more closely resembled species found in the tropical regions of that continent than species found in the temperate regions of England.

Biogeography can also to be used to explain the diversity and prevalence of endemic species of marsupials found on the Australian continent. When the continents existed as Pangaea, Australia was connected to South America and Antarctica. As the continents began to break apart and shift, Australia became isolated, preventing the migration of organisms to the other continents. As a result, the different groups of marsupials specifically isolated to Australia greatly outnumbered the presence of placental mammals, and therefore, they began to flourish and diversify. Although South America contains some marsupial species, the diversity and prevalence of these species are minimal when compared to the number of placental species found on the Australian continent.

Finally, biogeography can be used to construct the evolutionary history of different groups of organisms using fossils discovered across the various continents (Figure 10.15). A common example was the use of anatomical data from fossils collected across North America, South America, Europe, Asia, and Africa to construct the *Equus* fossil record. As discussed, the *Equus* fossil record is the most complete, well-preserved fossil record because of the large number of fossils discovered on each of these continents, especially North America. The *Equus* fossil record was used to construct the different evolutionary branches by evolutionary biologists who understood the geographical distribution of the fossils, as well as the history of continental shift. Using biogeography and the *Equus* fossil record enabled evolutionary biologists to determine that the genus *Equus* originated in North America about 5 million years ago. The use of these two tools proved not only beneficial in helping construct the evolutionary history of the modern-day horse, but they have been used to construct the evolutionary histories of other organisms as well, which only provides more evidence to further substantiate the theory of evolution.

Comparative Anatomy and Embryology

The use of comparative anatomy and embryology provides the third line of evidence to support the theory of evolution.

Comparative anatomy is the study of the similarities and differences of anatomical structures of living organisms and is one of the strongest pieces of evidence to support the idea of descent with modification. One of the first scientists to study comparative anatomy was Leonardo da Vinci, whose anatomical notes and drawings depicted different animals ranging from bears to horses, as well as humans. However, the founding of the modern-day study of comparative anatomy has been credited to English physician Edward Tyson, who focused on dissecting various animals for the sheer purpose of comparative analysis. Tyson was one of the first to suggest that porpoises were mammals, based on how the convoluted ridges of the brain were more like land mammals than fishes. In 1699, Tyson published *Orang-Outang, sive Homo Sylvestris: or, the Anatomy of a Pygmie Compared with that of a Monkey, an Ape, and a Man,* in which he summarized the internal and external anatomy of a chimpanzee and described the anatomical similarities and differences of the chimpanzee with those

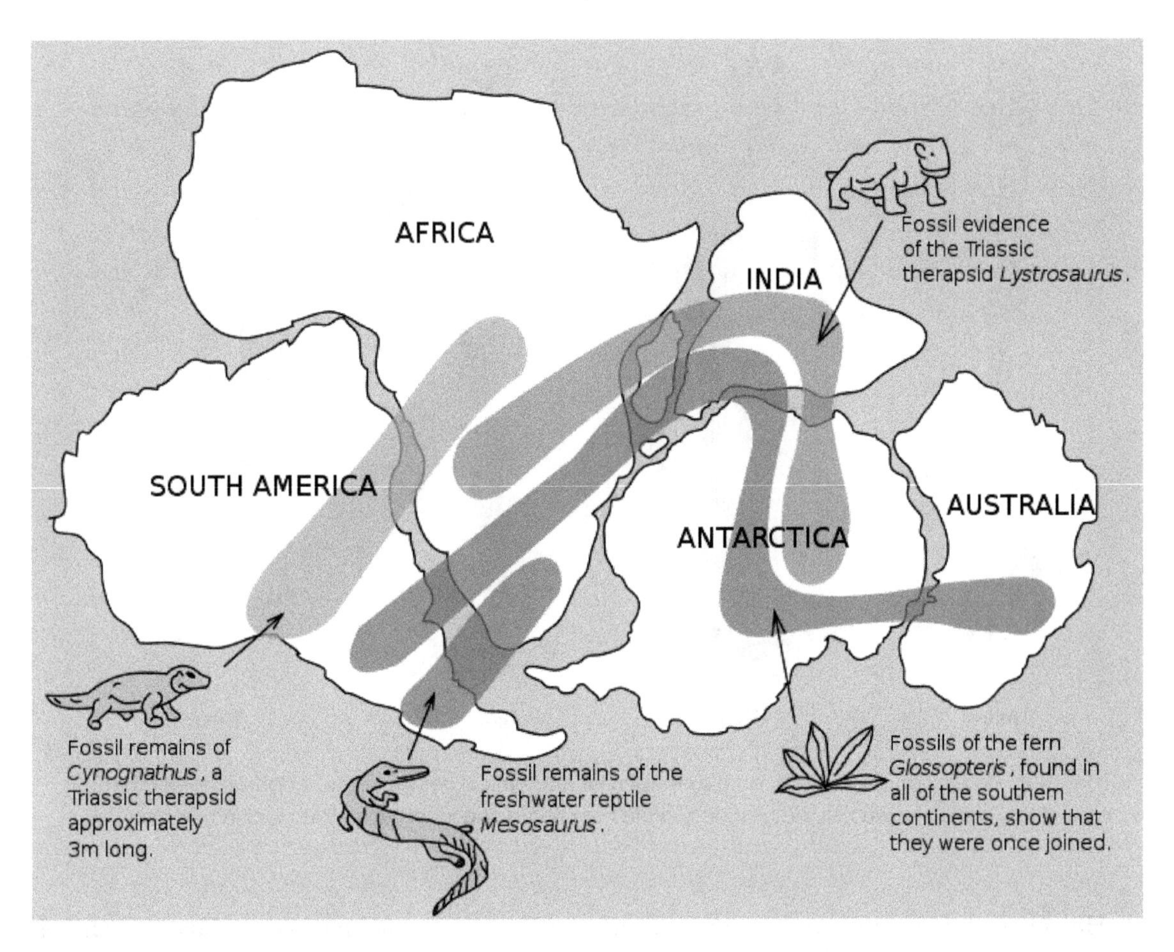

FIGURE 10.15 The biogeography of fossils is an important piece of evidence to support the theory of evolution. The distribution of fossils is useful in constructing the evolutionary history of different organisms and can be used to explain the diversity and prevalence of species found on Earth today.

of humans, apes, and monkeys. Tyson concluded that his anatomical comparison suggested that chimpanzees were more anatomically like humans than to monkeys or apes. Tyson's detailed work on comparative anatomy paved the way for future anatomists, including Richard Owen and Thomas Huxley, whose own comparative anatomy studies influenced Darwin's work on the theory of evolution.

Comparative anatomy provides compelling evidence to support the descent with modification idea, because organisms who are closely related reveal more anatomical similarities. Two different anatomical structures provide this evidence, including homologous and vestigial structures.

Homologous structures, first observed by Aristotle and later named by Richard Owen in 1843, are anatomically similar structures found in different organisms whose evolutionary origin and inheritance are traced back to a common ancestor. Although an anatomical design of a particular structure is recognizably similar between different organisms, its function may be remarkably dissimilar based on how natural selection modified the structure for a specific use. An example of homologous structures are vertebrate forelimbs, all of which contain the same arrangement of bones but are slightly modified for lifting in humans, walking in dogs, flight in birds, and swimming in whales (Figure 10.16).

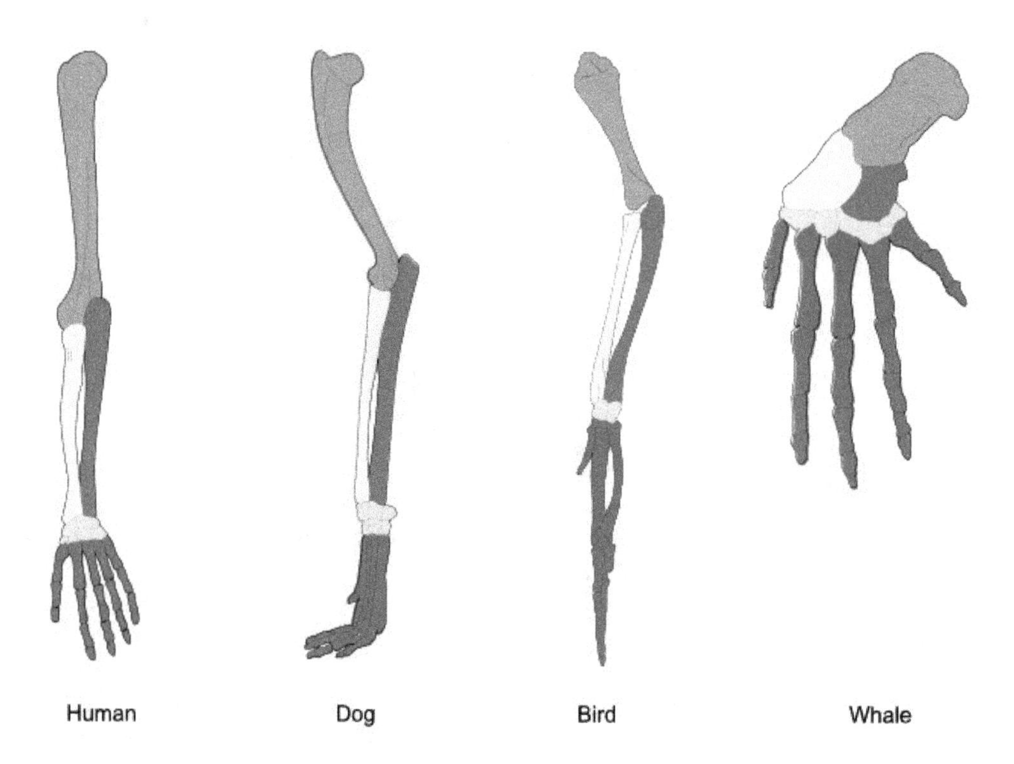

FIGURE 10.16 Homologous structures, such as the similar anatomical design of the vertebrate forelimbs of humans, dogs, birds, and whales, suggest an evolutionary origin and inheritance from a common ancestor.

Homologous structures, however, should not be confused with **analogous structures**, structures whose origin did not result from common ancestry but independently evolved due to **convergent evolution**. Convergent evolution is a process by which different organisms living in the same habitat adapt to similar environmental challenges, resulting in the development of similarly functioning structures. For example, the wings of bats and birds are homologous; however, the wings of insects are analogous to these structures (Figure 10.17). Although the wings of all three organisms appear superficially similar and perform the same overall function, the wings of bats and birds consist of forelimb bones modified by natural selection, while butterfly wings are extensions of its body's cuticle.

FIGURE 10.17 Analogous structures, such as the wings of insects, are the result of convergent evolution in which similar structures are found in different organisms like birds and bats.

Vestigial structures are remnants of once functioning structures in a common ancestor, whose function is reduced or nonexistent in a descendant. As with homologous structures, Aristotle was one of the first to describe the presence of vestigial structures in different organisms in *History of Animals*. Lamarck also described different vestigial structures in *Zoological Philosophy: An Exposition with Regard to the Natural History of Animals*, including the eyes of the blind mole rat (*Spalax typhlus*), stating it "has altogether lost the use of sight, so that it shows nothing more than vestiges of this organ." Along with Lamarck, Darwin also recognized the presence of vestigial structures, describing them as "rudimentary or atrophied organs ... imperfect and useless" (Darwin 1859, 453). Darwin expanded his discussion of vestigial organs in *The Descent of Man and Selection in Relation to Sex*, compiling a list of several vestigial features in man, including the appendix and coccyx. Although described by the likes of Aristotle, Lamarck, and Darwin, the term "vestigial" was never used by these contemporaries. The term was first used by German anatomist Robert Wiedersheim in his book *The Structure of Man* (1895), in which he compiled a long list of vestigial structures existing in man. Wiedersheim stated "the occurrence of them in certain organs, or parts of organs, now known as 'vestigial' by such organs are meant those which were formerly of greater physiological significance than at present" (Wiedersheim 1893, 2). The presence of vestigial structures further validates the idea of descent with modification, as their presence could have only occurred by means of inheritance from a common ancestor. For example, the presence of vestigial pelvic and leg bones in snakes and whales suggests these groups of organisms descended from ancestors whose use of legs was their primary means of mobility (Figure 10.18). As a result of natural selection, the lack of limbs provided a much better advantage for these modern organisms in each of their habitats, resulting in the presence of small pelvic and leg bone remnants in their skeletons.

Embryology is the study of embryonic development in organisms and provides yet another line of evidence to support the idea of descent with modification. Many different scientists were instrumental in studying and expanding on the ideas of embryology, but none more than German scientist Karl von Baer, who has been credited with founding the modern study of embryology, in the early 1800s. Von Baer's studies focused on the embryonic development of animals, which led to the discovery of the mammalian egg in a dog, later confirmed by findings in other mammalian species, including humans. Von Baer detailed the observations and descriptions of his findings in an 1827 pamphlet entitled "On the Genesis of the Ovum of Mammals and of Men," laying the foundation

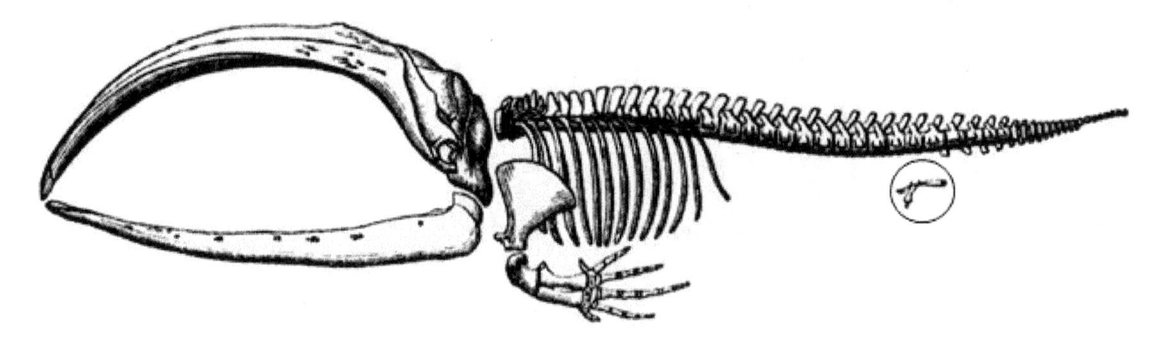

FIGURE 10.18 Vestigial structures are remnant structures that no longer serve a purpose in living organisms, such as the vestigial pelvic bones in whales (circled). The presence of vestigial structures suggests a practical use in a common ancestor.

for the study of modern embryology. During his studies, von Baer also observed structural similarities, such as tails and gill slits, in the embryonic development of many vertebrates, including turtles, fishes, chickens, and humans.

The work of von Baer inspired another German scientist, Ernst Haeckel, to publish a series of drawings demonstrating the various stages of embryonic development in different vertebrate species, with each stage of development showing varying degrees of resemblance to one another. Haeckel's drawings appeared for the first time in his 1868 publication, *Natural History of Creation*. He wrote later to a colleague who questioned his initial drawings that they "are completely exact, partly copied from nature, partly assembled from all illustrations of these early stages that have hitherto become known," and they were continually published in textbooks throughout the 20th century (Hopwood 2006, 272). Since the illustrations were first published in 1868, many critics of his time and today have accused Haeckel of fraud, suggesting he plagiarized the embryonic drawings of his contemporaries, inappropriately used artistic license to exaggerate similarities and omit distinct differences, and failed to draw the embryos to scale. Haeckel stated in his 1874 publication, *Anthropogenie* (later translated to *The Evolution of Man*), in which the next series of drawings appeared (Figure 10.19), that he had left out certain embryonic structures, such as the yolk sac or amnion, and the images had been reduced to the same size and oriented in a way to simplify the comparison of anatomical similarities.

Additional work by modern-day embryologists has discredited Haeckel's work further, as photographs taken of embryos at comparable stages to those of Haeckel's drawings reveal striking differences, suggesting that Haeckel misrepresented the state of embryonic development. However, more recent data and photographs show that embryos in early stages of development undeniably share anatomical similarities, providing further support to the idea of descent with modification.

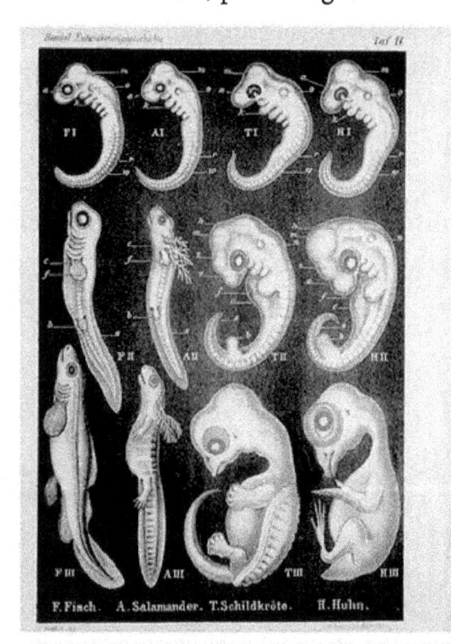

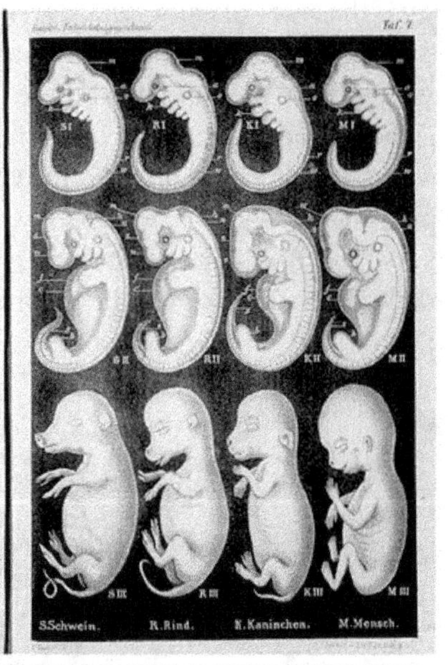

Key: (F) Fish (A) Salamander (T) Turtle (H) Chicken (S) Pig (R) Cow (K) Rabbit (M) Human

FIGURE 10.19 One of the first scientists to publish the observed embryonic similarities of different organisms was Ernst Haeckel. This figure represents his drawings from his 1874 publication, *Anthropogenie*, which showed simplified versions of the anatomical similarities in different embryos. The similarities observed in early embryonic stages provide further evidence to support descent with modification.

In evaluating the early embryonic stages of fishes, amphibians, reptiles, birds, and mammals, we observe that all display many common features that suggest descent with modification from a common ancestor. All embryos in the early stages of embryonic development have gill pouches on the side of their necks and a post-anal tail, but the embryos begin to look quite different as certain genes influence the future development of these structures. For example, certain genes are responsible for the development of gills from gill pouches in fishes, while genes active early in human development result in the loss of the post-anal tail.

Molecular Biology

For centuries, scientists have used fossils, biogeography, and comparative anatomy and embryology as the primary lines of evidence to support the idea of descent with modification. However, beginning with the discovery of the structure of DNA in 1953 and, several years later, the genetic code in 1966, molecular biology has become a revolutionary piece of evidence to support descent with modification. **Molecular biology** is the field of biology that focuses on the chemical structure and biological activity of molecules found in the cells of living organisms. The field of molecular biology has revealed that organisms use many of the same molecules, including DNA, RNA, and proteins, which has generated substantial evidence to verify relatedness between organisms.

Recall from chapter 6 that all organisms, from bacteria to humans, possess DNA arranged in a unique sequence of nucleotides that is responsible for identifying an organism's genetic information. Today, scientists compare the DNA nucleotide sequences of two organisms to determine how closely related or unrelated they are to one another. The comparison of DNA nucleotide sequences of different organisms began as early as the 1970s, when American molecular biologists Charles Sibley and Jon Ahlquist developed an innovative process known as DNA-DNA hybridization to determine the evolutionary relationship of birds. The technique involves unwinding the double-stranded DNA molecule of both organisms and mixing the resulting strands to form a DNA hybrid, a double-stranded molecule consisting of one strand from each organism. The relatedness of two organisms is determined by how quickly the strands form the DNA hybrid. The faster the strands from the two organisms reform into a DNA hybrid, the higher the number of shared nucleotides, suggesting that the organisms are closely related. In 1984, Sibley and Ahlquist used DNA-DNA hybridization to compare the DNA nucleotide sequences of different primate species, including humans, chimpanzees, gorillas, and orangutans. In comparing human DNA with the DNA of the other primate species, the scientists discovered a 1.5-percent difference to chimpanzees, a 2.4-perent difference to gorillas, and a 3.7-percent difference to orangutans, a finding that revealed that humans are more closely related to chimpanzees than to other primate species. A more precise method has replaced DNA-DNA hybridization, a process known as DNA sequencing, which is used to determine the specific sequence of DNA nucleotides in an organism. Upon sequencing and comparing human and chimpanzee DNA, the results revealed genetic relatedness; however, the process also showed slight variations in the nucleotide sequences of noncoding DNA. Scientists believe these variations, along with differences in gene expression, account for the various dissimilarities between the two primate species.

During the process of protein synthesis, DNA is used as a template to generate a specific strand of RNA known as mRNA, which consists of three-letter codons that specify a particular sequence of amino acids required to build a functional protein. These three-letter codons are translated into one of the possible 20 different amino acids using the genetic code, a tool discovered by Marshall Nirenberg and a group of scientists in 1966. Since all living organisms have DNA, they use the same genetic code to synthesize essential proteins from these same 20 amino acids, with some amino acid sequences of proteins differing slightly between various groups of organisms. For example, located between the inner and outer membranes of the mitochondria of all living eukaryotes and most single-celled organisms is a protein called cytochrome *c*. In all these organisms, cytochrome *c* functions in promoting cellular apoptosis and plays an integral part in the mitochondrial electron transport chain. Cytochrome *c* is an important molecule to use when determining evolutionary relationships (Figure 10.20), because it has evolved slowly over time through random mutations. In eukaryotic organisms, the protein is composed of 105 amino acids, of which about 33 percent have been conserved.

This conservatism is evident, as studies have revealed that when sequences of amino acids for cytochrome *c* are more closely identical between different eukaryotic organisms, it suggests the organisms are genealogically related and evolved from a more recent common ancestor. In 2016, scientists performed an amino acid sequence comparison of the cytochrome *c* protein of different eukaryotic organisms, ranging from humans to yeast. The results of the study revealed that the cytochrome *c* protein of humans and chimpanzees is identical, less one amino acid, while the amino acid sequence of the protein in humans compared to that of a mouse is 92-percent identical. The close similarity of amino acid sequences between the human and mouse suggests that the gene from

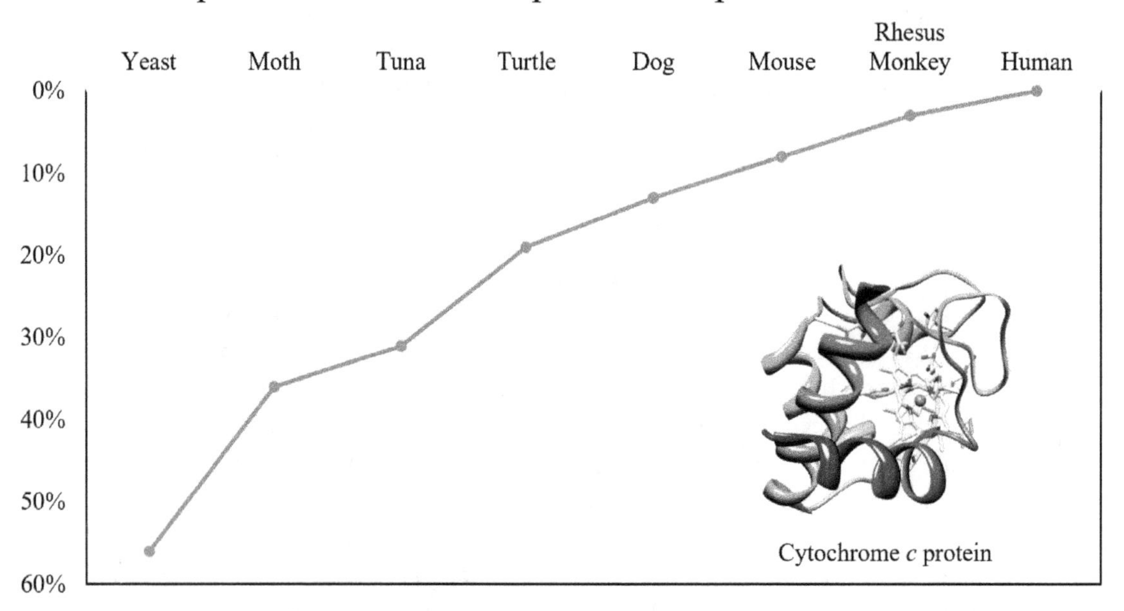

FIGURE 10.20 Molecular biology has become an important piece of evidence to support the theory of evolution. Relatedness among different species of organisms has become more evident through the comparison of DNA and in the comparison of amino acid sequences of proteins, such as cytochrome c. This figure shows the results of a study performed in 2016 and indicates the percent difference in amino acid sequences of cytochrome c in various organisms as compared to humans. *Data sourced from: Kulkarni Keya and Sundarrajan Priya. A Study of Phylogenetic Relationships and Homology of Cytochrome C using Bioinformatics.* Int. Res. Journal of Science & Engineering, *2016,4(3–4):65–75.*

which the cytochrome *c* protein is synthesized was inherited from a recent common ancestor but has undergone some minor random mutations as the species began to separately evolve. To differentiate, the same study compared the cytochrome *c* amino acid sequence of a human to yeast and discovered that the sequences were only 44-percent identical, suggesting the duration of evolution between the species occurred much longer ago, providing more time for random mutations to occur. The evidence generated from molecular biology studies, including the evidence obtained from DNA-DNA hybridization, DNA sequencing, and amino acid sequence comparisons, appear to correlate with and further emphasize Darwin's idea of descent with modification and the relatedness of organisms.

Field and Laboratory Studies

The final line of evidence providing support for the idea of descent with modification has been through direct observations made during field and laboratory studies. Although evolution can be a slow process, many scientists have committed a lifetime of research and experimentation to study the evolution of specific group of organisms, and the evidence they have discovered has been irrefutable.

Recall from earlier in this chapter Peter and Rosemary Grant, the evolutionary biologists whose field studies on the Galápagos Island finches revealed significant changes in the beak sizes of finches. Beginning in 1973, the pair of researchers would spend several months on Daphne Major, one of the Galápagos Islands, for the next 40 years, tagging and studying every aspect of the island's finches, including mating and production of offspring, recording their songs, and taking blood samples for genetic analysis. During their studies, the Grants also tracked how changes in the finches' morphology (body size and beak size) correlated to the harsh, environmental changes occurring on the island, which resulted in the fluctuation of food availability for the finches. The Grants observed that natural selection would oscillate directionally on the population of finches, based on how the environment of the island would transition from times of excessive rain to periods of extensive drought, influencing the availability of different-size seeds. Within the first few years of studying the finches of Daphne Major, the Grants made their first discovery as the drought of 1977 resulted in a directional shift to a larger beak size in the next generation of medium ground finches (*Geospiza fortis*) due to seed availability (Figure 10.21). (Please see *Directional Selection* under section 10-1 for a more thorough description).

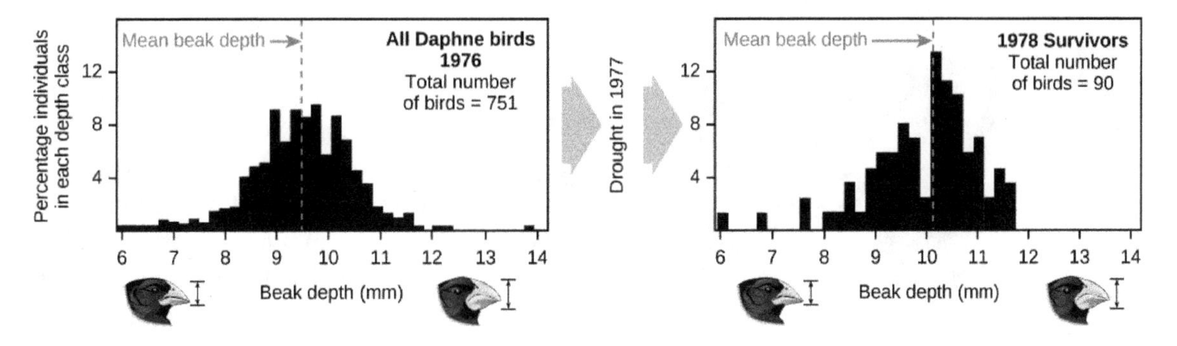

FIGURE 10.21 The following graphs show the directional shift of the medium ground finches (*Geospiza fortis*) on Daphne Major before and after the drought of 1977. As the graphs indicate, the directional shift reveals an increase in beak size related to seed availability during the drought.

Another directional shift in the finch population on Daphne Major occurred between 1982 and 1983, when a prolonged El Niño event produced steady rainfall and high temperatures, altering the island's vegetation, and shifting the size and availability of seeds. Small, soft seeds were more abundant in the year following the El Niño event, and medium ground finches with smaller beaks held a selective advantage during the drought of 1985, as larger birds experienced a higher mortality rate as the seed supply decreased. Over the course of 40 years, the Grants repeatedly observed directional shifts based on environmental changes occurring on Daphne Major, concluding that the morphology of the finches was constantly changing and considerably different from when they began their study in 1973.

In 2004, a team of researchers, including Peter and Rosemary Grant, performed a comparative analysis of expression patterns of different growth factors responsible for beak development in six *Geospiza* species. The focus of the study was to determine the growth factor responsible for producing the differences observed in the beak morphology of these finches. Upon analysis, the researchers discovered a correlation between beak morphology and different expression patterns of the gene *Bmp4*. When expressed, the relative levels and timing of *Bmp4* varied during certain stages of beak development. The researchers discovered that when *Bmp4* was expressed at higher levels and appeared much sooner in certain stages of beak development, the finches developed larger, deeper beaks when compared to the gene's expression patterns for other beak morphologies. The researchers concluded that the varied levels and timing of *Bmp4* expression were responsible for differences observed in beak morphology; therefore, these levels and time of expression have provided the necessary variation required for natural selection to act during beak evolution in the *Geospiza* species.

Another example of evolution in action has been observable antibiotic resistance in bacteria during lab studies. In 1928, an accidental discovery was made in a laboratory at St. Mary's Hospital in London, England, when culture plates inoculated with the bacterium *Staphylococcus aureus* were pushed aside and left unattended for more than a month on a laboratory bench. In July of 1928, Scottish physician Sir Alexander Fleming (Figure 10.22, left), who discovered lysozymes, a group of enzymes found in the saliva, tears, and mucus with antimicrobial properties, six years earlier, left for a family vacation during the middle of his studies on strain variation of *Staphylococcus*. While away, Fleming was appointed professor of bacteriology at St. Mary's Hospital Medical School in September, prompting his return to London. Fleming returned to the laboratory shortly after his new appointment and discovered that in a corner of one of his inoculated plates, of which the cover was off, a white, fluffy mass of mold was growing; however, no *Staphylococcus* was growing in the region. Fleming's

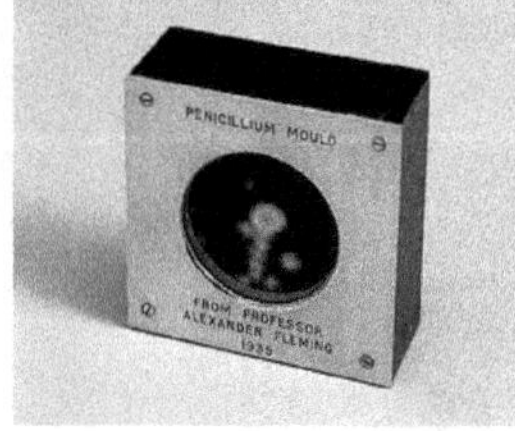

FIGURE 10.22 Alexander Fleming (left) discovered a strain of fungus, known as *Penicillium* (right), produced antimicrobial properties in 1928. This discovery led to the production of the first effective antibiotic, penicillin, in the early 1940s.

excitement over the unique discovery was unmet by his colleagues, who reacted with indifference and apathy. Regardless, Fleming, unfazed by the reactions of his contemporaries, began performing experiments with the mold, growing colonies of it, and testing its antimicrobial abilities, in late September.

Fleming discovered that the mold inhibited the growth of not only *Staphylococcus* but *Streptococcus* and *Corynebacterium diphtheriae* bacterial strains as well, determining that the mold must possess specific antimicrobial substances. Fleming began to experiment with different mold broth filtrates to isolate the antimicrobial substance, while also experimenting with different species of fungi, including *Aspergillus fumigatus, Cladosporium,* and eight strains of *Penicillium* to determine which species produced these similar properties. Fleming discovered that one of the eight strains of *Penicillium* produced the antimicrobial substances (Figure 10.22, right), and with little knowledge of mycology (the study of fungi), obtained only from his literature research, he believed the mold to be *Penicillium chrysogenum.* To verify his assertion, Fleming consulted Charles J. La Touche, an Irish mycologist working in a lab one floor below, who identified the mold as *Pencillium rubrum.* However, a comprehensive phylogenetic analysis performed in 2011 revealed the actual penicillin-producing strain as *Penicillium rubens,* not *P. rubrum* as La Touche identified.

Fleming continued his studies on the antimicrobial properties of *P. rubens,* noting its low toxicity to animals, including humans, and how it did not interfere with white blood cell function, making it the first-ever nontoxic antimicrobial drug ever discovered. Fleming published his findings, which received little recognition from the scientific community at the time, in May 1929 under the title "On the Antibacterial Action of Cultures of a Penicillium, with Special Reference to Their Use in the Isolation of B. influenza." Within the summary section of his publication, Fleming named the antimicrobial agent *penicillin,* later explaining that he chose the name in order "to avoid the repetition of the rather cumbersome phrase 'mould broth filtrate.'"

Fleming continued working with penicillin throughout the 1930s, performing promising clinical trials, but he found the lack of success producing it in large quantities disheartening. As Fleming contemplated retirement in early 1940, two scientists from the University of Oxford, Howard Florey and Ernst Chain, learned of Fleming's work and began researching large-scale production methods to isolate and purify penicillin, administering it for the first time to a patient in 1941. Shortly after World War II escalated in 1941, the United States and United Kingdom governments collaborated to begin mass production of penicillin. By 1943, the antibiotic became a widely used treatment to control bacterial infections among the Allied forces. For the discovery of penicillin and its development into a useful antibiotic, Fleming, Florey, and Chain all shared the 1945 Nobel Prize in Physiology or Medicine.

Penicillin was a highly effective antibiotic when introduced in the early 1940s, functioning by disrupting the bacterial enzymes responsible for synthesizing their cell walls. However, as early as 1942, clinical specimens of *Staphylococcus aureus* began to demonstrate antibiotic resistance, predicted to occur by many scientists, including Fleming, who stated, "the microbes are educated to resist penicillin and a host of penicillin-fast organisms is bred out" (Society for Healthcare Epidemiology of America 2012, 322). The next two decades showed increased bacterial resistance to penicillin, with roughly 80 percent of *Staphylococcus aureus* strains resistant by the 1960s, and today, more than 90 percent of strains are resistant.

The development of resistance is best explained by natural selection. In 1945, scientists discovered that resistant strains of *S. aureus* produced an enzyme known as penicillinase, which they believed either had arisen through a spontaneous mutation, resulting in new resistant variants, or had already been possessed or acquired by one or more bacterial cells through intraspecies or interspecies gene exchange. Regardless of the mechanism, penicillin-resistant strains were favored by natural selection when exposed to penicillin, causing the number of resistant strains to proliferate due to their short generation time. The increased prevalence of penicillin-resistant strains sparked an antibiotic revolution as scientists began to develop and produce new, stronger antibiotics. Two new antibiotics were introduced in the early 1950s, tetracycline and erythromycin, but by 1959, scientists had discovered bacterial resistance to tetracycline (*Shigella*), and in 1968, the first erythromycin-resistant bacterial strain (*Streptococcus*) emerged. A new antibiotic, methicillin, was introduced in 1960, and within two years, *S. aureus* was resistant. Between 1943 and 2015, several different antibiotics have been introduced, and by 2015, multiple bacterial strains, including *Staphylococcus*, *Streptococcus*, and several from the Enterobacteriaceae group of bacteria, were resistant to most if not all antibiotics (Figure 10.23). This resistance has led to deadly bacterial infections, such as methicillin-resistant *S. aureus* (MRSA), which is resistant to multiple antibiotics. Although teixobactin, a new antibiotic introduced in 2015, has shown promise in treating MRSA, it is only a matter of time before resistance could arise from prolonged use of the drug.

Antibiotic Introduced (Year)	First Antibiotic Resistant Strain(s) Identified (Year)
Penicillin (1941)	*Staphylococcus aureus* (1942)
Tetracycline (1950)	*Shigella* (1959)
Erythromycin (1953)	*Streptococcus pyogenes* (1968)
Vancomycin (1958)	*Enterococcus faecium* (1988)
Methicillin (1960)	*Staphylococcus aureus* (1962)
Cefotaxime (1980)	*Escherichia coli* (1983)
Azithromycin (1980)	*Neisseria gonorrhoeae* (2011)
Imipenem (1985)	*Klebsiella pneumoniae* (1996)
Ceftazidime (1985)	Enterobacteriaceae (inc. *Klebsiella, Escherichia, Shigella*) (1987)
Levofloxacin (1996)	*Streptococcus pneumoniae* (1996)
Linezolid (2000)	*Staphylococcus aureus* (2001)
Daptomycin (2003)	*Staphylococcus aureus* (2004)
Ceftaroline (2010)	*Staphylococcus aureus* (2011)
Ceftazidime-avibactam (2015)	*Klebsiella pneumoniae* (2015)

FIGURE 10.23 Antibiotic resistance is a common example of evolution in action. Since the introduction of penicillin in the 1940s, the efficacy of different types of antibiotics has greatly diminished due to the evolution of resistant strains of bacteria. Within 70 years since the initial development of antibiotics, many strains of bacteria are now resistant to most, if not all, antibiotics. *Data sourced from: Ventola C. L. (2015). The antibiotic resistance crisis: Part 1: Causes and threats.* P & T: a peer-reviewed journal for formulary management, *40(4), 277–283 & Centers for Disease Control and Prevention. (2020). Brief History of Resistance and Antibiotics. Select Germs Showing Resistance Over Time.*

Finally, evolution has also been observed using a combination of both field and laboratory studies. In 2017, Hurricanes Irma and Maria devastated the islands of Turks and Caicos, a study site for a group of scientists, led by American biologist Colin Donihue, investigating anole lizards (*Anolis scriptus*) (Figure 10.24, left). Donihue and his collaborators returned to the islands after weather conditions improved to determine whether the anoles had survived the devastation and to investigate potential pre- and postevolutionary effects the hurricanes had on the anole population, a study believed to be the first of its kind. The primary focus of the study was to obtain post-hurricane data on specific morphological traits, such as body size, toepad (which they use to grasp onto the branches of bushes and other vegetation) size, and forelimb and hindlimb length, and compare this with data collected prior to the hurricanes. The team discovered that the anoles that survived the hurricanes demonstrated significant phenotypic changes: smaller body sizes, increased toepad size (Figure 10.24, right) and forelimb length, and shorter hindlimbs.

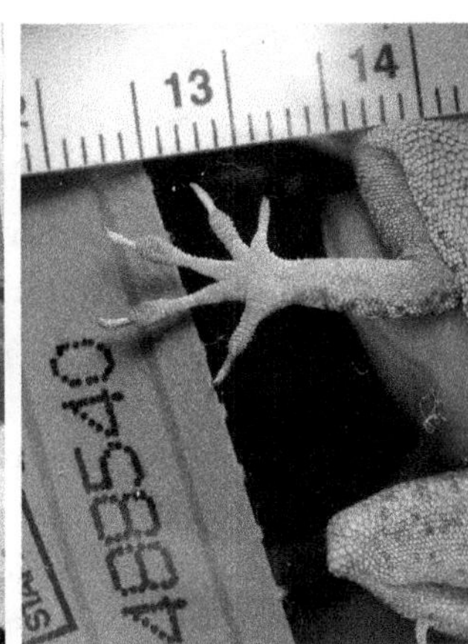

FIGURE 10.24 An organism that has undergone evolution due to weather conditions are anole lizards (*Anolis scriptus*) found on the islands of Turks and Caicos (left). In response to Hurricanes Irma and Maria devastating the islands in 2017, it was observed the lizards had developed significant phenotypic changes, including increased toepad size (right).

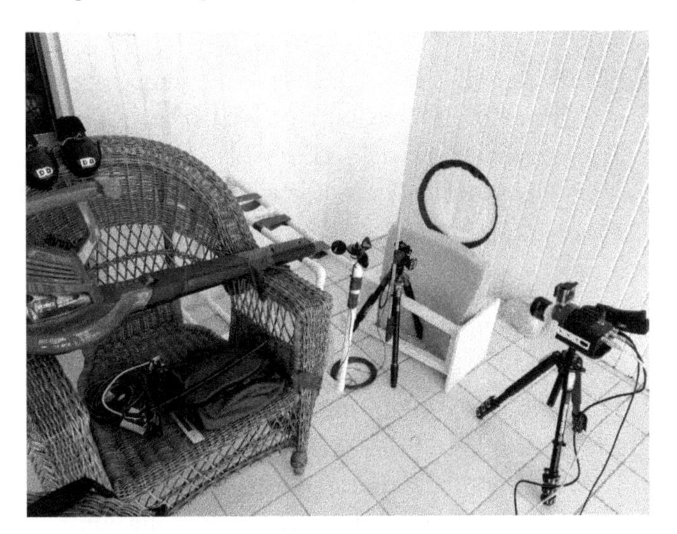

FIGURE 10.25 The laboratory setup used by scientists to investigate how the anole lizards of Turks and Caicos respond to high winds, which involved the use of a leaf blower.

The scientists decided to take their research a step further and performed a field laboratory experiment to determine how the anoles responded to high wind using a leaf blower (Figure 10.25). Once the lizards were perched, the scientists turned on the leaf blower and observed the lizards using their toepads to grasp while tucking their forelimbs under their body and extending their hindlimbs out behind them, until the wind resulted in them losing their grip. The researchers speculated that smaller body size, increased toepad size, and shorter hindlimbs provided an improved ability for clinging to branches during high-wind events, as in the cases of Hurricanes Irma and Maria. This is an example of natural selection intervening, as anoles without these unique characteristics perished because they were unable to hold on to branches when the hurricanes began. The scientists concluded the population experienced a directional shift, favoring individuals with traits better adapted to clinging onto vegetation in high winds.

As with most scientific research, the results of one study led to additional questions, which prompted researchers to design another series of experiments to discover the answers. Donihue and his collaborators questioned what effects natural selection had on the next generation of hurricane-surviving anoles and whether the selection event was only isolated to the anoles of Turks and Caicos or if other Caribbean anole species had been affected as well. The team of researchers returned to Turks and Caicos to answer the first question and discovered the toepad size of the offspring of hurricane survivors was larger than the pre-hurricane population but similar in size to their parents, suggesting the offspring inherited this adaptation. The second question was answered using 70 years of hurricane data, along with toepad measurements generated from pictures of 188 different anole species across the Caribbean. Donihue and the team discovered the same general pattern—species of anoles surviving more hurricanes have a larger relative toepad size—suggesting that hurricanes have the potential to shape the evolutionary path of populations living in certain regions of the world that are highly vulnerable to these meteorological events.

Chapter Summary

10-1. Mechanisms of Evolutionary Change

- Evolution is the change in allele frequencies of a population over time. A population is a group of interbreeding individuals that live in the same area and can produce fertile offspring. Population genetics is the study of allele frequencies within the gene pool of a population to determine how these alleles are inherited and influence genetic variation.
- The allele frequencies of a population can change through mutations, genetic drift, gene flow, or natural selection.
  - A mutation changes allele frequencies in a population as DNA base pairs are altered, either converting one allele to another or producing a new allele.
  - Genetic drift influences changes in allele frequencies due to random, chance events. Two ways in which genetic drift can affect allele frequencies of a population are the bottleneck effect and founder effect. The bottleneck effect occurs when changes in allele frequencies result from a severe reduction in the number of individuals in a population due to a catastrophic event, such as an earthquake or habitat destruction. The founder effect changes allele frequencies when a small number of individuals from the original population leaves or is geographically isolated.
  - Gene flow influences allele frequencies as individuals introduce their alleles as they migrate between populations.
  - Natural selection changes allele frequencies in a population as individuals with alleles more favorable to a particular environment have higher reproductive rates than individuals with less favorable alleles. The effects natural selection has on a population depends on four different factors. Heritable variation: natural selection works on individual variations within a population that arise through mutations and are inherited in subsequent generations. Competition for resources: as the availability of resources becomes limited within a population, individuals with favorable variations capable of outcompeting others

will survive and reproduce. Adaptation: traits that allow organisms to be better suited to their environment. Differential reproductive success and fitness: fitness is the measure of reproductive success of individuals with beneficial variations compared to individuals with less favorable variations. Fitness influences sexual selection as mates choose their partners based on traits that imply higher fitness. There are two types of sexual selection, including intrasexual (direct competition) and intersexual (female choice) selection. Natural selection favors particular phenotypes in three ways. Directional selection favors one extreme phenotype or another, resulting in a shift of individuals with that particular phenotype. Diversifying (disruptive) selection favors individuals with both extreme phenotypes over an intermediate phenotype. Stabilizing selection favors individuals with an intermediate phenotype over those with extreme phenotypes. Natural selection does not always produce perfectly adapted organisms for several different reasons. Natural selection only works on preexisting variations within the population. A fast-changing environment does not always give natural selection time to operate. Natural selection does not produce new anatomical structures but rather modifies existing structures based on its required adaptation. Evolutionary change is restricted as certain trait adaptations may require a trade-off with another.

10-2. The Hardy-Weinberg Principle

- G. H. Hardy and Wilhelm Weinberg independently proposed a mathematical model, known as the Hardy-Weinberg Principle, establishing a theoretical state of genetic equilibrium where the allele frequencies within a large, randomly mating population remain unchanged from one generation to the next. The Hardy-Weinberg Principle expresses the allele frequencies in a population using the formula $p + q = 1$, with p representing the dominant allele and q the recessive allele. The Hardy-Weinberg Principle expresses the genotypes frequencies in a population using the formula $p^2 + 2pq + q^2 = 1$. p^2 represents the homozygous dominant alleles; $2pq$ represents the heterozygous alleles; q^2 represents the homozygous recessive alleles.
- Hardy-Weinberg equilibrium is the state in which all allele frequencies and genotype frequencies remain unchanged in subsequent generations. For a population to remain in Hardy-Weinberg equilibrium, five conditions must be met: no mutations, random mating, no gene flow, large population size, and no natural selection. Any deviation from these conditions causes allele frequencies to change, resulting in evolution of the population. The allele frequencies in natural populations are continually changing; therefore, it is rare for any population to maintain Hardy-Weinberg equilibrium. The conditions upon which Hardy-Weinberg equilibrium is defined are useful in determining whether evolution is occurring within a population.

10-3. Lines of Evidence Supporting Evolution

- Many different lines of evidence support the theory of evolution.
  - The fossil record is a collection of evidence that provides a look into the history of an organism's life, including its basic structure, relatedness to other organisms, habitat, and

global events that influenced its extinction or evolution. Fossils are organismal remains, such as bones, teeth, or imprints, located within the earth's different stratified rock layers that paleontologists use to construct an organism's fossil record. The formation of fossils is rare given the limited number of structures that can produce fossils or the inability of some organisms to leave fossils; therefore, the fossil record is considered incomplete. Analysis of the fossil record has revealed progressive, intermediate stages, suggesting the accumulation of changes over time, supporting descent with modification. Transitional fossils are a group of fossils considered intermediaries of evolutionary change. Common examples of transitional fossils include *Archaeopteryx* (dinosaurs and birds), *Tiktaalik*, (fishes and land animals), *Seymouria* (amphibians and reptiles), and therapsids (reptiles and mammals). The most complete, well-preserved fossil record, consisting of several transitional fossils revealing subtle anatomical modifications to a changing environment, is the evolution of the modern-day horse from its doglike, forest-dwelling ancestor (*Hyracotherium*) to its current, grassland-grazing genus, *Equus*.

- Biogeography is the study of the distribution pattern of organisms living on Earth. The unique distribution pattern of organisms observed today can be explained by the theory of continental drift. The theory of continental drift states that all continents were at one time a large land mass, called Pangaea, which broke apart and shifted to their current locations. As one land mass, Pangaea enabled organisms to travel between different regions; however, when continental drift began, it restricted movement, isolating organisms to specific continents and regions. A common example of how continental drift influenced biogeography can be observed in the diversity of mammalian species found in Australia and South America, which were once connected when Pangaea existed. Australia is home to the most prevalent and diverse species of marsupials found on Earth in comparison to placental mammals, while placental mammals greatly outnumber marsupials in South America. Biogeography is also used to develop fossil records for certain organisms to determine their evolutionary history, as was the case of the genus *Equus*, which includes the modern-day horse.
- Comparative anatomy and embryology provide another line of evidence to support the theory of evolution. Comparative anatomy is the study of anatomical similarities and differences among living organisms. Two main types of anatomical structures are studied using comparative anatomy. Homologous structures are anatomically similar but functionally dissimilar structures found in related organisms who share a common ancestor. The anatomical arrangement of bones in vertebrate forelimbs is a common example. Analogous structures are superficially similar structures found in different organisms but have evolved due to convergent evolution, not common ancestry. The wings of birds and bats are analogous to butterfly wings. Vestigial structures are structures whose function is reduced or nonexistent but that were once functional in and inherited from a common ancestor. The presence of vestigial pelvic and leg bones in snakes and whales are examples. Embryology is the study of embryos to determine similarities and differences in structures between organisms at various developmental stages. The work of scientists Karl von Baer and Ernst Haeckel provided early evidence of similar developmental patterns among different organisms. Many similar features are shared among different organisms during

early embryonic development, including gill pouches and a post-anal tail, but certain genes influence the future development of these structures.

- Molecular biology is the study of the structure and activity of different molecules found in cells, including DNA, RNA, and proteins. DNA is composed of nucleotide sequences, which defines an organism's genetic information. Organisms that are closely related show minimal differences in their DNA nucleotide sequences. Laboratory techniques such as DNA-DNA hybridization and DNA sequencing have been used to compare the DNA nucleotide sequences of two organisms to determine their relatedness. Both techniques have revealed that humans are more closely related to chimpanzees than to other primate species. An RNA strand known as mRNA is an intermediary molecule produced by DNA to produce proteins. mRNA consists of three-letter codons translated into one of 20 amino acids based on the genetic code to produce a protein. All organisms have DNA; therefore, all organisms use the same genetic code and amino acids to synthesize proteins. Slight similarities and differences in the amino acid sequences of proteins can reveal the duration of evolution between species beginning with a common ancestor and their relatedness to one another. The cytochrome *c* protein is a common protein used to determine evolutionary relationships, because it has evolved slowly with minimal mutations. The amino acid sequence of cytochrome *c* in humans and chimpanzees is identical, but there is an 8-percent difference between humans and mice and a much greater difference between humans and yeast.
- Field and laboratory studies provide evidence of evolution through direct observations. Fieldwork performed in the Galápagos Islands by Peter and Rosemary Grant for 40 years has revealed evolution in action as they have observed and documented several directional shifts in the morphology in the population of finches on Daphne Major due to harsh environmental changes occurring on the island. Laboratory studies have demonstrated evolution in action as the prevalence of antibiotic resistance began upon Alexander Fleming's discovery of penicillin in 1928, the isolation and purification of penicillin for mass production in the early 1940s by Howard Florey and Ernst Chain, and the introduction of penicillin as an antibiotic treatment for the Allied Forces during World War II. As early as 1942, *Staphylococcus aureus* began developing antibiotic resistance to penicillin, and today more than 90 percent of *S. aureus* strains are resistant to the antibiotic, due to the presence of an enzyme known as penicillinase. From 1943 to 2010, 12 different antibiotics were introduced, and by 2011, several different bacterial strains were resistant to most, if not all, antibiotics, which has led to MRSA, a deadly bacterial infection. Evolution in action has also been observed during combined field work and laboratory studies performed by Colin Donihue, whose studies have revealed morphological changes in body size, limb length, and relative toepad size in the anole population of Turks and Caicos, as well as other anole populations in the Caribbean, in response to high winds generated from hurricanes.

End-of-Chapter Activities and Questions

Directions: Please refer back to what you learned in this chapter to complete the following activities.

Define Each Term in Your Own Words

1. Evolution
2. Fitness
3. Sexual Selection
4. Hardy-Weinberg Equilibrium
5. Transitional Fossils

Chapter Review

1. What are the differences between the bottleneck effect and the founder effect? How do they contribute to the overall mechanism of genetic drift?
2. Identify and describe the four components of natural selection, then apply each identified component to the results of the field studies performed on the finches of the Galápagos Islands by Peter and Rosemary Grant.
3. There exists a population of mice whose dominant fur color is brown (B) and whose recessive fur color is white (b). Assuming the mouse population is in Hardy-Weinberg equilibrium, calculate the allele frequency of the dominant allele (B) and the recessive allele (b) in the population given that 9 percent of the population is homozygous recessive for white fur (bb). Upon calculating the allele frequency of the dominant allele (B), what percentage of the population is homozygous dominant (BB) for brown fur color?
4. Identify and discuss the differences between the three types of structural evidence that support the idea of descent with modification.
5. Explain how the following scientists have contributed to a better understanding of the process of evolution: Gregor Mendel, Hermann von Meyer, Ernst Haeckel, Peter and Rosemary Grant, Hopi Hoekstra, and Colin Donihue.

Multiple Choice

1. Excessive hunting of northern elephant seals by humans in the late 19th century greatly reduced the original population to as few as 20 individuals. Although the population has significantly rebounded since the early 1900s due to the implementation of strict conservation initiatives, the genetic variation of the current population represents only a fraction of the variation once present in the original population. The lack of genetic variation in northern elephant seals today can be attributed to which of the following occurrences?

   a. Mutation

   b. genetic drift

   c. gene flow

   d. natural selection

2. In 18th- and 19th-century England, the frequency of light-colored peppered moths (*Biston betularia*) predominated the dark-colored moth variants. However, at the height of the Industrial Revolution, the frequency of dark-colored moths increased as pollution killed many of the light-colored lichens and soot emitted from the factories darkened the trees that the light-colored moths would use to

camouflage themselves against predators. This shift in phenotypes of the peppered moth due to environmental change would represent what mode of natural selection?

a. disruptive selection

b. stabilizing selection

c. directional selection

d. sexual selection

3. In what ways can new alleles be introduced into a population?

   a. genetic drift and mutations

   b. gene flow and natural selection

   c. natural selection and genetic drift

   d. mutations and gene flow

4. Using information obtained from *Chapter Review, question 3* above, regarding fur color in the mouse population, what percentage of individuals are heterozygous for fur color (Bb), assuming the mouse population is in Hardy-Weinberg equilibrium? Recall that 9 percent of the population is homozygous recessive for white fur (bb).

   a. 30%

   b. 42%

   c. 70%

   d. 91%

5. Which of the following is not considered a line of evidence supporting the theory of evolution?

   a. amino acid sequences

   b. biogeographical distribution

   c. species fitness

   d. homologous structures

References

Darwin, C. (1859). *The origin of species by means of natural selection, or the preservation of favoured races in the struggle for life* (1st ed.). John Murray.

Darwin, C. (1872). *The origin of species by means of natural selection, or the preservation of favoured races in the struggle for life* (2nd ed.). John Murray.

Hopwood, N. (2006). Pictures of evolution and charges of fraud: Ernst Haeckel's embryological illustrations. *Isis, 97*(2), 260–301.

Punnett, R. C., (1950). Early days of genetics. *Heredity, 4*, 1–10.

Society for Healthcare Epidemiology of America, Infectious Diseases Society of America, & Pediatric Infectious Diseases Society. (2012). Policy Statement on Antimicrobial Stewardship by the Society for

Healthcare Epidemiology of America (SHEA), the Infectious Diseases Society of America (IDSA), and the Pediatric Infectious Diseases Society (PIDS). *Infection Control and Hospital Epidemiology, 33*(4), 322–327.

Wiedersheim, R. (1895). *The structure of man: An index to his past history* (2nd ed.; H. and M. Bernard, Trans.). Macmillan and Co. (Original work published 1893)

Image Credits

Fig. 10.1: Source: https://www.sutori.com/item/wilhelm-weinberg-1862-1937-wilhelm-weinberg-a-german-physician-didn-t-collabor.

Fig. 10.2: Copyright © by OpenStax (CC BY 4.0) at https://commons.wikimedia.org/wiki/File:Genetic_drift_in_a_population_Figure_19_02_02.png.

Fig. 10.3: Copyright © by OpenStax (CC BY 4.0) at https://commons.wikimedia.org/wiki/File:Bottleneck_effect_Figure_19_02_03.jpg.

Fig. 10.4: Copyright © by Tsaneda (CC BY 3.0) at https://commons.wikimedia.org/wiki/File:Founder_effect_Illustration.jpg.

Fig. 10.5: Copyright © by Tsaneda (CC BY 3.0) at https://commons.wikimedia.org/wiki/File:Gene_flow.jpg.

Fig. 10.6: Copyright © by NicholasToal (CC BY-SA 4.0) at https://commons.wikimedia.org/wiki/File:Natural_Selection.png.

Fig. 10.7a: Source: https://commons.wikimedia.org/wiki/File:Peromyscus_polionotus_oldfield_mouse.jpg.

Fig. 10.7b: Source: https://commons.wikimedia.org/wiki/File:Peromyscus_polionotus_ammobates.jpg.

Fig. 10.8a: Copyright © by Jeffrey Pang (CC BY 2.0) at https://commons.wikimedia.org/wiki/File:Elk_Sparring_-_Jasper_National_Park.jpg.

Fig. 10.9a: Source: https://commons.wikimedia.org/wiki/File:Peromyscus_polionotus_oldfield_mouse.jpg.

Fig. 10.9b: Source: https://commons.wikimedia.org/wiki/File:Peromyscus_polionotus_ammobates.jpg.

Fig. 10.9b: Copyright © by Airwolfhound (CC BY-SA 2.0) at https://commons.wikimedia.org/wiki/File:Stoat_-_RSPB_Sandy_(28596785111).jpg.

Fig. 10.10: Copyright © by OpenStax (CC BY 4.0) at https://commons.wikimedia.org/wiki/File:Figure_19_03_01.png.

Fig. 10.11a: Source: https://commons.wikimedia.org/wiki/File:Godfrey_Harold_Hardy_1.jpg.

Fig. 10.11b: Source: https://www.sutori.com/item/wilhelm-weinberg-1862-1937-wilhelm-weinberg-a-german-physician-didn-t-collabor.

Fig. 10.12: Copyright © by OpenStax (CC BY 4.0) at https://commons.wikimedia.org/wiki/File:Figure_19_01_01.png.

Fig. 10.13a: Source: https://commons.wikimedia.org/wiki/File:Archaeopteryx_fossil.jpg.

Fig. 10.13b: Copyright © by Ghedoghedo (CC BY-SA 3.0) at https://commons.wikimedia.org/wiki/File:Tiktaalik_roseae.jpg.

Fig. 10.13c: Copyright © by Sanjay Acharya (CC BY-SA 4.0) at https://commons.wikimedia.org/wiki/File:Seymouria_Fossil.jpg.

Fig. 10.14: Copyright © by Mcy jerry (CC BY-SA 3.0) at https://commons.wikimedia.org/wiki/File:Horseevolution.png.

Fig. 10.15: Source: https://commons.wikimedia.org/wiki/File:Snider-Pellegrini_Wegener_fossil_map.svg.

Fig. 10.16: Copyright © by Волков Владислав Петрович (CC BY-SA 4.0) at https://commons.wikimedia.org/wiki/File:Homology_vertebrates-en.svg.

Fig. 10.17a: Source: https://commons.wikimedia.org/wiki/File:European_honey_bee_extracts_nectar.jpg.

Fig. 10.17b: Copyright © by Charles J. Sharp (CC BY 3.0) at https://commons.wikimedia.org/wiki/File:Purple-throated_carib_hummingbird_feeding.jpg.

Fig. 10.17c: Copyright © by Charles J. Sharp (CC BY-SA 4.0) at https://commons.wikimedia.org/wiki/File:Indian_flying_fox_(Pteropus_giganteus_giganteus)_in_flight.jpg.

Fig. 10.18: Source: https://commons.wikimedia.org/wiki/File:Whale_skeleton.png.

Fig. 10.19: Source: https://commons.wikimedia.org/wiki/File:Haeckel_Anthropogenie_1874.jpg.

Fig. 10.20a: from Kulkarni Keya and Sundarrajan Priya, "A Study of Phylogenic Relationships and Homology of Cytochrome C using Bioinformatics," International Research Journal of Science and Engineering, vol. 4, no. 3-4. Copyright © 2016 by Kulkarni Keya and Sundarrajan Priya.

Fig. 10.20b: Copyright © by Cytochrome c (CC BY-SA 3.0) at https://commons.wikimedia.org/wiki/File:Cytochromec.png.

Fig. 10.21: Copyright © by OpenStax (CC BY 4.0) at https://openstax.org/books/concepts-biology/pages/11-1-discovering-how-populations-change.

Fig. 10.22a: Source: https://commons.wikimedia.org/wiki/File:Alexander_Fleming.jpg.

Fig. 10.22b: Copyright © by Alexander Flemming (CC BY-SA 2.0) at https://commons.wikimedia.org/wiki/File:Sample_of_penicillin_mould_presented_by_Alexander_Fleming_to_Douglas_Macleod,_1935_(9672239344).jpg.

Fig. 10.24a: Copyright © by Colin Donihue. Reprinted with permission.

Fig. 10.24b: Copyright © by Colin Donihue. Reprinted with permission.

Fig. 10.25: Copyright © by Colin Donihue. Reprinted with permission.

Answers to End-of-Chapter Multiple Choice Questions

Chapter 1. Science and Biology

1. Which of the following occurs in the correct order from simplest to most complex?

    b. macromolecule, cell, organism, ecosystem

2. What is the best example of irritability?

    a. rabbit escaping a pursuing fox

3. All of the following are examples of common biological theories, except __________.

    b. thermodynamics

4. Which of the following scientists is not correctly paired to the proper number of kingdoms in his proposed classification system?

    c. Haeckel—five kingdoms

5. What macromolecule is the basis for the currently used three-domain, six-kingdom classification system?

    d. ribosomal RNA

Chapter 2. Chemistry

1. Isotopes are atoms of the same element with the same number of _______, but a different number of_______.

    b. protons, neutrons

2. Which of the following elements was the first element synthetically discovered in a laboratory?

    a. technetium

3. Oxygen has six valence electrons. How many of these valence electrons are available to bond with hydrogen atoms to make water? (Hint: Draw a Lewis diagram.)

   b. two

4. Which of the following properties of water is (are) important in the transport of water from roots to leaves in plants?

   c. adhesion and cohesion

5. An important pH buffering system in the human body is shown below:

   $$\_\_\_\_\_\_\_ + H^+ \leftrightarrow \_\_\_\_\_\_\_ \leftrightarrow CO_2 + H_2O$$

   Which molecules are missing from the equation?

   c. HCO_3^-; H_2CO_3

Chapter 3. Macromolecules

1. The chemistry of the carbon atom is unique in that it __________.

   b. has four valence electrons

2. What type of bond is found between two monosaccharides to form a disaccharide?

   a. glycosidic bond

3. Which of the following best describes a triglyceride?

   b. glycerol head bonded to three fatty acid tails

4. An amino acid consists of all the following functional groups except:

   c. phosphate group

5. Which of the following nitrogenous bases pertains to an RNA molecule and not to a DNA molecule?

   c. uracil

Chapter 4. Cell Structure and Function

1. The cell theory was proposed by which group of scientists?

   b. Matthias Schleiden, Theodor Schwann, Rudolf Virchow

2. In what ways do archaeal cells differ from bacterial cells?

   d. all the choices are correct

3. Which of the following structures is only unique to eukaryotic cells?

   b. nucleus

4. The endomembrane system of eukaryotic cells contains all the following except ________.

   a. mitochondria

5. What structural evidence supports the bacterial origin of mitochondria and chloroplasts as outlined by the endosymbiotic theory?

   d. all the choices are correct

Chapter 5. Metabolism and Energy

1. The metabolic pathways of both anabolism and catabolism require which of the following?

   b. enzymes

2. The majority of the energy located within an ATP molecule is found between the ________.

   a. three phosphate groups

3. Oxygen production is derived from the splitting of __________ molecules during photosynthesis.

   d. water

4. Which of the following molecules is not matched with the correct metabolic reaction of cellular respiration?

   c. oxygen—Krebs cycle

5. Who was the scientist credited with introducing the idea that chemical reactions are involved in the process of fermentation?

   a. Eduard Buchner

Chapter 6. DNA and DNA Technology

1. The DNA molecule consists of two strands running in alternating directions, with each strand differing from the other because of the location of a phosphate group and sugar molecule. As a result, the DNA strand is said to exhibit a(n) __________ arrangement.

   b. antiparallel

2. Which of the following enzymes is not associated with the process of DNA replication?

   a. restriction enzyme

3. During the transcription process, RNA polymerase comes to an exposed adenine (A) on the DNA strand. The resulting complementary nitrogenous base would be __________.

   d. uracil

4. A DNA sequence contains a mutation that results in a STOP codon after being transcribed and translated. Which of the following diseases may result from this type of mutation?

   c. cystic fibrosis

5. Which of the following enzymes is used both during the process of DNA replication and in genetic engineering?

   a. DNA ligase

Chapter 7. Chromosomes and Cellular Division

1. What is the major difference between a prokaryotic and eukaryotic chromosome?

   b. prokaryotic chromosome is circular; eukaryotic chromosome is linear

2. Why is binary fission considered a disadvantage to prokaryotic cells?

   a. reduces genetic diversity of the population

3. What structure forms along the old metaphase plate during cytokinesis in plant cells?

   b. cell plate

4. Upon the completion of meiosis, what immediate structure(s) is (are) produced?

   c. gametes

5. What enzyme enables cancer cells to become immortal and ignore the cellular signals triggering apoptosis?

   c. telomerase

Chapter 8. Genetics and Inheritance

1. Which of the following plants did Gregor Mendel use during the course of experiments?

   b. *Pisum sativum*

2. Which of the following conclusions from Mendel's experiments laid the foundation for his law of segregation?

   a. alleles segregate during gamete formation

3. Why are males more likely than females to inherit sex-linked traits?

   b. males only have one X chromosome

4. If two individuals with wavy hair mate, what is the probability their offspring will have curly hair? (Hint: this is an example of incomplete dominance.)

   c. 25%

5. As a result of nondisjunction, an egg without either X chromosome (an "O" egg) was fertilized by a sperm cell carrying a Y chromosome. Which of the following genetic disorders would result upon fertilization?

   d. none of the choices are correct

Chapter 9. The Origins of the Theory of Evolution

1. Which pre-Darwinian scientist is correctly paired with his early thoughts on evolutionary change?

   b. Georges-Louis Leclerc—descent with modification

2. Which of the following observations made by Darwin during the HMS *Beagle* voyage is best paired with an idea proposed by an early thinker of evolutionary change?

    c. modified beak structures in finches—inheritance of acquired characteristics

3. The species considered the closest living relative of the Galápagos finches is __________.

    d. *Tiaris obscura*

4. What was the name of Alfred Russel Wallace's essay that prompted Charles Darwin to publish his "species theory"?

    a. "On the Tendency of Varieties to Depart Indefinitely from the Original Type"

5. Who is credited with coining the term *survival of the fittest*?

    a. Herbert Spencer

Chapter 10. Mechanisms of and Evidence for Evolution

1. Excessive hunting of northern elephant seals by humans in the late 19th century greatly reduced the original population to as few as 20 individuals. Although the population has significantly rebounded since the early 1900s due to the implementation of strict conservation initiatives, the genetic variation of the current population represents only a fraction of the variation once present in the original population. The lack of genetic variation in northern elephant seals today can be attributed to which of the following occurrences?

    b. genetic drift

2. In 18th- and 19th-century England, the frequency of light-colored peppered moths (*Biston betularia*) predominated the dark-colored moth variants. However, at the height of the Industrial Revolution, the frequency of dark-colored moths increased as pollution killed many of the light-colored lichens and soot emitted from the factories darkened the trees that the light-colored moths would use to camouflage themselves against predators. This shift in phenotypes of the peppered moth due to environmental change would represent what mode of natural selection?

    c. directional selection

3. In what ways can new alleles be introduced into a population?

    d. mutations and gene flow

4. Using information obtained from *Chapter Review, question 3* above, regarding fur color in the mouse population, what percentage of individuals are heterozygous for fur color (Bb), assuming the mouse population is in Hardy-Weinberg equilibrium? Recall that 9 percent of the population is homozygous recessive for white fur (bb).

    b. 42%

5. Which of the following is not considered a line of evidence supporting the theory of evolution?

    c. species fitness

Printed in the USA
CPSIA information can be obtained
at www.ICGtesting.com
LVHW081104080923
757442LV00014B/45

9 781516 589609